Selected Titles in This Series

35 **James F. Davis and Paul Kirk,** Lecture notes in algebraic topology, 2001
34 **Sigurdur Helgason,** Differential geometry, Lie groups, and symmetric spaces, 2001
33 **Dmitri Burago, Yuri Burago, and Sergei Ivanov,** A course in metric geometry, 2001
32 **Robert G. Bartle,** A modern theory of integration, 2001
31 **Ralf Korn and Elke Korn,** Option pricing and portfolio optimization: Modern methods of financial mathematics, 2001
30 **J. C. McConnell and J. C. Robson,** Noncommutative Noetherian rings, 2001
29 **Javier Duoandikoetxea,** Fourier analysis, 2001
28 **Liviu I. Nicolaescu,** Notes on Seiberg-Witten theory, 2000
27 **Thierry Aubin,** A course in differential geometry, 2001
26 **Rolf Berndt,** An introduction to symplectic geometry, 2001
25 **Thomas Friedrich,** Dirac operators in Riemannian geometry, 2000
24 **Helmut Koch,** Number theory: Algebraic numbers and functions, 2000
23 **Alberto Candel and Lawrence Conlon,** Foliations I, 2000
22 **Günter R. Krause and Thomas H. Lenagan,** Growth of algebras and Gelfand-Kirillov dimension, 2000
21 **John B. Conway,** A course in operator theory, 2000
20 **Robert E. Gompf and András I. Stipsicz,** 4-manifolds and Kirby calculus, 1999
19 **Lawrence C. Evans,** Partial differential equations, 1998
18 **Winfried Just and Martin Weese,** Discovering modern set theory. II: Set-theoretic tools for every mathematician, 1997
17 **Henryk Iwaniec,** Topics in classical automorphic forms, 1997
16 **Richard V. Kadison and John R. Ringrose,** Fundamentals of the theory of operator algebras. Volume II: Advanced theory, 1997
15 **Richard V. Kadison and John R. Ringrose,** Fundamentals of the theory of operator algebras. Volume I: Elementary theory, 1997
14 **Elliott H. Lieb and Michael Loss,** Analysis, 1997
13 **Paul C. Shields,** The ergodic theory of discrete sample paths, 1996
12 **N. V. Krylov,** Lectures on elliptic and parabolic equations in Hölder spaces, 1996
11 **Jacques Dixmier,** Enveloping algebras, 1996 Printing
10 **Barry Simon,** Representations of finite and compact groups, 1996
9 **Dino Lorenzini,** An invitation to arithmetic geometry, 1996
8 **Winfried Just and Martin Weese,** Discovering modern set theory. I: The basics, 1996
7 **Gerald J. Janusz,** Algebraic number fields, second edition, 1996
6 **Jens Carsten Jantzen,** Lectures on quantum groups, 1996
5 **Rick Miranda,** Algebraic curves and Riemann surfaces, 1995
4 **Russell A. Gordon,** The integrals of Lebesgue, Denjoy, Perron, and Henstock, 1994
3 **William W. Adams and Philippe Loustaunau,** An introduction to Gröbner bases, 1994
2 **Jack Graver, Brigitte Servatius, and Herman Servatius,** Combinatorial rigidity, 1993
1 **Ethan Akin,** The general topology of dynamical systems, 1993

Lecture Notes in Algebraic Topology

James F. Davis
Paul Kirk

Graduate Studies in Mathematics
Volume 35

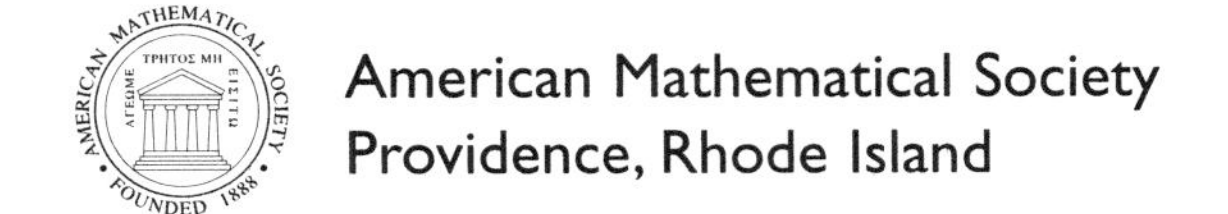

American Mathematical Society
Providence, Rhode Island

2000 *Mathematics Subject Classification.* Primary 55–01, 57–01.

ABSTRACT. This is a textbook for a second-year course on algebraic topology. Some familiarity with homology and the fundamental group is assumed. This book covers the basics of homotopy theory, including fibrations, cofibrations, homology with local coefficients, obstruction theory, and applications of spectral sequences. The intent of the book is to bridge the gap between algebraic and geometric topology, both by providing the algebraic tools that a geometric topologist needs and by concentrating on those areas of algebraic topology which are geometrically motivated.

Library of Congress Cataloging-in-Publication Data

Davis, James F. (James Frederic), 1955–
Lecture notes in algebraic topology / James F. Davis, Paul Kirk.
p. cm. — (Graduate studies in mathematics, ISSN 1065-7339 ; v. 35)
Includes bibliographical references and index.
ISBN 0-8218-2160-1 (alk. paper)
1. Algebraic topology. I. Kirk, P. (Paul) II. Title. III. Series.

QA612.D39 2001
514′.2—dc21 2001022609

---

10 9 8 7 6 5 4 3 2 1 06 05 04 03 02 01

Contents

Preface

To paraphrase a comment in the introduction to a classic point-set topology text, this book might have been titled *What Every Young Topologist Should Know.* It grew from lecture notes we wrote while teaching second–year algebraic topology at Indiana University.

The amount of algebraic topology a student of topology must learn can be intimidating. Moreover, by their second year of graduate studies students must make the transition from understanding simple proofs line-by-line to understanding the overall structure of proofs of difficult theorems.

To help our students make this transition, the material in these notes is presented in an increasingly sophisticated manner. Moreover, we found success with the approach of having the students meet an extra session per week during which they took turns presenting proofs of substantial theorems and writing lecture notes to accompany their explanations. The responsibility of preparing and giving these lectures forced them to grapple with "the big picture" and also gave them the opportunity to learn how to give mathematical lectures, preparing for their participation in research seminars. We have collated a number of topics for the students to explore in these sessions; they are listed as projects in the table of contents and are enumerated below.

Our perspective in writing this book was to provide the topology graduate students at Indiana University (who tend to write theses in geometric topology) with the tools of algebraic topology they will need in their work, to give them a sufficient background to be able to interact with and appreciate the work of their homotopy theory cousins, and also to make sure that they are exposed to the critical advances in mathematics which came about with the development of topology in the years 1950-1980. The topics discussed in varying detail include homological algebra, differential topology, algebraic K-theory, and homotopy theory. Familiarity with these topics is important not just for a topology student but any student of pure mathematics, including the student moving towards research in geometry, algebra, or analysis.

The prerequisites for a course based on this book include a working knowledge of basic point-set topology, the definition of CW-complexes, fundamental group/covering space theory, and the construction of singular homology including the Eilenberg-Steenrod axioms. In Chapter 8, familiarity with the basic results of differential topology is helpful. In addition, a command of basic algebra is required. The student should be familiar with the notions of R-modules for a commutative ring R (in particular the definition of tensor products of two R-modules) as well as the structure theorem for modules over a principal ideal domain. Furthermore, in studying non simply-connected spaces it is necessary to work with tensor products over (in general non-commutative) group rings, so the student should know the definition of a right or left module over such a ring and their tensor products. Basic terminology from category theory is used (sometimes casually), such as category, functor, and natural transformation. For example, if a theorem asserts that some map is natural, the student should express this statement in categorical language.

In a standard first-year course in topology, students might also learn some basic homological algebra, including the universal coefficient theorem, the cellular chain complex of a CW-complex, and perhaps the ring structure on cohomology. We have included some of this material in Chapters 1, 2, and 3 to make the book more self-contained and because we will often have to refer to the results. Depending on the pace of a first-year course, a course based on this book could start with the material of Chapter 2 (Homological Algebra), Chapter 3 (Products), or Chapter 4 (Fiber Bundles).

Chapter 6 (Fibrations, Cofibrations and Homotopy Groups) and Chapter 9 (Spectral Sequences) form the core of the material; any second-year course should cover this material. Geometric topologists must understand how to work with non simply-connected spaces, and so Chapter 5 (Homology with Local Coefficients) is fundamental in this regard. The material in Chapters 7 (Obstruction Theory and Eilenberg-MacLane Spaces) and 8 (Bordism, Spectra, and Generalized Homology) introduces the student to the modern perspective in algebraic topology. In Chapter 10 (Further Applications of Spectral Sequences) many of the fruits of the hard labor that preceded this chapter are harvested. Chapter 11 (Simple-Homotopy theory) introduces the ideas which lead to the subject of algebraic K-theory and to the s-cobordism theorem. This material has taken a prominent role in research in topology, and although we cover only a few of the topics in this area (K_1, the Whitehead group, and Reidemeister torsion), it serves as good preparation for more advanced courses.

These notes are meant to be used in the classroom, freeing the student from copying everything from the chalkboard and hopefully leaving more time to think about the material. There are a number of exercises in the text; these are usually routine and are meant to be worked out when the student studies. In many cases, the exercises fill in a detail of a proof or provide a useful generalization of some result. Of course, this subject, like any subject in mathematics, cannot be learned without thinking through some exercises. Working out these exercises as the course progresses is one way to keep up with the material. The student should keep in mind that, perhaps in contrast to some areas in mathematics, topology is an example driven subject, and so working through examples is the best way to appreciate the value of a theorem.

We will omit giving a diagram of the interdependence of various chapters, or suggestions on which topics could be skipped, on the grounds that teachers of topology will have their own opinion based on their experience and the interests of the students. (In any case, every topic covered in this book is related in some way to every other topic.) We have attempted (and possibly even succeeded) to organize the material in such a way as to avoid the use of technical facts from one chapter to another, and hence to minimize the need to shuffle pages back and forth when reading the book. This is to maximize its usefulness as a textbook, as well as to ensure that the student with a command of the concepts presented can learn new material smoothly and the teacher can present the material in a less technical manner. Moreover, we have not taken the view of trying to present the most elementary approach to any topic, but rather we feel that the student is best served by learning the high-tech approach, since this ultimately is faster and more useful in research. For example, we do not shrink from using spectral sequences to prove basic theorems in algebraic topology.

Some standard references on the material covered in this course include the books [**14**], [**36**], [**43**], [**9**], [**17**] [**31**], and [**7**]. A large part of the material in these notes was distilled from these books. Moreover, one can find some of the material covered in much greater generality and detail in these tomes. Our intention is not to try to replace these wonderful books, but rather to offer a textbook to accompany a course in which this material is taught.

We recommend that students look at the article "Fifty years of homotopy theory" by G. Whitehead [**44**] for an overview of algebraic topology, and look back over this article every few weeks as they are reading this book. The books a student should read after finishing this course (or in conjunction with this course) are Milnor and Stasheff, *Characteristic Classes* [**30**] (every mathematician should read this book), and Adams, *Algebraic Topology: A Student's Guide* [**1**].

The authors would like to thank Eva-Marie Elliot and Mary Jane Wilcox for typing early versions of the manuscript. Special thanks are due to our colleagues Ayelet Lindenstrauss and Allan Edmonds for their careful proofreading of our manuscripts; all remaining mistakes and typographical errors are entirely the authors' fault. The second author would like to thank John Klein for teaching him algebraic topology while they were in graduate school. Special thanks to Marcia and Beth.

Projects

The following is a list of topics to be covered in the extra meetings and lectured on by the students. They do not always match the material of the corresponding chapter but are usually either related to the chapter material or preliminary to the next chapter. Sometimes they form interesting subjects which could reasonably be skipped. Some projects are quite involved (e.g. "state and prove the Hurewicz theorem"), and the students and instructor should confer to decide how deeply to cover each topic. In some cases (e.g. the Hopf degree theorem, the Hurewicz theorem, and the Freudenthal suspension theorem) proofs are given in later chapters using more advanced methods.

- **Chapter 1.**
    1. The cellular approximation theorem.
    2. Singular homology theory.
- **Chapter 2.**
    1. The acyclic models theorem and the Eilenberg-Zilber map.
- **Chapter 3.**
    1. Algebraic limits and the Poincaré duality theorem.
    2. Exercises on intersection forms.
- **Chapter 4.**
    1. Fiber bundles over paracompact bases are fibrations.
    2. Classifying spaces.
- **Chapter 5.**
    1. The Hopf degree theorem.
    2. Colimits and limits.

- **Chapter 6.**
  1. The Hurewicz theorem.
  2. The Freudenthal suspension theorem.
- **Chapter 7.**
  1. Postnikov systems.
- **Chapter 8.**
  1. Basic notions from differential topology.
  2. Definition of K-theory.
  3. Spanier-Whitehead duality.
- **Chapter 9.**
  1. Construction of the Leray-Serre-Atiyah-Hirzebruch spectral sequence.
- **Chapter 10.**
  1. Unstable homotopy theory.
- **Chapter 11.**
  1. Handlebody theory and torsion for manifolds.

Chain Complexes, Homology, and Cohomology

1.1. Chain complexes associated to a space

A *chain complex* is a sequence of homomorphisms

$$\cdots \to C_n \xrightarrow{\partial} C_{n-1} \xrightarrow{\partial} C_{n-2} \to \cdots$$

so that the double composite $\partial \circ \partial : C_n \to C_{n-2}$ is zero for all n.

Here are 3 examples of chain complexes associated to a space:

1. The singular chain complex $(S_q(X), \partial)$ of a topological space X.
2. The cellular chain complex $(C_q(X), \partial)$ of a CW-complex X.
3. The simplicial chain complex $(\Delta_q(K), \partial)$ of a simplicial complex K.

Their constructions are as follows.

1.1.1. Construction of the singular chain complex.

The *(geometric) q-simplex* Δ^q is defined by

$$\Delta^q = \left\{(t_0, t_1, \ldots, t_q) \in \mathbf{R}^{q+1} \mid \Sigma t_i = 1,\ t_i \geq 0 \text{ for all } i\right\}.$$

The *face maps* are the functions

$$f_m^q : \Delta^{q-1} \to \Delta^q$$

defined by

$$(t_0, t_1, \ldots, t_{q-1}) \mapsto \quad (t_0, \ldots, t_{m-1}, 0, t_m, \ldots, t_{q-1})$$
$$\uparrow m^{\text{th}}\text{coordinate}$$

A *singular q-simplex in a space X* is a continuous map $\sigma : \Delta^q \to X$.

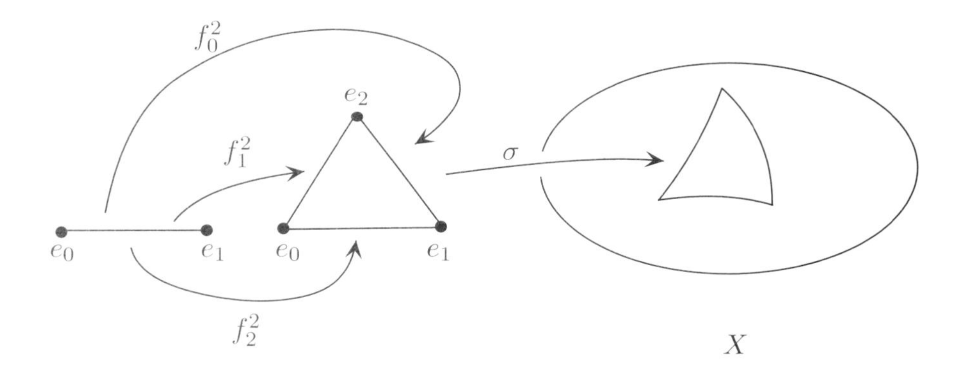

Let R be a *commutative ring* with unit. Denote by $S_q(X; R)$ the free R-module with basis the singular q-simplices $\{\sigma : \Delta^q \to X\}$ and define the differential $\partial : S_q(X; R) \to S_{q-1}(X; R)$ to be the R-linear map defined on a singular simplex σ by

$$\partial(\sigma) = \sum_{m=0}^{q} (-1)^m \sigma \circ f_m^q.$$

Thus, on a chain $\sum_{i=1}^{\ell} r_i \sigma_i$, ∂ has the formula

$$\partial\left(\sum_{i=1}^{\ell} r_i \sigma_i\right) = \sum_{i=1}^{\ell} r_i \left(\sum_{m=0}^{q} (-1)^m \sigma_i \circ f_m^q\right).$$

One calculates that $\partial^2 = 0$.

Recall that $S_*(-; R)$ is a covariant functor; that is, a continuous map $f : X \to Y$ induces a homomorphism of free modules $f_* : S_*(X; R) \to S_*(Y; R)$. One defines f_* by

$$f_*(\sigma) = f \circ \sigma$$

on a singular simplex σ and extends by linearity.

Definition 1.1. The complex $(S_*(X; R), \partial)$ is called the *singular chain complex of X with coefficients in R*. Its homology is defined by

$$H_q(X; R) = \frac{\ker \partial : S_q(X; R) \to S_{q-1}(X; R)}{\operatorname{im} \partial : S_{q+1}(X; R) \to S_q(X; R)}$$

and is called the *singular homology of X with coefficients in R*.

When $R = \mathbf{Z}$, we omit the reference to R. Homology is a covariant functor from the category of spaces to the category of R-modules.

To say that (X, A) is a *pair of spaces* means that A is a subspace of a topological space X. We next recall the definition of the relative singular chain complex of a pair of spaces. If $A \subset X$, define

$$S_q(X, A; R) = \frac{S_q(X; R)}{S_q(A; R)}.$$

Thus $S_q(X, A; R)$ is a free R-module; a basis element is represented by a singular q-simplex in X whose image is not contained in A. One obtains a commutative diagram with exact rows

$$\begin{array}{ccccccccc} 0 & \longrightarrow & S_q(A;R) & \longrightarrow & S_q(X;R) & \longrightarrow & S_q(X,A;R) & \longrightarrow & 0 \\ & & \downarrow{\partial} & & \downarrow{\partial} & & \downarrow{\partial} & & \\ 0 & \longrightarrow & S_{q-1}(A;R) & \longrightarrow & S_{q-1}(X;R) & \longrightarrow & S_{q-1}(X,A;R) & \longrightarrow & 0 \end{array}$$

Exercise 1. Show that $\partial : S_q(X, A; R) \to S_{q-1}(X, A; R)$ is well defined and $\partial^2 = 0$.

The complex $(S_q(X, A; R), \partial)$ is called the *singular chain complex for the pair* (X, A) *with coefficients in* R.

Its homology is defined by

$$H_q(X, A; R) = \frac{\ker \partial : S_q(X, A; R) \to S_{q-1}(X, A; R)}{\operatorname{im} \partial : S_{q+1}(X, A; R) \to S_q(X, A; R)}.$$

One can show there is a exact sequence

$$\cdots \to H_q(A; R) \to H_q(X; R) \to H_q(X, A; R) \xrightarrow{\partial} H_{q-1}(A; R) \to \cdots$$
$$\cdots \to H_1(X, A; R) \xrightarrow{\partial} H_0(A; R) \to H_0(X; R) \to H_0(X, A; R) \to 0.$$

1.1.2. Definition of a CW-complex. CW-complexes form a nice collection of topological spaces which include most spaces of interest in geometric and algebraic topology.

Definition 1.2. Let X be a topological space and let $A \subset X$ be a closed subspace. We say X is *obtained from* A *by adjoining* n*-cells* $\{e_i^n\}_{i \in I}$ if:

1. For each index $i \in I$, e_i^n is a subset of X, called an n*-cell.*
2. $X = A \cup_i e_i^n$.
3. Letting ∂e_i^n denote the intersection of e_i^n with A, we require $e_i^n - \partial e_i^n$ to be disjoint from $e_j^n - \partial e_j^n$ for $i \neq j$.
4. For each $i \in I$ there exists a continuous surjective map

$$\phi_i^n : (D^n, S^{n-1}) \to (e_i^n, \partial e_i^n),$$

called a *characteristic map for the cell* e_i^n, so that the restriction of ϕ_i^n to the interior $\text{int}(D^n)$ is a homeomorphism onto $e_i^n - \partial e_i^n$.

5. A subset $B \subset X$ is closed in X if and only if $B \cap A$ is closed in A and $(\phi_i^n)^{-1}(B)$ is closed in D^n for all i, n.

The characteristic maps themselves are not part of the definition; only their existence is. The restriction of the characteristic map to S^{n-1} is called the *attaching map* for the n-cell. Notice that adjoining 0-cells to a space A is the same as taking a disjoint union of A with a discrete set.

There is another point of view on the above. Given a continuous map

$$f : \amalg\, S^{n-1} \to A$$

from a disjoint union of $(n-1)$-spheres to a space A, one can form the space

$$X = A \cup_f (\amalg\, D^n) = \frac{A\ \amalg (\amalg\, D^n)}{\sim}$$

where $f(x) \sim x$ for $x \in \amalg\, S^{n-1}$. Then X is obtained from A by adjoining n-cells.

Definition 1.3. A *relative CW-complex* (X, A) is a topological space X, a closed subspace A, and a sequence of closed subspaces $X^n, n = -1, 0, 1, 2, \cdots$ called the *relative n-skeleton* so that

1. $X^{-1} = A$ and X^n is obtained from X^{n-1} by adjoining n-cells.
2. $X = \cup_{i=-1}^{\infty} X^i$.
3. A subset $B \subset X$ is closed in X if and only if $B \cap X^n$ is closed in X^n for each n.

The largest n so that $X = X^n$, if it exists, is called the *dimension* of (X, A). If such an n does not exist, we say (X, A) is infinite-dimensional. If A is itself a CW-complex, we call (X, A) a *CW-pair.*

If A is empty, X is called an *(absolute) CW-complex* and X^n is called the *(absolute) n-skeleton.*

1.1.3. Definition of the cellular chain complex of a CW-complex.

If X is a CW-complex and $X^n \subset X$ is the n-skeleton of X, denote by $C_q(X; R)$ the relative homology group

$$C_q(X; R) = H_q(X^q, X^{q-1}; R).$$

If (X, A) is a *CW*-pair, then you have seen an argument (using excision for singular homology) which shows that $H_q(X, A; R) \cong H_q(X/A, A/A; R)$. Applying this to (X^q, X^{q-1}), one finds that $C_q(X; R)$ is a free R-module with a basis corresponding to the q-cells of X.

Exercise 2. What is the excision argument? How can one use characteristic maps $\phi_i : D^q \to X^q$ for cells to give basis elements? To what extent do the basis elements depend on the choice of characteristic maps? Despite the ambiguity in sign it is traditional to denote the basis element by e_i^n.

The differential $\partial : C_q(X; R) \to C_{q-1}(X; R)$ can be defined in two ways. The first is purely algebraic; the second is geometric and involves the notion of the *degree* of a map $f : S^n \to S^n$. If you don't know what the degree of such a map is, look it up. If you know the definition of degree, then look up the differential-topological definition of degree for a smooth map $f : S^n \to S^n$.

First definition of ∂. Take ∂ to be the composite

$$H_q(X^q, X^{q-1}; R) \xrightarrow{\delta} H_{q-1}(X^{q-1}; R) \xrightarrow{i} H_{q-1}(X^{q-1}, X^{q-2}; R)$$

where δ is the connecting homomorphism in the long exact sequence in singular homology for the pair (X^q, X^{q-1}) and i is induced by the inclusion $(X^{q-1}, \phi) \hookrightarrow (X^{q-1}, X^{q-2})$.

Second definition of ∂. The quotient X^{q-1}/X^{q-2} is homeomorphic to a one-point union of $(q-1)$-spheres, one for each $(q-1)$-cell of X, since the boundary of each $(q-1)$-cell has been collapsed to a point, and a $(q-1)$-cell with its boundary collapsed to a point is a $(q-1)$-sphere. For each q-cell e^q consider the attaching map $S^{q-1} \to X^{q-1}$. The composite of this map with the quotient map to X^{q-1}/X^{q-2} defines a map from a $(q-1)$-sphere to a one-point union of $(q-1)$-spheres. Taking the degree of this map in the factor corresponding to a $(q-1)$-cell e_i^{q-1} gives an integer denoted by $[e^q : e_i^{q-1}]$.

Now define the differential $\partial : C_q(X; R) \to C_{q-1}(X; R)$ on a q-cell e^q by

$$\partial(e^q) = \sum [e^q : e_i^{q-1}] e_i^{q-1}$$

(the sum is over all $(q-1)$-cells e_i^{q-1}), and extend to all chains by linearity.

Exercise 3. Prove that the two definitions of the differential

$$\partial : C_q(X; R) \to C_{q-1}(X; R)$$

are the same. (Hint: $H_n(D^n, S^{n-1}) \xrightarrow{\delta} H_{n-1}(S^{n-1})$ is an isomorphism.) Use the first definition to show that $\partial^2 = 0$.

The cellular homology groups are the homology groups of the complex $(C_*(X; R), \partial)$:

$$H_q(X, R) = \frac{\ker \partial : C_q(X) \to C_{q-1}(X)}{\operatorname{im} \partial : C_{q+1}(X) \to C_q(X)}.$$

Definition 1.4. A *cellular map* $f : X \to Y$ is a continuous function between CW-complexes so that $f(X^q) \subset Y^q$ for all q.

A cellular map induces a chain map $f_* : C_*(X; R) \to C_*(Y; R)$, since f restricts to a map of pairs $f : (X^q, X^{q-1}) \to (Y^q, Y^{q-1})$. Thus for every q, cellular homology is a functor

$$\{\text{CW-complexes, cellular maps}\} \to \{\text{abelian groups, homomorphisms}\}.$$

The proof of the following theorem can be found in any standard first-year algebraic topology textbook.

Theorem 1.5. *The cellular and singular homology of a CW-complex are naturally isomorphic.* □

So, for example, the circle S^1 has a cell structure with one 0-cell and one 1-cell. The boundary map is trivial, so $H_1(S^1) \cong \mathbf{Z}$. A generator $[S^1] \in H_1(S^1)$ is specified by taking the 1-cell which parameterizes the circle in a counterclockwise fashion. We can use this to define the *Hurewicz map*

$$\rho : \pi_1(X, x_0) \to H_1(X; \mathbf{Z})$$

by

$$\rho(\alpha) = \alpha_*([S^1]).$$

Here $\pi_1(X, x_0)$ denotes the *fundamental group of* X *based at* $x_0 \in X$

$$\pi_1(X, x_0) = \left\{ \begin{matrix} \text{Homotopy classes of maps} \\ (S^1, 1) \to (X, x_0) \end{matrix} \right\}.$$

Recall that the fundamental group of a space is a non-abelian group in general. The Hurewicz theorem for the fundamental group is the following.

Theorem 1.6. *Suppose that X is path-connected. Then the Hurewicz map $\rho : \pi_1(X, x_0) \to H_1(X; \mathbf{Z})$ is a surjection with kernel the commutator subgroup of $\pi_1(X, x_0)$. Hence $H_1(X; \mathbf{Z})$ is isomorphic to the abelianization of $\pi_1(X, x_0)$.* □

1.1.4. Construction of the simplicial chain complex of a simplicial complex. For both historical reasons and to keep current with simplicial techniques in algebraic topology, it is worthwhile to have some familiarity with simplicial homology.

Definition 1.7. An *(abstract) simplicial complex* K is a pair (V, S) where V is a set and S is a collection of non-empty finite subsets of V satisfying:

1. If $v \in V$, then $\{v\} \in S$.
2. If $\tau \subset \sigma \in S$ and τ is non-empty, then $\tau \in S$.

Elements of V are called *vertices*. Elements of S are called *simplices*. A *q-simplex* is an element of S with $q+1$ vertices. If $\sigma \in S$ is a q-simplex, we say $\dim(\sigma) = q$.

Put a (total) ordering on the vertices V.

Define the simplicial q-chains $\Delta_q(K;R)$ to be the free R-module with basis the q-simplices of K. Denote a q-simplex by $\langle\sigma\rangle = \langle v_0, v_1, \dots, v_q\rangle$ where the vertices are listed in increasing order. Define the differential $\partial : \Delta_q(K;R) \to \Delta_{q-1}(K;R)$ on a q-simplex by

$$\partial\langle v_0, v_1, \dots, v_q\rangle = \sum_{m=0}^{q} (-1)^m \langle v_0, v_1, \dots, \widehat{v_m}, \dots, v_q\rangle,$$

where $\widehat{v_m}$ means omit the m-th vertex, and then extend by R-linearity, i.e.

$$\partial\left(\sum_{i=1}^{\ell} r_i\langle\sigma_i\rangle\right) = \sum_{i=1}^{\ell} r_i\left(\sum_{m=0}^{q}(-1)^m\partial\langle\sigma_i\rangle\right).$$

The homology of this chain complex is denoted $H_*(K;R)$. Notice that these definitions are purely combinatorial; the notions of topological space and continuity are not used. The connection with topology is given by the next definition.

Definition 1.8. The *geometric realization* of a simplicial complex K is the quotient space

$$|K| = \frac{\amalg_{\sigma\in S}\Delta^{\dim(\sigma)}}{\sim}.$$

In other words, we take a geometric q-simplex for each abstract q-simplex of K and glue them together. The identifications are given as follows: if $\sigma = \langle v_0, v_1, \dots, v_q\rangle \in S$ and if $\tau = \langle v_0, v_1, \dots, \widehat{v_m}, \dots, v_q\rangle$ is a face of σ, then identify the geometric simplex $\Delta^{\dim(\tau)}$ corresponding to τ with a face of the geometric simplex $\Delta^{\dim(\sigma)}$ corresponding to σ using the m-th face map f_m^q. In other words, the equivalence relation is generated by $x \sim f_m^q(x)$ for $x \in \langle v_0, v_1, \dots, \widehat{v_m}, \dots, v_q\rangle$. (Look at [**26**] for a deeper perspective on this construction.)

If one chooses another ordering of the vertices, then one can, with some fuss about orientation, define a canonical isomorphism between the simplicial chain complex (or geometric realization) defined using one ordering to the simplicial chain complex (or geometric realization) defined using the other ordering.

A *triangulation* of a topological space X is a homeomorphism from the geometric realization of a simplicial complex to X.

Exercise 4. Find a triangulation of $\mathbf{R}P^2$ and compute its simplicial homology.

The homology $H_*(K;R)$ of an abstract simplicial complex K is isomorphic to $H_*(|K|;R)$, the singular homology of its geometric realization. This can be seen by noting that $|K|$ is naturally a CW-complex, the q-skeleton is the union of simplices of dimension $\leq q$, the q-cells are the the q-simplices. The cellular chain complex of $|K|$ is isomorphic to the simplicial chain complex of K.

Another construction of homology uses the cubical singular complex (this is the point of view taken in Massey's book [**23**]). This gives yet another chain complex associated to a topological space. It is not hard, using the acyclic models theorem, to show that the simplicial and cubical singular homology functors are naturally isomorphic.

1.2. Tensor products, adjoint functors, and Hom

1.2.1. Tensor products. Let A and B be modules over a commutative ring R.

Definition 1.9. The *tensor product of* A *and* B is the R-module $A \otimes_R B$ defined as the quotient

$$\frac{F(A \times B)}{R(A \times B)}$$

where $F(A \times B)$ is the free R-module with basis $A \times B$ and $R(A \times B)$ the submodule generated by

1. $(a_1 + a_2, b) - (a_1, b) - (a_2, b)$
2. $(a, b_1 + b_2) - (a, b_1) - (a, b_1)$
3. $r(a, b) - (ra, b)$
4. $r(a, b) - (a, rb)$.

One denotes the image of a basis element (a, b) in $A \otimes_R B$ by $a \otimes b$. Note that one has the relations

1. $(a_1 + a_2) \otimes b = a_1 \otimes b + a_2 \otimes b$
2. $a \otimes (b_1 + b_2) = a \otimes b_1 + a \otimes b_2$
3. $(ra \otimes b) = r(a \otimes b) = (a \otimes rb)$.

Informally, $A \otimes_R B$ is the largest R-module generated by the set of symbols $\{a \otimes b\}_{a \in A, b \in B}$ satisfying the above "product type relations". Any element of $A \otimes B$ can be expressed as a finite sum $\sum_{i=1}^{n} a_i \otimes b_i$, but it may not be possible to take $n = 1$, nor is the representation as a sum unique.

Recall that a function $\phi : A \times B \to M$ is *R-bilinear* if M is an R-module and

1. $\phi(a_1 + a_2, b) = \phi(a_1, b) + \phi(a_2, b)$
2. $\phi(a, b_1 + b_2) = \phi(a, b_1) + \phi(a, b_2)$
3. $\phi(ra, b) = r\phi(a, b) = \phi(a, rb)$.

For example, the map $\pi : A \times B \to A \otimes_R B$, $(a, b) \mapsto a \otimes b$ is R-bilinear. The universal property of the tensor product is that this map π is initial in the category of bilinear maps with domain $A \times B$.

Proposition 1.10. *Given an R-bilinear map $\phi : A \times B \to M$, there is a unique R-module map $\overline{\phi} : A \otimes_R B \to M$ so that $\overline{\phi} \circ \pi = \phi$.*

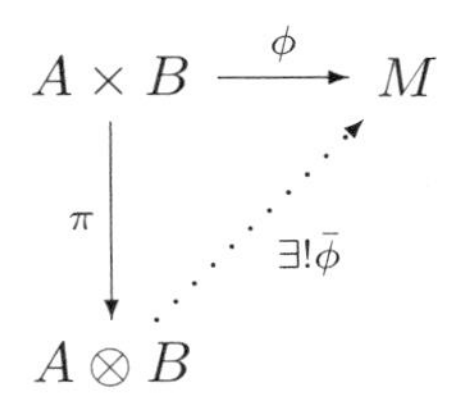

Proof. If $\overline{\phi}$ exists, then $\overline{\phi}(\sum a_i \otimes b_i) = \sum \overline{\phi}(a_i \otimes b_i) = \sum \overline{\phi} \circ \pi(a_i \otimes b_i) = \sum \phi(a_i, b_i)$. Thus uniqueness is clear. For existence, define $\hat{\phi} : F(A \times B) \to M$ on basis elements by $(a, b) \mapsto \phi(a, b)$ and extend by R-linearity. The bilinearity of ϕ implies $\hat{\phi}(R(A \times B)) = 0$, so $\hat{\phi}$ induces $\overline{\phi} : A \otimes_R B \to M$ by the universal property of quotients. □

Proposition 1.10 is useful for defining maps *out* of tensor products, and the following exercise indicates that this is the defining property of tensor products.

Exercise 5. Suppose $p : A \times B \to T$ is an R-bilinear map so that for any R-bilinear map $\psi : A \times B \to M$, there is a unique R-module map $\overline{\psi} : T \to M$ so that $\overline{\psi} \circ p = \psi$. Then $T \cong A \otimes_R B$.

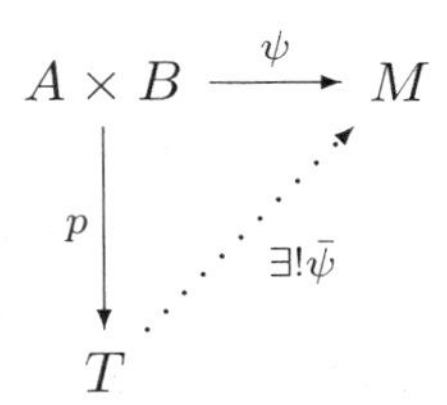

For the rest of this section, we will omit the subscript R from the tensor product. The basic properties of the tensor product are given by the next theorem.

Theorem 1.11.

1. *$A \otimes B \cong B \otimes A$.*
2. *$R \otimes B \cong B$.*
3. *$(A \otimes B) \otimes C \cong A \otimes (B \otimes C)$.*
4. *$(\oplus_\alpha A_\alpha) \otimes B \cong \oplus_\alpha (A_\alpha \otimes B)$.*
5. *Given R-module maps $f : A \to C$ and $g : B \to D$, there is an R-module map $f \otimes g : A \otimes B \to C \otimes D$ so that $a \otimes b \mapsto f(a) \otimes g(b)$.*
6. *The functor $- \otimes M$ is right exact. That is, given an R-module M, and an exact sequence*
$$A \xrightarrow{f} B \xrightarrow{g} C \to 0,$$
the sequence
$$A \otimes M \xrightarrow{f \otimes \mathrm{Id}} B \otimes M \xrightarrow{g \otimes \mathrm{Id}} C \otimes M \to 0$$
is exact.

Proof.

1. There is a map $A \otimes B \to B \otimes A$ with $a \otimes b \mapsto b \otimes a$. More formally, the map $A \times B \to B \otimes A$, $(a, b) \mapsto b \otimes a$ is bilinear; for example, one sees $(a_1 + a_1, b) \mapsto b \otimes (a_1 + a_2) = b \otimes a_1 + b \otimes a_2$. By the universal property there is a map $A \otimes B \to B \otimes A$ with $\sum a_i \otimes b_i \mapsto \sum b_i \otimes a_i$. The inverse map is clear.
2. Define $R \otimes B \to B$ by $r \otimes b \mapsto rb$ and $B \to R \otimes B$ by $b \mapsto 1 \otimes b$.
3. $(a \otimes b) \otimes c \leftrightarrow a \otimes (b \otimes c)$.
4. $(\oplus a_\alpha) \otimes b \leftrightarrow \oplus\, (a_\alpha \otimes b)$.
5. $A \times B \to C \otimes D$, $(a, b) \mapsto f(a) \otimes g(b)$ is R-bilinear.
6. We explicitly define an isomorphism
$$\overline{g \otimes \mathrm{Id}} : \frac{B \otimes M}{(f \otimes \mathrm{Id})(A \otimes M)} \to C \otimes M.$$
Since $(g \otimes \mathrm{Id}) \circ (f \otimes \mathrm{Id}) = (g \circ f) \otimes \mathrm{Id} = 0$, the map $g \otimes \mathrm{Id}$ descends to the map $\overline{g \otimes \mathrm{Id}}$ by the universal property of quotients. The inverse map is given by defining an R-bilinear map $C \times M \to \frac{B \otimes M}{(f \otimes \mathrm{Id})(A \otimes M)}$ by $(c, m) \mapsto [\hat{c} \otimes m]$ where $g(\hat{c}) = c$. Note that the map is independent of the choice of lift $\hat{c}$, indeed if $\hat{c}'$ is another lift, then $\hat{c} - \hat{c}' \in \ker g = \operatorname{im} f$, so $[\hat{c} \otimes m] - [\hat{c}' \otimes m] = 0$.

$\square$

Example 1.12. Let M be an abelian group. Applying properties 5 and 2 of Theorem 1.11 we see that if we tensor the short exact sequence

$$0 \to \mathbf{Z} \xrightarrow{\times n} \mathbf{Z} \to \mathbf{Z}/n \to 0$$

by M, we obtain the exact sequence

$$M \xrightarrow{\times n} M \to \mathbf{Z}/n \otimes_{\mathbf{Z}} M \to 0.$$

Notice that $\mathbf{Z}/n \otimes_{\mathbf{Z}} M \cong M/nM$ and that the sequence is not short exact if M has torsion whose order is not relatively prime to n. Thus $- \otimes M$ is not left exact.

Example 1.13. If V and W are vector spaces over $\mathbf{R}$ with bases $\{e_i\}$ and $\{f_j\}$ respectively, then $V \otimes_{\mathbf{R}} W$ has basis $\{e_i \otimes f_j\}$; thus $\dim(V \otimes_{\mathbf{R}} W) = (\dim V)(\dim W)$.

Exercise 6. Compute $A \otimes_{\mathbf{Z}} B$ for any finitely generated groups A and B.

1.2.2. Adjoint functors. Note that an R-bilinear map $\beta : A \times B \to C$ is the same as an element of $\mathrm{Hom}_R(A, \mathrm{Hom}_R(B, C))$. The universal property of the tensor product can be rephrased as follows.

Proposition 1.14 (Adjoint Property of Tensor Products). *There is an isomorphism of R-modules*

$$\mathrm{Hom}_R(A \otimes_R B, C) \cong \mathrm{Hom}_R(A, \mathrm{Hom}_R(B, C)),$$

natural in A, B, C given by $\phi \leftrightarrow (a \mapsto (b \mapsto \phi(a \otimes b)))$. □

This is more elegant than the universal property for three reasons: It is a statement in terms of the category of R-modules, it gives a reason for the duality between tensor product and Hom, and it leads us to the notion of adjoint functor.

Definition 1.15. (Covariant) functors $F : \mathcal{C} \to \mathcal{D}$ and $G : \mathcal{D} \to \mathcal{C}$ form an *adjoint pair* if there is a 1-1 correspondence $\mathrm{Mor}_{\mathcal{D}}(Fc, d) \longleftrightarrow \mathrm{Mor}_{\mathcal{C}}(c, Gd)$, for all $c \in \mathrm{Ob}\,\mathcal{C}$ and $d \in \mathrm{Ob}\,\mathcal{D}$, natural in c and d. The functor F is said to be the left adjoint of G and G is the right adjoint of F.

The adjoint property says that for any R-module B, the functors

$$- \otimes_R B : R\text{-MOD} \to R\text{-MOD}$$

and

$$\mathrm{Hom}_R(B, -) : R\text{-MOD} \to R\text{-MOD}$$

form an adjoint pair. Here R-MOD is the category whose objects are R-modules and whose morphisms are R-maps.

It turns out that the right exactness of $- \otimes_R B$ and the left exactness of $\mathrm{Hom}_R(B, -)$ are formal consequences of being an adjoint pair, but we won't pursue this. A random functor may not have a left (or right) adjoint, but if it does, the adjoint is unique up to natural isomorphism.

Exercise 7. Let SETS be the category whose objects are sets and whose morphisms are functions. The *forgetful functor* R-MOD $\to$ SETS takes a module to its underlying set. Find an adjoint for the forgetful functor. Find another adjoint pair of your own. "Adjoints are everywhere."

1.2.3. Hom.

Exercise 8. For any finitely generated abelian groups A and B, compute $\mathrm{Hom}_{\mathbf{Z}}(A, B)$, the group of all homomorphisms from A to B.

For an R-module A, define $A^* = \mathrm{Hom}_R(A, R)$. The module A^* is often called the *dual* of A. For an R-module map $f : A \to B$, the dual map $f^* : B^* \to A^*$ is defined by $f^*(\varphi) = \varphi \circ f$. Hence taking duals defines a contravariant functor from the category of R-modules to itself.

More generally, for R-modules A and M, $\mathrm{Hom}_R(A, M)$ is the R-module of homomorphisms from A to M. It is contravariant in its first variable and covariant in its second variable. For an R-map $f : A \to B$, we have $\mathrm{Hom}_R(f, M) : \mathrm{Hom}_R(B, M) \to \mathrm{Hom}_R(A, M)$, defined by $\varphi \mapsto \varphi \circ f$. Usually we write f^* for $\mathrm{Hom}_R(f, M)$. The following computational facts may help with Exercise 8.

1. $\mathrm{Hom}_R(R, M) \cong M$.
2. $\mathrm{Hom}_R(\oplus_\alpha A_\alpha, M) \cong \prod_\alpha \mathrm{Hom}_R(A_\alpha, M)$.
3. $\mathrm{Hom}_R(A, \prod_\alpha M_\alpha) \cong \prod_\alpha \mathrm{Hom}_R(A, M_\alpha)$.

The distinction between direct sum and direct product in the category of modules is relevant only when the indexing set is infinite, in which case the direct sum allows only a finite number of nonzero coordinates.

1.3. Tensor and Hom functors on chain complexes

We will turn our attention now to the algebraic study of (abstract) chain complexes (C_*, ∂) of R-modules. We do not assume that the chain groups are free R-modules, although they are for the three geometric examples of Section 1.1.

The starting observation is that the singular (or cellular) homology functor is a composite of two functors: the singular complex functor

$$S_* : \{ \text{ spaces, continuous maps } \} \to \{ \text{ chain complexes, chain maps } \}$$

and the homology functor

H_*:{chain complexes, chain maps}→{graded R-modules, homomorphisms}.

A useful strategy is to place interesting *algebraic* constructions between S_* and H_*, i.e. to use functors

$$\{\text{chain complexes}\} \to \{\text{chain complexes}\}$$

to construct new homology invariants of spaces. The two families of functors we use are the following.

1. Forming the tensor product of a chain complex and an R-module M.

This is the functor

$$(C_*, \partial) \to (C_* \otimes M,\ \partial \otimes \mathrm{Id})$$

with

$$(\partial \otimes \mathrm{Id})\left(\sum c_i \otimes m_i\right) = \sum (\partial c_i) \otimes m_i.$$

Since $(\partial \otimes \mathrm{Id})^2 = 0$, $(C_* \otimes M,\ \partial \otimes \mathrm{Id})$ is a chain complex. You should show that this is a *covariant functor;* i.e. write down the formula for the map $C_* \otimes M \to C'_* \otimes M$ induced by a chain map $C_* \to C'_*$ and check that it is a chain map.

Definition 1.16. Taking the homology of $C_* \otimes M$ yields the *homology of C_* with coefficients in M*:

$$H_*(C_*; M) = \frac{\ker \partial : C_* \otimes M \to C_* \otimes M}{\operatorname{im} \partial : C_* \otimes M \to C_* \otimes M}.$$

Applying this to the singular complex of a space leads to the following definition.

Definition 1.17. The homology of $S_*(X; R) \otimes M$ is called the *singular homology of X with coefficients in the R-module M* and is denoted by $H_*(X; M)$.

In the same way one can define $H_*(X, A; M)$, the relative singular homology with coefficients in M. A similar construction applies to the cellular complex to give the cellular homology with coefficients.

2. Taking the R-linear maps from C_* to an R-module M.

This is the functor

$$(C_*, \partial) \to (\mathrm{Hom}_R(C_*, M), \delta)$$

where the differential δ is the dual to ∂; i.e. $\delta = \mathrm{Hom}_R(\partial, M)$ (sometimes denoted by ∂^*). Explicitly $\delta : \mathrm{Hom}_R(C_*, M) \to \mathrm{Hom}_R(C_*, M)$ is defined by

$$(\delta f)(c) = f(\partial c).$$

Then $\delta^2 = 0$ since $(\delta^2 f)(c) = (\delta f)(\partial c) = f(\partial^2 c) = 0$. There are two important facts to notice about this chain complex:

1. The differential δ of this complex has degree $+1$, and not -1 as in the three examples we encountered before. What this means is that
$$\delta : \mathrm{Hom}_R(C_q, M) \to \mathrm{Hom}_R(C_{q+1}, M).$$
2. The functor is *contravariant*, in contrast to the tensor product functor we considered above. The point here is that for any R–module M the functor $\mathrm{Hom}(-, M)$ is a contravariant functor.

A useful terminology is to call a chain complex with degree $+1$ differential a *cochain complex.*

This leads us to the definition of the cohomology of a chain complex with coefficients in M:

Definition 1.18.
$$H^*(C_*; M) = \frac{\ker \delta : \mathrm{Hom}(C_*, M) \to \mathrm{Hom}(C_*, M)}{\mathrm{im}\, \delta : \mathrm{Hom}(C_*, M) \to \mathrm{Hom}(C_*, M)}$$
is the *cohomology of* (C_*, ∂) *with coefficients in* M.

1.4. Singular cohomology

Applying the construction of the previous section to a space X, one obtains the *singular cochain complex of* X *with coefficients in the* R*-module* M
$$(\mathrm{Hom}_R(S_q(X; R), M), \delta).$$
This cochain complex is denoted by $S^*(X; M)$. (Note that we can take $M = R$.) Similar constructions apply to the cellular chain complex and the simplicial chain complex.

Exercise 9. For a useful way to think about singular cochains, show that
$$S^q(X; R) = \mathrm{Hom}_{\mathbf{Z}}(S_q(X; \mathbf{Z}), R)$$
and also show that
$$S^q(X; R) = \text{functions}(\{\text{singular simplexes}\}, R).$$

Definition 1.19. The *singular cohomology of* X *with coefficients in the* R*-module* M is
$$H^q(X; M) = \frac{\ker \delta : \mathrm{Hom}_R(S_q(X, R), M) \to \mathrm{Hom}_R(S_{q+1}(X, R), M)}{\mathrm{im}\, \delta : \mathrm{Hom}_R(S_{q-1}(X, R), M) \to \mathrm{Hom}_R(S_q(X, R), M)}.$$

Similarly one defines the cellular cohomology of a CW-complex and the simplicial cohomology of a simplicial complex.

Exercise 10. What is $H^0(X; \mathbf{Z})$? Show that $H_0(\mathbf{Q}; \mathbf{Z}) \not\cong H^0(\mathbf{Q}; \mathbf{Z})$, where we consider the rational numbers $\mathbf{Q}$ as a subspace of the real numbers $\mathbf{R}$.

The primary motivation for introducing cohomology comes from the fact that $H^*(X; R)$ admits a ring structure, while homology does not. This will be discussed in Chapter 3.

Recall that for a chain complex (C_*, ∂), a *cycle* is an element of $\ker \partial$ and a *boundary* is an element of $\operatorname{im} \partial$. The terminology for cochain complexes is obtained by using the "co" prefix:

Definition 1.20. A *cocycle* is an element in the kernel of δ and a *coboundary* is an element in the image of δ.

Exercise 11. Show that a cocycle applied to a boundary is zero and a coboundary applied to a cycle is zero. Deduce that there is a bilinear pairing (the *Kronecker pairing*)

$$H^n(C_*; R) \times H_n(C_*; R) \to R$$

given by the formula

$$\langle [\varphi], [\alpha] \rangle = \varphi(\alpha).$$

Deduce by taking adjoints that the Kronecker pairing defines a map

$$H^n(C_*; R) \to \operatorname{Hom}_R(H_n(C_*; R), R).$$

The Kronecker pairing on the homology and cohomology of a space should be thought of as an analogue (in fact a generalization) of integrating a differential n-form along an n-dimensional submanifold. (See the paragraph on the DeRham complex on page 16.)

We will study the Kronecker pairing in detail for R a principal ideal domain (PID) in Section 2.6. It is important to note that cohomology is not the dual of homology in general. The map $H^n(C_*; R) \to \operatorname{Hom}_R(H_n(C_*; R), R)$ need not be injective nor surjective. The following example illustrates this. A precise relationship between the dual of homology and cohomology is provided by the universal coefficient theorem (Theorem 2.29) when R is a PID.

The cellular chain complex of $\mathbf{R}P^2$ is

$$\begin{array}{ccccccccccc}
\cdots & \xrightarrow{\partial} & C_3 & \xrightarrow{\partial} & C_2 & \xrightarrow{\partial} & C_1 & \xrightarrow{\partial} & C_0 & \xrightarrow{\partial} & 0 \\
 & & \updownarrow\cong & & \updownarrow\cong & & \updownarrow\cong & & \updownarrow\cong & & \\
\cdots & \longrightarrow & 0 & \longrightarrow & \mathbf{Z} & \xrightarrow[\times 2]{} & \mathbf{Z} & \xrightarrow[\times 0]{} & \mathbf{Z} & \longrightarrow & 0
\end{array}$$

so

$$H_0(\mathbf{R}P^2; \mathbf{Z}) = \mathbf{Z},\ H_1(\mathbf{R}P^2; \mathbf{Z}) = \mathbf{Z}/2, \text{ and } H_2(\mathbf{R}P^2; \mathbf{Z}) = 0.$$

The corresponding cochain complex is

$$\begin{array}{ccccccccc}
\cdots \xleftarrow{\delta} & C^3 & \xleftarrow{\delta} & C^2 & \xleftarrow{\delta} & C^1 & \xleftarrow{\delta} & C^0 & \longleftarrow 0 \\
& \updownarrow\cong & & \updownarrow\cong & & \updownarrow\cong & & \updownarrow\cong & \\
\cdots \longleftarrow & \mathrm{Hom}(C_3,\mathbf{Z}) & \leftarrow & \mathrm{Hom}(C_2,\mathbf{Z}) & \leftarrow & \mathrm{Hom}(C_1,\mathbf{Z}) & \leftarrow & \mathrm{Hom}(C_0,\mathbf{Z}) & \longleftarrow 0 \\
& \updownarrow\cong & & \updownarrow\cong & & \updownarrow\cong & & \updownarrow\cong & \\
\cdots \longleftarrow & 0 & \longleftarrow & \mathrm{Hom}(\mathbf{Z},\mathbf{Z}) & \leftarrow & \mathrm{Hom}(\mathbf{Z},\mathbf{Z}) & \leftarrow & \mathrm{Hom}(\mathbf{Z},\mathbf{Z}) & \longleftarrow 0 \\
& \updownarrow\cong & & \updownarrow\cong & & \updownarrow\cong & & \updownarrow\cong & \\
\cdots \longleftarrow & 0 & \longleftarrow & \mathbf{Z} & \xleftarrow[\times 2]{} & \mathbf{Z} & \xleftarrow[\times 0]{} & \mathbf{Z} & \longleftarrow 0
\end{array}$$

Thus

$$H^0(\mathbf{R}P^2;\mathbf{Z}) = \mathbf{Z}$$
$$H^1(\mathbf{R}P^2;\mathbf{Z}) = 0$$
$$H^2(\mathbf{R}P^2;\mathbf{Z}) = \mathbf{Z}/2.$$

In particular $H^2(\mathbf{R}P^2;\mathbf{Z}) \neq \mathrm{Hom}_{\mathbf{Z}}(H_2(\mathbf{R}P^2;\mathbf{Z}),\mathbf{Z})$. Hence the Kronecker pairing is singular.

Exercise 12. We will show that if R is a field, then homology and cohomology are dual. Verify this for $\mathbf{R}P^2$ and $R = \mathbf{Z}/2$.

Recall that if $f : X \to Y$ is continuous, then $f_* : S_q(X) \to S_q(Y)$ is defined via $f_*(\Sigma r_i\sigma_i) = \Sigma r_i f \circ \sigma_i$. Since f_* is a chain map its dual $f^* : S^q(Y) \to S^q(X)$ is a (co)chain map. Hence the singular cochain functor is contravariant. This implies the following theorem.

Theorem 1.21. *Singular cohomology is a contravariant functor*

$$\{\text{ spaces, continuous maps }\} \to \{\text{ graded } R\text{-modules, homomorphisms }\}.$$

□

Remark for those readers who know about differential forms. Suppose X is a smooth manifold. Let $\Omega^q(X)$ be the vector space of differential q-forms on a manifold. Let $d : \Omega^q(X) \to \Omega^{q+1}(X)$ be the exterior derivative. Then $(\Omega^*(X), d)$ is an $\mathbf{R}$-cochain complex whose cohomology is denoted by $H^*_{\mathrm{DR}}(\Omega^*(X), d)$ and is called the DeRham cohomology of X. This gives geometric analogues: q-form and q-cochain, d and δ, closed form and cocycle, exact form and coboundary.

DeRham's theorem states that the DeRham cohomology of a manifold X is isomorphic to the singular cohomology $H^*(X;\mathbf{R})$. More precisely, let $S_q^{\mathrm{smooth}}(X;\mathbf{R})$ be the free $\mathbf{R}$-module generated by smooth singular simplices $\sigma : \Delta^q \to X$. There is the chain map

$$S_*^{\mathrm{smooth}}(X;\mathbf{R}) \to S_*(X;\mathbf{R})$$

given by inclusion and the cochain map

$$\Omega^*(X) \to S_*^{\text{smooth}}(X; \mathbf{R})^*$$

given by integrating a q-form along a q-chain. DeRham's theorem follows from the fact that both maps are chain homotopy equivalences; i.e. they have inverses up to chain homotopy.

1.4.1. Relative cohomology. Recall that the relative singular chain complex of a pair (X, A) is defined by taking the chain groups $S_q(X, A) = S_q(X)/S_q(A)$. Similarly, let M be an R-module and define the relative singular cochain complex by

$$S^q(X, A; M) = \mathrm{Hom}_R(S_q(X, A; R), M)$$

$$\delta = \mathrm{Hom}_R(\partial, M), \ \delta(\varphi) = \varphi \circ \partial.$$

Theorem 1.22. *The diagram*

$$\begin{array}{ccccccccc} 0 & \longrightarrow & S^q(X, A; M) & \longrightarrow & S^q(X; M) & \longrightarrow & S^q(A; M) & \longrightarrow & 0 \\ & & \downarrow \delta & & \downarrow \delta & & \downarrow \delta & & \\ 0 & \longrightarrow & S^{q+1}(X, A; M) & \longrightarrow & S^{q+1}(X; M) & \longrightarrow & S^q(A; M) & \longrightarrow & 0 \end{array}$$

commutes and the horizontal rows are exact.

The proof will depend on a few exercises.

Exercise 13. The diagram commutes.

Exercise 14. Given any short exact sequence of R-modules

$$0 \longrightarrow A \longrightarrow B \longrightarrow C \longrightarrow 0$$

show that

$$0 \to \mathrm{Hom}(C, M) \to \mathrm{Hom}(B, M) \to \mathrm{Hom}(A, M)$$

and

$$0 \to \mathrm{Hom}(M, A) \to \mathrm{Hom}(M, B) \to \mathrm{Hom}(M, C)$$

are exact.

We recall what it means for homomorphisms to split.

Definition 1.23.

1. An injection $0 \longrightarrow A \xrightarrow{\alpha} B$ is said to *split* if there is a map $\delta : B \to A$ so that $\delta \circ \alpha = \mathrm{Id}_A$. The map δ is called a *splitting*.
2. A surjection $B \xrightarrow{\beta} C \longrightarrow 0$ *splits* if there is a map $\gamma : C \to A$ so that $\beta \circ \gamma = \mathrm{Id}_C$.

A surjection $B \to C \to 0$ splits if C is free (prove this basic fact). In general, for an injection $0 \longrightarrow A \xrightarrow{\alpha} B$ the dual $\mathrm{Hom}_R(B, M) \to \mathrm{Hom}_R(A, M)$ need not be a surjection (find an example!), but if α is split by δ, then the dual map is a split surjection with splitting map $\mathrm{Hom}_R(\delta, M)$.

Lemma 1.24. *Given a short exact sequence of R-modules*

$$0 \longrightarrow A \xrightarrow{\alpha} B \xrightarrow{\beta} C \longrightarrow 0,$$

show that α splits if and only if β splits. (If either of these possibilities occur, we say the short exact sequence splits.) Show that in this case $B \cong A \oplus C$.

Exercise 15. Prove this lemma.

Corollary 1.25. *If $0 \longrightarrow A \xrightarrow{\alpha} B \xrightarrow{\beta} C \longrightarrow 0$ is a short exact sequence of R-modules which splits, then*

$$0 \to \mathrm{Hom}(C, M) \to \mathrm{Hom}(B, M) \to \mathrm{Hom}(A, M) \to 0$$

is exact and splits.

Exercise 16. $S_q(X, A; R)$ is a free R-module with basis

$$\{\sigma : \Delta^q \to X | \sigma(\Delta^q) \not\subset A\}.$$

Theorem 1.22 now follows from Corollary 1.25 and Exercise 16. □

Applying the zig-zag lemma (a short exact sequence of (co)chain complexes gives a long exact sequence in (co)homology) immediately implies the following corollary.

Corollary 1.26. *To a pair (X, A) of spaces there corresponds a long exact sequence in singular cohomology*

$$0 \to H^0(X, A; M) \to H^0(X; M) \to H^0(A; M) \xrightarrow{\delta} H^1(X, A; M) \to$$
$$\cdots \to H^{q-1}(A; M) \xrightarrow{\delta} H^q(X, A; M) \to H^q(X; M) \to \cdots.$$

□

Note that the connecting homomorphism δ has degree $+1$, in contrast to the homology connecting homomorphism ∂ in homology which has degree -1.

Exercise 17. Prove the zig-zag lemma, which says that a short exact sequence of (co)chain complexes yields a long exact sequence in (co)homology, to remind yourself how to carry out diagram chase arguments.

Exercise 18. Using the facts that $S_*(X;R)$ and $S_*(X,A;R)$ are *free* chain complexes with bases consisting of singular simplices (see Exercise 16), show that

1. $S_q(X;M) = S_q(X;R) \otimes_R M$ can be expressed as the set of all sums $\left\{\sum_{i=1}^{\ell} \sigma_i \otimes m_i | m_i \in M, \sigma_i \text{ a } q\text{-simplex}\right\}$ with $\partial(\sigma \otimes m) = \partial(\sigma) \otimes m$. What is the corresponding statement for $S_q(X,A;M)$?
2. $S^q(X;M) = \mathrm{Hom}_R(S_q(X;R),M)$ is in 1-1 correspondence with the set of functions from the set of singular q-simplices to M. Under this identification, given a cochain
$$\alpha : \{\text{Singular } q\text{-simplices}\} \to M,$$
its differential $\delta\alpha : \{\text{Singular } (q+1)\text{-simplices}\} \to M$ corresponds to the map
$$\delta\alpha(\sigma) = \Sigma(-1)^i \alpha(\sigma \circ f_i^{q+1})$$
and $S^q(X,A;M) \subset S^q(X;M)$ corresponds to those functions which vanish on the q-simplices entirely contained in A.

Exercise 19. Define and identify the cellular cochain complex in two different ways: as the dual of the cellular chain complex and in terms of relative cohomology of the skeleta. (This will be easier after you have learned the universal coefficient theorem.)

1.5. The Eilenberg-Steenrod axioms

An important conceptual advance took place in algebraic topology when Eilenberg and Steenrod "axiomatized" homology and cohomology.

Definition 1.27. An *(ordinary) homology theory* is a covariant functor
$$H_* : \{\text{ pairs of spaces, continuous maps of pairs }\} \to$$
$$\{\text{ graded } R\text{-modules, homomorphisms }\},$$
in other words a collection of covariant functors H_q for each non-negative integer q which assign an R-module $H_q(X,A)$ to a pair (X,A) of topological spaces and a homomorphism $f_* : H_q(X,A) \to H_q(Y,B)$ to every continuous map of pairs $f : (X,A) \to (Y,B)$. These are required to satisfy the following axioms:

1. There exist *natural* connecting homomorphisms
$$\partial : H_q(X,A) \to H_{q-1}(A)$$
for each pair (X,A) and each integer q so that the sequence
$$\cdots \to H_q(A) \to H_q(X) \to H_q(X,A) \xrightarrow{\partial} H_{q-1}(A) \to \cdots$$
$$\cdots \to H_1(X,A) \xrightarrow{\partial} H_0(A) \to H_0(X) \to H_0(X,A) \to 0$$

is exact.

(*Long exact sequence of a pair*)

2. If $f, g : (X, A) \to (Y, B)$ are homotopic maps, then the induced maps on homology are equal, $g_* = f_* : H_q(X, A) \to H_q(Y, B)$.

   (*Homotopy invariance*)

3. If $U \subset X$, $\overline{U} \subset \operatorname{Int} A$, then $H_q(X - U, A - U) \to H_q(X, A)$ is an isomorphism for all q.

   (*Excision*)

4. If "pt" denotes the one-point space, then $H_q(\mathrm{pt}) = 0$ when $q \neq 0$.

   (*Dimension Axiom*)

Theorem 1.28 (Existence). *For any R-module M, there is a homology theory H_* with $H_0(\mathrm{pt}) = M$.* □

In fact, existence is shown by using singular homology with coefficients in M.

Definition 1.29. An *(ordinary) cohomology theory* is a contravariant functor

$$H^* : \{ \text{ pairs of spaces, continuous maps of pairs } \} \to \{ \text{ graded } R\text{-modules, homomorphisms } \},$$

in other words a collection of contravariant functors H^q, one for each nonnegative integer q, which assign an R-module $H^q(X, A)$ to a pair (X, A) of topological spaces and a homomorphism $f^* : H^q(Y, B) \to H^q(X, A)$ to every continuous function of pairs $f : (X, A) \to (Y, B)$. These are required to satisfy the following axioms:

1. There exist *natural* connecting homomorphisms

   $$\delta : H^q(A) \to H^{q+1}(X, A)$$

   for each pair (X, A) so that the sequence of Corollary 1.26 is exact.

   (*Long exact sequence of a pair*)

2. If $f, g : (X, A) \to (Y, B)$ are homotopic maps, then $g^* = f^* : H^q(Y, B) \to H^q(X, A)$.

   (*Homotopy invariance*)

3. If $U \subset X$, $\overline{U} \subset \operatorname{Int} A$, then $H^q(X, A) \to H^q(X - U, A - U)$ is an isomorphism.

   (*Excision*)

4. If pt is a point, $H^q(\mathrm{pt}) = 0$ when $q \neq 0$.

   (*Dimension Axiom*)

Theorem 1.30 (Existence). *For any R-module M, there is a cohomology theory with $H^0(\mathrm{pt}) = M$.* □

There are many different approaches to constructing homology and cohomology theories; the choice of method is often dictated by the kind of problem one is attacking. For example, the singular homology and cohomology are defined for all spaces. The abstract definition simplifies the proofs of many theorems and is makes it easy to see that (co)homology is a homeomorphism invariant, but the singular complex is too large to be effective for computations. DeRham cohomology is defined for smooth manifolds and has many nice properties, including direct relationships to solutions of differential equations on manifolds. There exist some extensions of DeRham theory to more general spaces; these tend to be technical. Cellular homology is often the most useful for computing, but of course applies only to CW-complexes.

Čech cohomology theory is another theory that satisfies the axioms (at least for the subcategory of pairs of compact spaces), but the Čech cohomology of the topologist's sine curve is not isomorphic to the singular cohomology. Thus the axioms do not determine the cohomology of all spaces. However they do for finite CW-complexes. An informal way of saying this is that the proof that cellular cohomology equals singular cohomology uses only the axioms. A more precise version is the following theorem, whose proof we omit. We state it for cohomology, but the dual result holds for homology also.

Theorem 1.31 (Uniqueness). *Let H^* and $\hat{H}^*$ be contravariant functors from the category {pairs of finite CW-complexes, cellular maps} to {graded R-modules, homomorphisms} satisfying the Eilenberg-Steenrod Axioms. Let* pt *be a point.*

1. *Given a homomorphism $H^0(\mathrm{pt}) \to \hat{H}^0(\mathrm{pt})$, there is a natural transformation of functors $H^* \to \hat{H}^*$ inducing the given map.*
2. *Any natural transformation $H^* \to \hat{H}^*$ inducing an isomorphism for a point is an isomorphism for all finite CW-complexes.*

In fact the excision axiom can be replaced by the weaker axiom that for all finite CW-pairs $H^*(X/A, A/A) \to H^*(X, A)$ is an isomorphism. This shows that cellular cohomology and singular cohomology coincide for finite CW-complexes.

In light of this theorem, one can do all computations of cohomology groups of finite CW-complexes using the axioms, i.e. without resorting to the definition of the singular or cellular cochain complex. This is not always the best way to proceed, but usually in doing cohomology computations one makes repeated use of the axioms and a few basic computations.

There are also many functors from spaces to R-modules for which the dimension axiom of Eilenberg and Steenrod does not hold. These are called generalized (co)homology theories. They will be introduced in Chapter 8.

1.6. Projects for Chapter 1

1.6.1. Cellular approximation theorem. Recall that a cellular map $f : X \to Y$ is a map between CW-complexes which satisfies $f(X^n) \subset Y^n$ for all n. The cellular approximation theorem says that any map between CW-complexes is homotopic to a cellular map. Prove the cellular approximation theorem and its relative version. This is stated as Theorem 6.47. Give applications to homotopy groups. A good reference is [**13**].

1.6.2. Singular homology theory. Give an outline of the proof that singular homology theory satisfies the Eilenberg-Steenrod axioms, concentrating on the excision axiom. State the Mayer-Vietoris exact sequence, give a computational example of its use, and show how it follows from the Eilenberg-Steenrod axioms.

Chapter 2

Homological Algebra

2.1. Axioms for Tor and Ext; projective resolutions

Definition 2.1. An *exact functor* R-MOD $\to$ R-MOD is a functor which takes short exact sequences to short exact sequences.

More generally, a covariant functor $F : R\text{-MOD} \to R\text{-MOD}$ is called *right exact* (resp. *left exact*) if $F(A) \to F(B) \to F(C) \to 0$ is exact (resp. $0 \to F(A) \to F(B) \to F(C)$ is exact) whenever $0 \to A \to B \to C \to 0$ is a short exact sequence. Similarly a contravariant functor is called *right exact* (resp. *left exact*) if $F(C) \to F(B) \to F(A) \to 0$ is exact (resp. $0 \to F(C) \to F(B) \to F(A)$ is exact) whenever $0 \to A \to B \to C \to 0$ is a short exact sequence.

We have already seen that the functors $-\otimes_R M$, $\mathrm{Hom}_R(M, -)$, and $\mathrm{Hom}_R(-, M)$ are not exact in general. For example, taking $R = \mathbf{Z}$, $M = \mathbf{Z}/2$, and the short exact sequence

$$0 \to \mathbf{Z} \xrightarrow{\times 2} \mathbf{Z} \to \mathbf{Z}/2 \to 0,$$

we obtain

$$\begin{array}{ccccccc}
\mathbf{Z}\otimes\mathbf{Z}/2 & \longrightarrow & \mathbf{Z}\otimes\mathbf{Z}/2 & \longrightarrow & \mathbf{Z}/2\otimes\mathbf{Z}/2 & \longrightarrow & 0 \\
\downarrow\cong & & \downarrow\cong & & \downarrow\cong & & \\
\mathbf{Z}/2 & \xrightarrow{\times 2} & \mathbf{Z}/2 & \xrightarrow{\mathrm{Id}} & \mathbf{Z}/2 & \longrightarrow & 0
\end{array}$$

$$\begin{array}{ccccccc}
0 & \longrightarrow & \mathrm{Hom}(\mathbf{Z}/2,\mathbf{Z}) & \longrightarrow & \mathrm{Hom}(\mathbf{Z}/2,\mathbf{Z}) & \longrightarrow & \mathrm{Hom}(\mathbf{Z}/2,\mathbf{Z}) \\
 & & \downarrow \cong & & \downarrow \cong & & \downarrow \cong \\
0 & \longrightarrow & 0 & \longrightarrow & 0 & \longrightarrow & \mathbf{Z}/2
\end{array}$$

and

$$\begin{array}{ccccccc}
0 & \longrightarrow & \mathrm{Hom}(\mathbf{Z}/2,\mathbf{Z}/2) & \longrightarrow & \mathrm{Hom}(\mathbf{Z},\mathbf{Z}/2) & \longrightarrow & \mathrm{Hom}(\mathbf{Z},\mathbf{Z}/2) \\
 & & \downarrow \cong & & \downarrow \cong & & \downarrow \cong \\
0 & \longrightarrow & \mathbf{Z}/2 & \longrightarrow & \mathbf{Z}/2 & \xrightarrow{\times 2} & \mathbf{Z}/2
\end{array}$$

However, we have seen in Theorem 1.11 that $-\otimes_R M$ is right exact and in Exercise 14 that $\mathrm{Hom}_R(M,-)$ and $\mathrm{Hom}_R(-,M)$ are left exact.

Exercise 20. If F is a free module, show that $-\otimes_R F$ and $\mathrm{Hom}_R(F,-)$ are exact functors. Show by example that $\mathrm{Hom}_R(-,F)$ need not be exact.

The idea of homological algebra is to find *natural* functors which measure the failure of a functor to preserve short exact sequences. (A first stab at this for $-\otimes_R M$ might be to take the kernel of $A\otimes M \to B\otimes M$ as the value of this functor. Unfortunately, this does not behave nicely with respect to morphisms.) To construct these functors the only things we will use are the left/right exactness properties, the above exercise and the observation that for any module M there is a surjective map from a free module to M.

Theorem 2.2 (Existence).

1. *There exist functors*

$$\mathrm{Tor}_n^R : R\text{-}MOD \times R\text{-}MOD \to R\text{-}MOD \quad \textit{for all} \;\; n = 0,1,2,\ldots$$

$(M_1, M_2) \mapsto \mathrm{Tor}_n^R(M_1,M_2)$ *covariant in* M_1 *and* M_2 *satisfying the following axioms:*

T1) $\mathrm{Tor}_0^R(M_1,M_2) = M_1 \otimes_R M_2$.

T2) *If* $0 \to A \to B \to C \to 0$ *is any short exact sequence of* R*-modules and* M *is any* R*-module, then there is a natural long exact sequence*

$$\cdots \to \mathrm{Tor}_n^R(A,M) \to \mathrm{Tor}_n^R(B,M) \to \mathrm{Tor}_n^R(C,M) \to \mathrm{Tor}_{n-1}^R(A,M) \to \cdots$$
$$\cdots \to \mathrm{Tor}_1^R(C,M) \to A\otimes_R M \to B\otimes_R M \to C\otimes_R M \to 0.$$

T3) $\mathrm{Tor}_n^R(F,M) = 0$ *if* F *is a free module and* $n > 0$.

The functor $\mathrm{Tor}_n^R(-,M)$ *is called the* n^{th} derived functor *of the functor* $-\otimes_R M$.

2. *There exist functors*

$$\mathrm{Ext}^n_R : R\text{-}MOD \times R\text{-}MOD \to R\text{-}MOD \quad \textit{for all} \ \ n = 0, 1, 2, \dots$$

$(M_1, M_2) \mapsto \mathrm{Ext}^n_R(M_1, M_2)$ *contravariant in* M_1 *and covariant in* M_2 *satisfying the following axioms:*

E1) $\mathrm{Ext}^0_R(M_1, M_2) = \mathrm{Hom}_R(M_1, M_2)$.

E2) *If* $0 \to A \to B \to C \to 0$ *is any short exact sequence of* R-*modules and* M *is any* R-*module, then there is a natural long exact sequence*

$$0 \to \mathrm{Hom}_R(C, M) \to \mathrm{Hom}_R(B, M) \to \mathrm{Hom}_R(A, M) \to \mathrm{Ext}^1_R(C, M) \to \cdots$$
$$\cdots \to \mathrm{Ext}^q_R(B, M) \to \mathrm{Ext}^q_R(A, M) \to \mathrm{Ext}^{q+1}_R(C, M) \to \cdots$$

E3) $\mathrm{Ext}^n_R(F, M) = 0$ *if* F *is a free module and* $n > 0$.

The functor $\mathrm{Ext}^n_R(-, M)$ *is called the* n^{th} derived functor *of the functor* $\mathrm{Hom}_R(-, M)$.

Before we embark on the proof of this theorem, we prove that these axioms characterize the functors Tor and Ext.

Theorem 2.3 (Uniqueness). *Any two functors satisfying T1), T2), and T3) are naturally isomorphic. Any two functors satisfying E1), E2), and E3) are naturally isomorphic.*

Proof. We will show that values of $\mathrm{Tor}^R_n(M_1, M_2)$ are determined by the axioms by induction on n. This is true for $n = 0$ by T1). Next note that for any module M_1, there is a surjection $F \xrightarrow{\phi} M_1 \to 0$ where F is a free module. For example, let $S \subset M_1$ be a set which generates M_1 as an R-module (e.g. $S = M_1$), and let $F = F(S)$ be the free module with basis S. There is an obvious surjection ϕ. Let $K = \ker \phi$. Apply T2) to the short exact sequence

$$0 \to K \to F \to M_1 \to 0.$$

Then by T2) and T3), one has

$$\mathrm{Tor}^R_1(M_1, M_2) \cong \ker(K \otimes_R M_2 \to F \otimes_R M_2)$$

and

$$\mathrm{Tor}^R_n(M_1, M_2) \cong \mathrm{Tor}^R_{n-1}(K, M_2) \quad \text{for} \quad n > 1.$$

The values of Tor^R_{n-1} are known by induction. The proof for Ext is similar. □

The technique of the above proof is called dimension shifting, and it can be useful for computations. For example, if F is a free module and

$$0 \to K \to F' \to M \to 0$$

is a short exact sequence with F' free, then

$$\operatorname{Tor}_1^R(M, F) \cong \ker(K \otimes F \to F' \otimes F),$$

but this is zero by Exercise 20. Thus $\operatorname{Tor}_1^R(-, F)$ is identically zero. But $\operatorname{Tor}_n^R(M, F) \cong \operatorname{Tor}_{n-1}^R(K, F)$ for $n > 1$, so inductively we see $\operatorname{Tor}_n^R(-, F)$ is zero for $n > 0$. To compute $\operatorname{Ext}^1_{\mathbf{Z}}(\mathbf{Z}/2, \mathbf{Z})$, we apply E2) to the exact sequence $0 \to \mathbf{Z} \xrightarrow{\times 2} \mathbf{Z} \to \mathbf{Z}/2 \to 0$ to get the exact sequence

$$\begin{array}{ccccccc}
\operatorname{Hom}(\mathbf{Z}, \mathbf{Z}) & \xrightarrow{(\times 2)^*} & \operatorname{Hom}(\mathbf{Z}, \mathbf{Z}) & \longrightarrow & \operatorname{Ext}^1(\mathbf{Z}/2, \mathbf{Z}) & \longrightarrow & \operatorname{Ext}^1(\mathbf{Z}, \mathbf{Z}) \\
\downarrow \cong & & \downarrow \cong & & \downarrow \cong & & \downarrow \cong \\
\mathbf{Z} & \xrightarrow{\times 2} & \mathbf{Z} & \longrightarrow & \operatorname{Ext}^1(\mathbf{Z}/2, \mathbf{Z}) & \longrightarrow & 0
\end{array}$$

so $\operatorname{Ext}^1_{\mathbf{Z}}(\mathbf{Z}/2, \mathbf{Z}) \cong \mathbf{Z}/2$.

The following proposition gives some simple but useful computations. This result should be memorized. (The subscript or superscript R is often omitted when the choice of the ring R is clear from context.)

Proposition 2.4. *Let R be a commutative ring and $a \in R$ a non-zero-divisor (i.e. $ab = 0$ implies $b = 0$). Let M be an R-module. Let $M/a = M/aM$ and ${}_aM = \{m \in M | am = 0\}$. Then*

1. $R/a \otimes M \cong M/a$,
2. $\operatorname{Tor}_1(R/a, M) \cong {}_aM$,
3. $\operatorname{Hom}(R/a, M) \cong {}_aM$,
4. $\operatorname{Ext}^1(R/a, M) \cong M/a$.

Proof. Since a is not a divisor of zero, there is a short exact sequence

$$0 \to R \xrightarrow{\times a} R \to R/A \to 0.$$

Apply the functors $-\otimes M$ and $\operatorname{Hom}(-, M)$ to the above short exact sequence. By the axioms we have exact sequences

$$0 \to \operatorname{Tor}_1(R/a, M) \to R \otimes M \to R \otimes M \to R/a \otimes M \to 0$$

and

$$0 \to \operatorname{Hom}(R/a, M) \to \operatorname{Hom}(R, M) \to \operatorname{Hom}(R, M) \to \operatorname{Ext}^1(R/a, M) \to 0.$$

The middle maps in the exact sequence above can be identified with

$$M \xrightarrow{\times a} M,$$

which has kernel ${}_aM$ and cokernel M/a. □

In particular if n is a nonzero integer and $R = \mathbf{Z}$, the four functors $\mathrm{Tor}_1, \otimes, \mathrm{Hom}$, and Ext^1 applied to the pair $(\mathbf{Z}/n, \mathbf{Z}/n)$ are all isomorphic to $\mathbf{Z}/n$. If m and n are relatively prime integers, then applied to the pair $(\mathbf{Z}/m, \mathbf{Z}/n)$ they are all zero.

Proposition 2.5.

1. *If R is a field, then $\mathrm{Tor}_n^R(-,-)$ and $\mathrm{Ext}_R^n(-,-)$ are zero for $n > 0$.*
2. *If R is a PID, then $\mathrm{Tor}_n^R(-,-)$ and $\mathrm{Ext}_R^n(-,-)$ are zero for $n > 1$.*

Proof. 1. All modules over a field are free so this follows from axioms T3) and E3).

2. A submodule of a free module over a PID is free, so for any module M there is a short exact sequence

$$0 \to F_1 \to F_0 \to M \to 0$$

with F_1 and F_0 free. Then by T2), T3), E2), and E3), for $n > 1$, $\mathrm{Tor}_n^R(M,-)$ and $\mathrm{Ext}_n^R(M,-)$ sit in long exact sequences flanked by zero, and hence must vanish. □

The functors $\mathrm{Tor}_1^{\mathbf{Z}}$ and $\mathrm{Ext}_{\mathbf{Z}}^1$ are typically abbreviated Tor and Ext.

Exercise 21. Using the axioms, compute $\mathrm{Tor}(A,B)$ and $\mathrm{Ext}(A,B)$ for all finitely generated abelian groups.

A couple of natural questions must have occurred to you. What is the behavior of these functors with respect to exact sequences in the second variable? Is $\mathrm{Tor}_n(A,B) \cong \mathrm{Tor}_n(B,A)$? This seems likely since $A \otimes B \cong B \otimes A$. (Since $\mathrm{Hom}(A,B) \not\cong \mathrm{Hom}(B,A)$ the corresponding question for Ext could not have possibly occurred to you!) Your questions are answered by the following theorem.

Theorem 2.6 (Existence′).

1. *The functors*

   $$\mathrm{Tor}_n^R : R\text{-}MOD \times R\text{-}MOD \to R\text{-}MOD \quad \textit{for all} \;\; n = 0,1,2,\ldots$$

   satisfy the following axioms.

   T1′) $\mathrm{Tor}_0^R(M_1, M_2) = M_1 \otimes_R M_2$.

   T2′) *If $0 \to A \to B \to C \to 0$ is any short exact sequence of R-modules and M is any R-module, then there is a natural long exact sequence*

$$\cdots \to \mathrm{Tor}_n^R(M,A) \to \mathrm{Tor}_n^R(M,B) \to \mathrm{Tor}_n^R(M,C) \to \mathrm{Tor}_{n-1}^R(M,A) \to \cdots$$
$$\cdots \to \mathrm{Tor}_1^R(M,C) \to M \otimes_R A \to M \otimes_R B \to M \otimes_R C \to 0.$$

   T3′) $\mathrm{Tor}_n^R(M,F) = 0$ *if F is a free module and $n > 0$.*

2. *The functors*

$$\operatorname{Ext}_R^n : R\text{-}MOD \times R\text{-}MOD \to R\text{-}MOD \quad \text{for all } n = 0, 1, 2, \ldots$$

satisfy the following axioms:

E1′) $\operatorname{Ext}_R^0(M_1, M_2) = \operatorname{Hom}_R(M_1, M_2)$.

E2′) *If* $0 \to A \to B \to C \to 0$ *is any short exact sequence of R-modules and M is any R-module, then there is a natural long exact sequence*

$$0 \to \operatorname{Hom}_R(M, A) \to \operatorname{Hom}_R(M, B) \to \operatorname{Hom}_R(M, C) \to \operatorname{Ext}_R^1(M, A) \to \cdots$$
$$\cdots \to \operatorname{Ext}_R^q(M, B) \to \operatorname{Ext}_R^q(M, C) \to \operatorname{Ext}_R^{q+1}(M, A) \to \cdots$$

E3′) $\operatorname{Ext}_R^n(M, I) = 0$ *if I is an injective module (see Definition 2.12) and* $n > 0$.

We postpone the proof of Theorem 2.6 until the next section.

Corollary 2.7. *The functors* $\operatorname{Tor}_n^R(A, B)$ *and* $\operatorname{Tor}_n^R(B, A)$ *are naturally isomorphic.*

Proof. By Theorem 2.6, the functor $(A, B) \mapsto \operatorname{Tor}_n^R(B, A)$ satisfies the axioms T1), T2), and T3) and thus by the uniqueness theorem, Theorem 2.3, it must be naturally isomorphic to $(A, B) \mapsto \operatorname{Tor}_n^R(A, B)$. □

Tor and Ext are higher derived versions of $\otimes_R$ and Hom, so they have analogous properties. For example we offer without proof:

1. $\operatorname{Tor}_n^R(\oplus_\alpha A_\alpha, B) \cong \oplus_\alpha \operatorname{Tor}_n^R(A_\alpha, B)$,
2. $\operatorname{Ext}_R^n(\oplus_\alpha A_\alpha, B) \cong \prod_\alpha \operatorname{Ext}_R^n(A_\alpha, B)$, and
3. $\operatorname{Ext}_R^n(A, \prod_\alpha B_\alpha) \cong \prod_\alpha \operatorname{Ext}_R^n(A, B_\alpha)$.

The proofs of Theorems 2.2 and 2.6 are carried out using *projective modules* and *projective resolutions.* The functors Ext_R^n can also be defined using *injective resolutions.* We will carry out the details in the projective case over the next few sections and sketch the approach to Ext using injective resolutions.

Much of what we say can be done in the more general setting of *abelian categories*; these are categories where the concept of exact sequence makes sense (for example the category of sheaves or the category of representations of a Lie algebra) provided there are "enough projectives" or "enough injectives" in the category.

2.2. Projective and injective modules

Recall, if F is a free module over R; A, B are R-modules; and

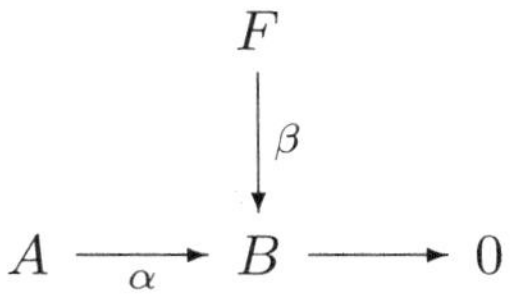

$$\begin{array}{ccccc} & & F & & \\ & & \downarrow{\scriptstyle\beta} & & \\ A & \xrightarrow[\alpha]{} & B & \longrightarrow & 0 \end{array}$$

is a diagram with α *onto*, then there exists a $\gamma : F \to A$ so that

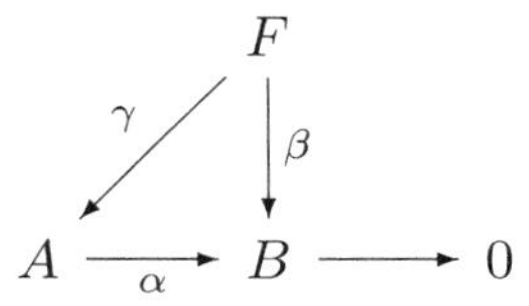

commutes. We say

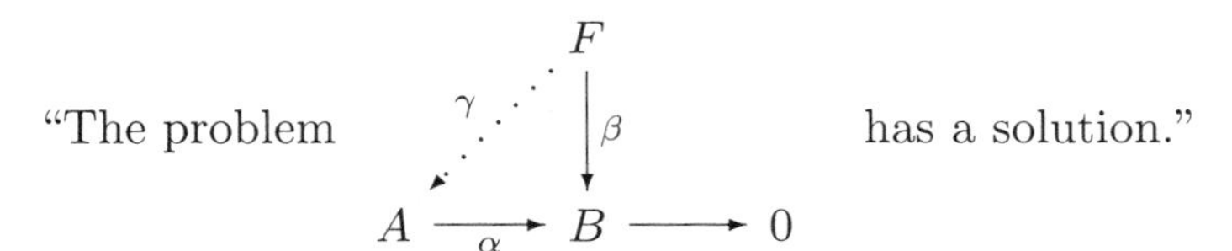

Remark. In general, whenever a commutative diagram is given with one dotted arrow, we will consider it as a problem whose solution is a map which can be substituted for the dashed arrow to give a commutative diagram. This makes sense in any category; we will use it mostly in the categories R-MOD and the category of topological spaces.

We make the following definition which encapsulates the basic property of free modules.

Definition 2.8. An R-module P is called *projective* if for any A, B, α, β with α onto, the problem

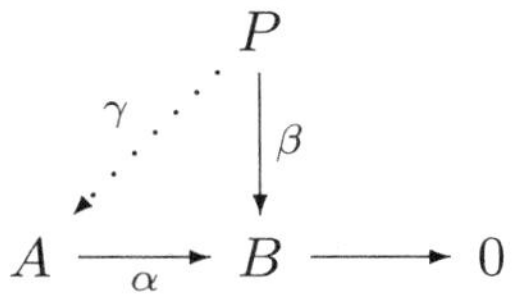

$$\begin{array}{ccccc} & & P & & \\ & \swarrow{\scriptstyle\gamma} & \downarrow{\scriptstyle\beta} & & \\ A & \xrightarrow[\alpha]{} & B & \longrightarrow & 0 \end{array}$$

has a solution γ.

Lemma 2.9. *An R-module P is projective if and only if there exists an R-module Q so that $P \oplus Q$ is a free R-module.*

Proof. If P is projective, choose F free and $F \to P$ an epimorphism. Let $Q = \ker(F \to P)$, so

$$0 \to Q \to F \to P \to 0$$

is exact. Since P is projective, the sequence splits, as one sees by considering the problem

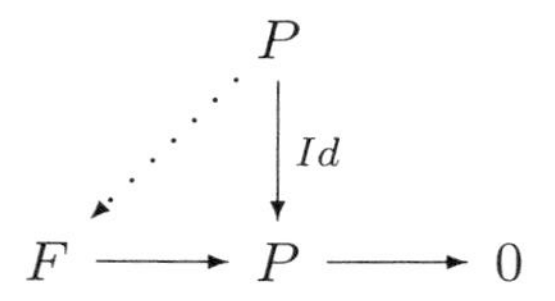

so $F = P \oplus Q$.

Conversely, if there exists an R-module Q so that $P \oplus Q$ is free, extend

$$\begin{array}{ccc} & P & \\ & \downarrow \beta & \\ A \longrightarrow & B & \longrightarrow 0 \end{array} \qquad \text{to} \qquad \begin{array}{ccc} & P \oplus Q & \\ & \downarrow \beta\oplus 0 & \\ A \longrightarrow & B & \longrightarrow 0 \end{array}$$

Since $P \oplus Q$ is free, there exists a solution $f : P \oplus Q \to A$ to

$$\begin{array}{ccccc} & & P \oplus Q & & \\ & \swarrow f & \downarrow \beta\oplus 0 & & \\ A & \longrightarrow & B & \longrightarrow & 0 \end{array}$$

But then let $\overline{f} = f \circ i$ where $i : P \to P \oplus Q$ is given by $p \mapsto (p, 0)$. Then $\overline{f}$ solves the problem

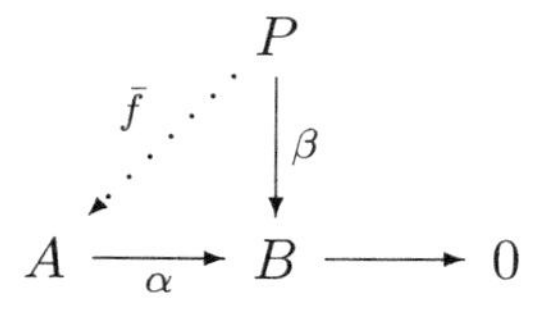

□

Thus projective modules generalize free modules by isolating one of the main properties of free modules. Furthermore the definition of a projective module is purely in terms of arrows in R-MOD, and hence is more elegant than the definition of a free module. On the other hand they are less familiar.

Exercise 22. Let P be a projective module.

1. Any short exact sequence $0 \to A \to B \to P \to 0$ is split.
2. If P is finitely generated, there is a finitely generated Q so that $P \oplus Q$ is free.

Proposition 2.10.

1. *Any module over a field is projective.*
2. *Any projective module over a PID is free.*

Proof. All modules over a field are free, hence projective. A projective module P is a submodule of the free module $P \oplus Q$, and for PIDs submodules of free modules are free. □

There are many examples of *non-free* projective modules over rings R. Note that R must be complicated, i.e. not a field nor a PID. For example, if $R = \mathbf{Z}/6$, then $P = \mathbf{Z}/2$ is a projective module. (To see this, use the Chinese remainder theorem $\mathbf{Z}/6 = \mathbf{Z}/2 \times \mathbf{Z}/3$.)

Here is a more interesting example, related to K-theory. Let R be the ring of continuous functions on the circle, $R = C^0(S^1, \mathbf{R})$. Let $E \to S^1$ be the twisted real line bundle over S^1 (so E = open Möbius band). Then as vector bundles $E \not\cong S^1 \times \mathbf{R}$, but $E \oplus E \cong S^1 \times \mathbf{R}^2$. So, if $M = C^0(E)$ (continuous sections of E), M is *not free* (why?), but $M \oplus M \cong C^0(S^1, \mathbf{R}) \oplus C^0(S^1, \mathbf{R}) = R \oplus R$. Thus M is projective.

Exercise 23. Show that the following are examples of projectives which are not free.

1. Let R be the ring of 2-by-2 matrices with real entries. Let $P = \mathbf{R}^2$ where the action of R on P is by multiplying a matrix by a vector. (Hint: Think of P as 2-by-2 matrices with the second column all zeroes.)
2. Let $R = \mathbf{R} \times \mathbf{R}$ (addition and multiplication are component-wise) and $P = \mathbf{R} \times \{0\}$.

One of the quantities measured by the functor K_0 of *algebraic K-theory* is the difference between projective and free modules over a ring. See Chapter 11 for another aspect of algebraic K-theory, namely the geometric meaning of the functor K_1.

As far as Tor and Ext are concerned, observe that

$$\operatorname{Tor}_n^R(A \oplus B, M) \cong \operatorname{Tor}_n^R(A, M) \oplus \operatorname{Tor}_n^R(B, M).$$

This is because $A \oplus B$ fits into the split exact sequence

$$0 \to A \to A \oplus B \to B \to 0.$$

Functoriality and Axiom T2) put $\operatorname{Tor}_n^R(A \oplus B, M)$ in a corresponding split exact sequence. Applying this to $P \oplus Q \cong$ free (and applying a similar argument to Ext), one obtains the following result.

Corollary 2.11. *For a projective module P, for $n > 0$, and for any module M, both $\operatorname{Tor}_n^R(P, M)$ and $\operatorname{Ext}_R^n(P, M)$ vanish.* □

Thus for purposes of computing Tor and Ext (e.g. dimension shifting), projective modules work just as well as free modules.

In the categorical framework in which we find ourselves, something interesting usually happens if one reverses all arrows. Reversing the arrows in the definition of projective modules leads to the definition of *injective* modules.

Definition 2.12. An R-module M is called *injective* if

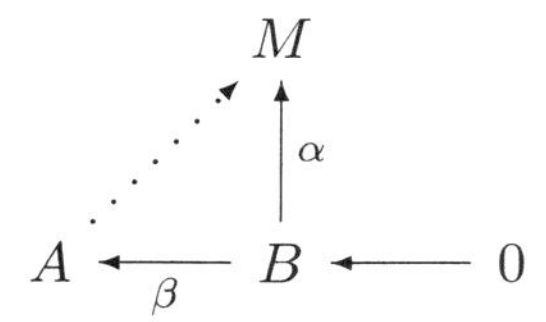

has a solution for all A, B, α, β (with β injective).

We will define Ext using projective modules instead of injective modules, so we omit most details about injective modules. See Rotman [**33**] or MacLane [**22**] for more. We list here a few results.

Theorem 2.13. *An abelian group A is injective if and only if A is divisible (i.e. the equation $nx = a$ has a solution $x \in A$ for each $n \in \mathbf{Z}$, $a \in A$).* □

Thus some examples of injective abelian groups are $\mathbf{Q}$ and $\mathbf{Q}/\mathbf{Z}$. (Note that a quotient of a divisible group is divisible, hence injective.)

Theorem 2.14.

1. *Given any R-module M, there exists a projective R-module P and an epimorphism $P \to M \to 0$.*
2. *Given any R-module M, there exists an injective R-module I and a monomorphism $0 \to M \to I$.*

Proof. We have already proved (1), by taking P to be a free module on a set of generators of M. The proof of (2) is more involved. One proves it first for an abelian group. Here is one way. Express $M = (\oplus \mathbf{Z})/K$. This injects to $D = (\oplus \mathbf{Q})/K$ which is divisible and hence injective.

Now suppose M is an R-module. Then, considered as an *abelian group*, there is an injection $\varphi : M \to D$ where D is divisible. One can show the map $M \to \mathrm{Hom}_{\mathbf{Z}}(R, D)$, $m \mapsto (r \mapsto \varphi(rm))$ is an injective R-module map and that $\mathrm{Hom}_{\mathbf{Z}}(R, D)$ is an injective R-module when D is divisible. □

2.3. Resolutions

We begin with the definition of projective and injective resolutions of an R-module.

Definition 2.15.

1. A *projective resolution* of an R-module M is a sequence (possibly infinitely long)

$$\cdots \to P_n \to P_{n-1} \to \cdots \to P_0 \to M \to 0$$

   where
   (a) the sequence is exact, and
   (b) each P_i is a projective R-module.

2. An *injective resolution* of M is a sequence

$$0 \to M \to I_0 \to I_1 \to I_2 \to \cdots \to I_n \to \cdots$$

   where
   (a) the sequence is exact, and
   (b) each I_n is an injective R-module.

Definition 2.16. Given a projective resolution, define the *deleted resolution* to be

$$\cdots \to P_n \to P_{n-1} \to \cdots \to P_0 \to 0.$$

We will use the notation $\mathbf{P}_*$ or $\mathbf{P}_M$. Note that $H_q(\mathbf{P}_M)$ is zero for $q \neq 0$ and is isomorphic to M for $q = 0$.

Theorem 2.17. *Every R-module M has (many) projective and injective resolutions.*

Proof. Choose a surjection $P_0 \to M$ with P_0 projective. Assume by induction that you have an exact sequence

$$P_n \xrightarrow{d_n} P_{n-1} \to \cdots \to P_0 \to M \to 0.$$

Let $K_n = \ker d_n$. Using the previous theorem, choose a projective module P_{n+1} which surjects to K_n. Then splice

$$P_{n+1} \to K_n \to 0 \qquad \text{to} \qquad 0 \to K_n \to P_n \to \cdots \to P_0 \to M$$

to get

$$P_{n+1} \to P_n \to \cdots \to P_0 \to M.$$

The proof for injective resolutions is obtained by rewriting the proof for projective resolutions but turning the arrows around.

To see that projective resolutions are not unique, notice that if

$$\to P_n \xrightarrow{d_n} P_{n-1} \to \cdots \to P_0 \to M \to 0$$

is a projective resolution and Q is projective, then

$$\to P_{n+1} \to P_n \oplus Q \xrightarrow{d_n \oplus Id} P_{n-1} \oplus Q \to P_{n-2} \to \cdots \to P_0 \to M$$

is also a projective resolution. □

If at any stage in the above construction the kernel K_n is projective, then one may stop there since

$$\cdots \to 0 \to 0 \to K_n \to P_{n-1} \to \cdots \to P_0 \to M \to 0$$

is a projective resolution. We omit typing the 0's.

We also record the following lemma which we used in constructing resolutions.

Lemma 2.18 (splicing lemma). *If the sequences* $A \to B \xrightarrow{\alpha} C \to 0$ *and* $0 \to C \xrightarrow{\beta} D \to E$ *are exact, then* $A \to B \xrightarrow{\beta \circ \alpha} D \to E$ *is exact.* □

Exercise 24. Prove the splicing lemma.

Theorem 2.19.

1. *If R is a field and M is any R-module, then*

$$0 \to M \xrightarrow{\text{Id}} M \to 0$$

is a projective resolution. In other words, every module over a field has a length 0 projective resolution. (It stops at P_0.)

2. *Every module over a PID has a length 1 projective resolution*

$$0 \to P_1 \to P_0 \to M \to 0.$$

3. *Every abelian group ($R = \mathbf{Z}$) has a length 1 injective resolution*

$$0 \to M \to I_0 \to I_1 \to 0.$$

Proof. 1. This is clear.

2. Every submodule of a free module over a PID is free. Thus if P_0 is a free module surjecting to M, and P_1 is its kernel,

$$0 \to P_1 \to P_0 \to M \to 0$$

is a projective (in fact free) resolution of M.

3. If $0 \to M \to D_0$ is an injection with D_0 divisible, then D_0/M is divisible, since the quotient of any divisible group is divisible. Thus $0 \to M \to D_0 \to D_0/M \to 0$ is an injective resolution. □

Comment about Commutative Algebra. A *Dedekind Domain* is a commutative domain (no zero divisors) in which every module has a projective resolution of length 1. Equivalently submodules of projective modules are projective. A PID is a Dedekind domain. From the point of view of category theory, they are perhaps more natural than PIDs. If $\zeta_n = e^{2\pi i/n}$ is a primitive n-th root of unity, then $\mathbf{Z}[\zeta_n]$ is a Dedekind domain. Projective modules (in fact ideals) which are not free first arise at $n = 23$. Non-free ideals are what make Fermat's Last Theorem so hard to prove.

A commutative Noetherian ring R has *height* equal to n ($\text{ht}(R) = n$) if the longest chain of nontrivial prime ideals in R has length n:

$$0 \subset P_1 \subset \cdots \subset P_n \subset R.$$

The *homological dimension of* R, $\text{hdim}(R)$, is the least upper bound on the length of projective resolutions for all finitely generated modules over R. The homological dimension of a field is 0 and a Dedekind domain is 1. If a ring has homological dimension n, then any module M has a projective resolution with $P_k = 0$ for $k > n$. The numbers $\text{ht}(R)$ and $\text{hdim}(R)$ are related. For a large class of rings (regular rings) they are *equal.*

2.4. Definition of Tor and Ext - existence

In this section we will complete the proof of Theorem 2.2.

Let M, N be R-modules. Let $\cdots \to P_n \to \cdots \xrightarrow{d_2} P_1 \xrightarrow{d_1} P_0 \xrightarrow{\epsilon} M \to 0$ be a projective resolution of M. Applying $- \otimes_R N$ to a deleted resolution $\mathbf{P}_M$, one obtains the sequence $\mathbf{P}_M \otimes_R N$

$$\mathbf{P}_M \otimes_R N = \{\cdots \to P_n \otimes N \to \cdots \xrightarrow{d_2 \otimes \text{Id}} P_1 \otimes N \xrightarrow{d_1 \otimes \text{Id}} P_0 \otimes N \to 0\}.$$

Note that $\mathbf{P}_M \otimes N$ is a chain complex (since $(d_{n-1} \otimes \text{Id}) \circ (d_n \otimes \text{Id}) = d_{n-1} \circ d_n \otimes \text{Id} = 0$), and by right exactness of $- \otimes N$, the 0th homology is $M \otimes N$. However, since $- \otimes N$ need not be an exact functor in general, $\mathbf{P}_M \otimes N$ might not be exact.

Similarly, by applying the functor $\text{Hom}_R(-, N)$ to a deleted projective resolution $\mathbf{P}_M$, one obtains the cochain complex

$$\text{Hom}(\mathbf{P}_M, N) = \{0 \to \text{Hom}_R(P_0, N) \xrightarrow{\alpha_1^*} \text{Hom}_R(P_1, N) \xrightarrow{\alpha_2^*} \text{Hom}_R(P_2, N) \to \cdots\}.$$

Exercise 25. Show that $\text{Hom}_R(\mathbf{P}_M, N)$ forms a cochain complex.

We will eventually define $\text{Tor}_n^R(M, N)$ as $H_n(\mathbf{P}_M \otimes_R N)$ and we will define $\text{Ext}_R^n(M, N)$ as $H^n(\text{Hom}(\mathbf{P}_M, N))$. (We could have also defined Tor and Ext as $H_*(M \otimes_R \mathbf{P}_N)$ and $H^*(\text{Hom}(M, \mathbf{I}_N))$.) For now we record some obvious facts. What is not obvious is that this will lead to well defined functors, i.e. independent of the choice of resolutions.

Theorem 2.20. *Let M and N be R-modules and let $\mathbf{P}_M$ be a deleted projective resolution of M.*

1. *For $n = 0, 1, 2, \ldots$ the assignment $(M, N) \mapsto H_n(\mathbf{P}_M \otimes_R N)$ is a covariant functor in N and satisfies Axioms T1′), T2′), and T3′) of Theorem 2.6. Furthermore, if M is free (or just projective), one can choose the resolution so that axiom T3) of Theorem 2.2 is satisfied.*
2. *For $n = 0, 1, 2, \ldots$ the assignment $(M, N) \mapsto H^n(\mathrm{Hom}_R(\mathbf{P}_M, N))$ is a covariant functor in N and satisfies Axioms E1′), E2′), and E3′) of Theorem 2.6. Furthermore, if M is free (or just projective), one can choose the resolution so that axiom E3) of Theorem 2.2 is satisfied.*

Exercise 26. Prove this theorem.

Exercise 27. An R-module F is called *flat* if $- \otimes_R F$ is exact. A free module is flat, and clearly a summand of a flat module is flat, so projectives are flat. There are modules which are flat, but not projective; show that $\mathbf{Q}$ is a flat but not projective $\mathbf{Z}$-module. In fact over a PID a module is flat if and only if it is torsion free.

Tor can be computed using a flat resolution rather than a projective one. Assume this and compute $\mathrm{Tor}(\mathbf{Q}/\mathbf{Z}, A) = H_1(P_{\mathbf{Q}/\mathbf{Z}} \otimes A)$ for any abelian group A.

2.5. The fundamental lemma of homological algebra

Taking inventory, we still need to show that our candidates for Tor and Ext are well defined, functorial in the first variable, and that short exact sequences in the first variable give long exact sequences in Tor and Ext. The well-definition and functoriality will follow from the fundamental lemma of homological algebra; the long exact sequences will follow from the horseshoe lemma.

Definition 2.21. A *projective chain complex*

$$P_* = \{\cdots \to P_2 \to P_1 \to P_0\}$$

is a chain complex where all the modules P_i are projective. An *acyclic chain complex*

$$C_* = \{\cdots \to C_2 \to C_1 \to C_0\}$$

is a chain complex where $H_i(C_*) = 0$ for all $i > 0$ (i.e. C_* is an exact sequence).

Theorem 2.22 (Fundamental lemma of homological algebra). *Let P_* be a projective chain complex and C_* be an acyclic chain complex over a ring R. Then given a homomorphism $\varphi : H_0(P_*) \to H_0(C_*)$, there is a chain map*

$f_* : P_* \to C_*$ inducing φ on H_0. Furthermore, any two such chain maps are chain homotopic.

We derive a few corollaries before turning to the proof.

Corollary 2.23. *Any two deleted projective resolutions of M are chain homotopy equivalent.*

Proof. Let $\mathbf{P}_M$ and $\mathbf{P}'_M$ be deleted projective resolutions of M. They are both projective and acyclic and they have $H_0 = M$. The existence part of the fundamental lemma gives chain maps $f_* : \mathbf{P}_M \to \mathbf{P}'_M$ and $g_* : \mathbf{P}'_M \to \mathbf{P}_M$ inducing the identity on H_0. The uniqueness part of the fundamental lemma gives a chain homotopy equivalence between $g_* \circ f_*$ and Id since they are both chain maps $\mathbf{P}_M \to \mathbf{P}_M$ inducing the identity map on H_0. Likewise $f_* \circ g_*$ is chain homotopy to Id. □

Corollary 2.24. *The assignments $(M, N) \mapsto H_n(\mathbf{P}_M \otimes_R N)$ and $(M, N) \mapsto H_n(\operatorname{Hom}_R(\mathbf{P}_M, N))$ do not depend on the choice of projective resolution and are functorial in M.*

Proof. Corollary 2.23 gives chain homotopy equivalences $\mathbf{P}_M \to \mathbf{P}'_M$ and hence a chain homotopy equivalence $\mathbf{P}_M \otimes_R N \to \mathbf{P}'_M \otimes_R N$ and similarly for Hom. This gives the independence of the resolution. Given a map $M \to M'$, the fundamental lemma gives a map $\mathbf{P}_M \to \mathbf{P}'_{M'}$, unique up to chain homotopy. This gives functoriality. □

Corollary 2.25. *The assignments $(M, N) \mapsto H_n(\mathbf{P}_M \otimes_R N)$ and $(M, N) \mapsto H^n(\operatorname{Hom}_R(\mathbf{P}_M, N))$ are functorial in both variables and satisfy the ' axioms of Theorem 2.6.* □

Proof of the fundamental lemma, Theorem 2.22. Let $M = H_0(P_*)$ and $M' = H_0(C_*)$. Then since $H_0(P_*) = P_0/\operatorname{im}(P_1 \to P_0)$, there is a surjection $P_0 \to M$ and likewise a surjection $C_0 \to M'$. So we wish to solve the following problem (i.e. fill in the dotted arrows so that the diagram commutes).

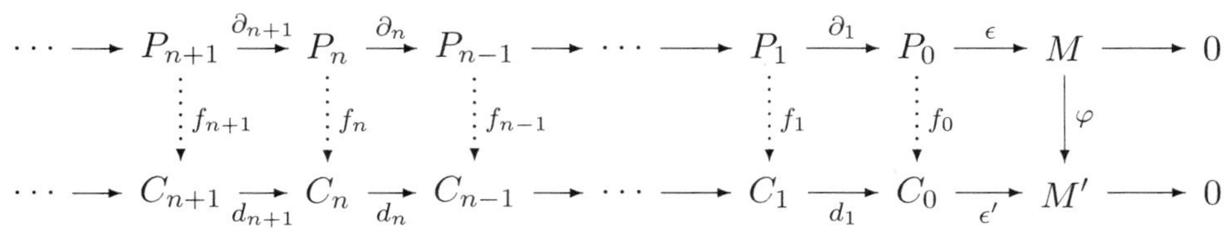

$$\begin{array}{ccccccccccccccc}
\cdots \longrightarrow & P_{n+1} & \xrightarrow{\partial_{n+1}} & P_n & \xrightarrow{\partial_n} & P_{n-1} & \longrightarrow \cdots \longrightarrow & P_1 & \xrightarrow{\partial_1} & P_0 & \xrightarrow{\epsilon} & M & \longrightarrow & 0 \\
& \vdots{\scriptstyle f_{n+1}} & & \vdots{\scriptstyle f_n} & & \vdots{\scriptstyle f_{n-1}} & & \vdots{\scriptstyle f_1} & & \vdots{\scriptstyle f_0} & & \downarrow{\scriptstyle \varphi} & & \\
\cdots \longrightarrow & C_{n+1} & \xrightarrow[d_{n+1}]{} & C_n & \xrightarrow[d_n]{} & C_{n-1} & \longrightarrow \cdots \longrightarrow & C_1 & \xrightarrow[d_1]{} & C_0 & \xrightarrow[\epsilon']{} & M' & \longrightarrow & 0
\end{array}$$

Here the P_i are projective, ϵ is onto and the bottom sequence is exact. We construct f_i by induction.

Step 0. The map f_0 exists since P_0 is projective:

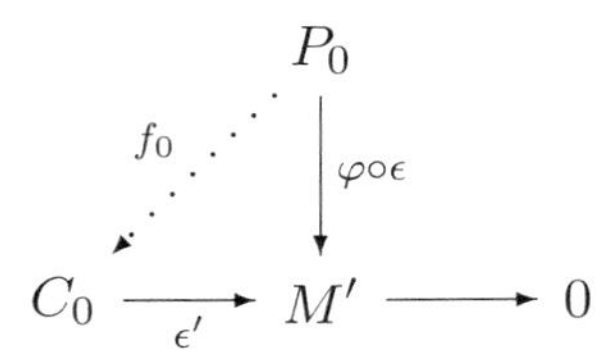

Step n. Suppose we have constructed $f_0, f_1, \cdots, f_{n-1}$. The problem

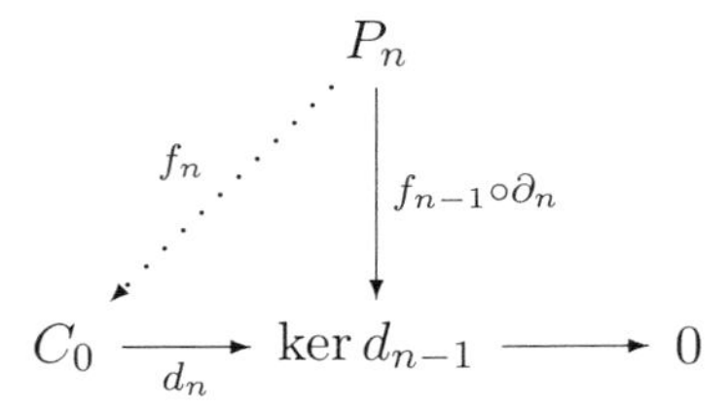

makes sense since $d_{n-1} \circ f_{n-1} \circ \partial_n = f_{n-2} \circ \partial_{n-1} \circ \partial_n = 0$. Furthermore $\ker d_{n-1} = \operatorname{im} d_n$ since C_* is acyclic, so the bottom map is onto. Then f_n exists since P_n is projective.

This completes the existence part of the fundamental lemma; we switch now to uniqueness up to chain homotopy. Suppose f_*, g_* are two choices of chain maps which induce φ on H_0.

$$\begin{array}{ccccccccccccc}
\longrightarrow & P_n & \xrightarrow{\partial_n} & P_{n-1} & \longrightarrow & \cdots & \longrightarrow & P_1 & \xrightarrow{\partial_1} & P_0 & \xrightarrow{\epsilon} & M & \longrightarrow 0 \\
 & \downarrow{\scriptstyle f_n-g_n} & \swarrow{\scriptstyle s_{n-1}} & \downarrow{\scriptstyle f_{n-1}-g_{n-1}} & & & & \downarrow{\scriptstyle f_1-g_1} & \swarrow{\scriptstyle s_0} & \downarrow{\scriptstyle f_0-g_0} & & \downarrow{\scriptstyle \varphi} & \\
\longrightarrow & C_n & \xrightarrow[d_n]{} & C_{n-1} & \longrightarrow & \cdots & \longrightarrow & C_1 & \xrightarrow[d_1]{} & C_0 & \xrightarrow[\epsilon']{} & M' & \longrightarrow 0
\end{array}$$

Here we want to define maps $s_n : P_n \to C_{n+1}$, but contrary to our usual convention, we don't want the diagram to commute, but instead we want s to be a chain homotopy, i.e. $f_0 - g_0 = d_1 \circ s_0$ and $f_n - g_n = d_{n+1} \circ s_n + s_{n-1} \circ \partial_n$ for $n > 0$.

We will construct a *chain homotopy* by induction on n.

Step 0. Since $\epsilon' \circ (f_0 - g_0) = (\varphi - \varphi) \circ \epsilon = 0 : P_0 \to C_0$, $\operatorname{Im}(f_0 - g_0) \subset \ker \epsilon' : C_0 \to M' = \operatorname{im} d_1 : C_1 \to C_0$.

Then s_0 exists since P_0 is projective

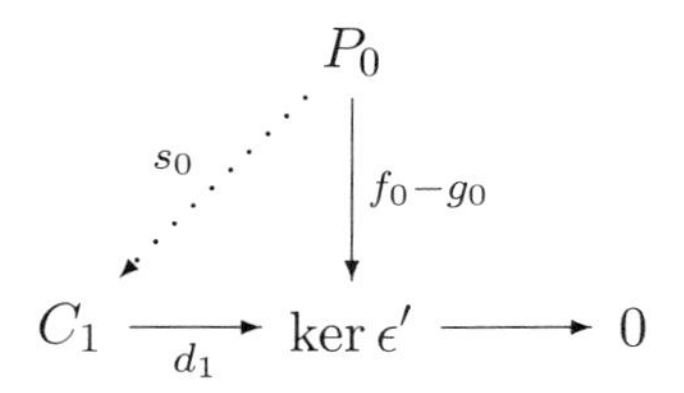

Step n. Suppose we have defined

$$s_q : P_q \to C_{q+1} \quad \text{for} \quad q = 0, \cdots, n-1$$

satisfying $f_q - g_q = d_{q+1}s_q + s_{q-1}\partial_q$ for each $q = 0, \cdots, n-1 \quad (s_{-1} = 0)$. Then the problem

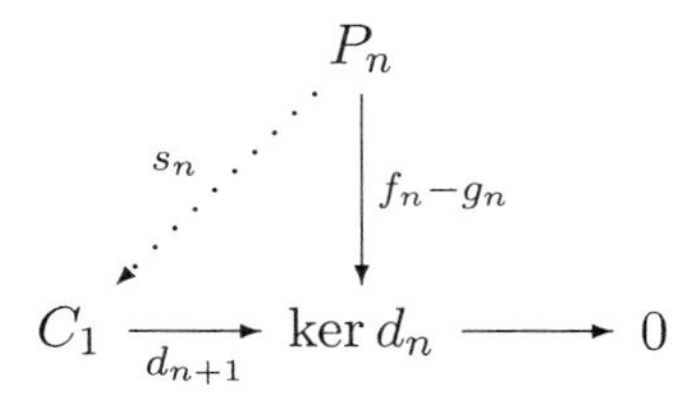

makes sense, since

$$\begin{aligned} d_n(f_n - g_n - s_{n-1}\partial_n) &= (f_{n-1} - g_{n-1})\partial_n - d_n s_{n-1}\partial_n \\ &= (d_n s_{n-1} + s_{n-2}\partial_{n-1})\partial_n - d_n s_{n-1}\partial_n \\ &= s_{n-2}\partial_{n-1}\partial_n = 0. \end{aligned}$$

Therefore $\operatorname{im}(f_n - g_n - s_{n-1}\partial_n) \subset \ker d_n = \operatorname{im}(d_{n+1} : C_{n+1} \to C_n)$. Thus $d_{n+1}s_n = f_n - g_n - s_{n-1}\partial_n$, proving the induction step.

This finishes the proof of the fundamental lemma. □

To show that our functors satisfy the remaining axioms, we need the following lemma.

Lemma 2.26 (Horseshoe lemma). *Let $0 \to A \to B \to C \to 0$ be a short exact sequence of R-modules. Let $\mathbf{P}_A$ and $\mathbf{P}_C$ be deleted projective resolutions of A and C. Then there exists a deleted projective resolution $\mathbf{P}_B$ of B, fitting into a short exact sequence of chain complexes $0 \to \mathbf{P}_A \to \mathbf{P}_B \to \mathbf{P}_C \to 0$ which induces the original sequence on H_0.*

Proof. We are given the following "horseshoe" diagram

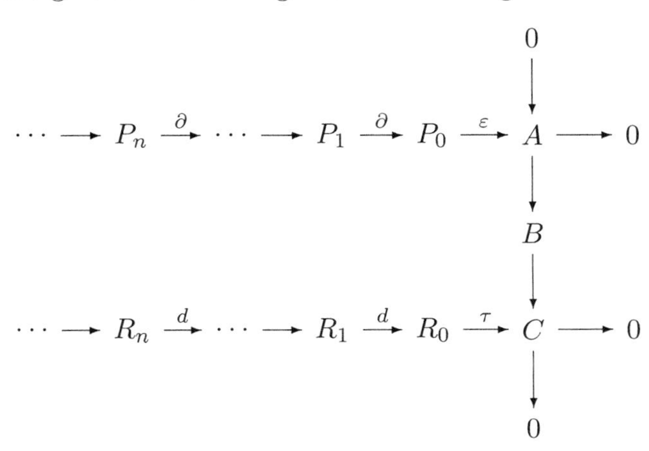

where the horizontal rows are projective resolutions. We want to add a middle row of projective modules to obtain a commutative diagram with exact rows and short exact columns. Since Q_n is projective, the columns will split, and so $Q_n = P_n \oplus R_n$ must go in the n-th slot in the middle. Furthermore we may assume that the maps $P_n \to Q_n$ and $Q_n \to R_n$ are the inclusion and projection maps, but the horizontal maps are yet unclear.

Step 0. The problem

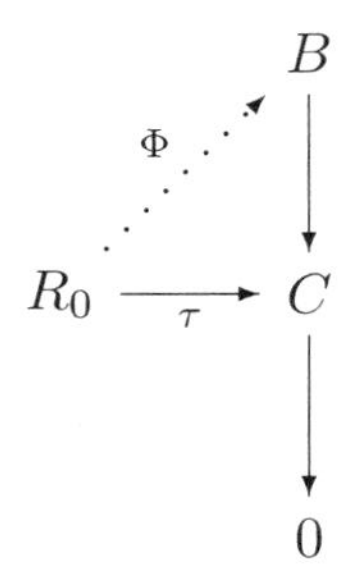

has a solution Φ since R_0 is projective. Let $\gamma : Q_0 \to B$ be $\gamma(p, r) = i\varepsilon(p) - \Phi(r)$ where $(p, r) \in Q_0 = P_0 \oplus R_0$ and $i : A \to B$. A diagram chase shows γ is *onto*. Thus we have the commutative diagram

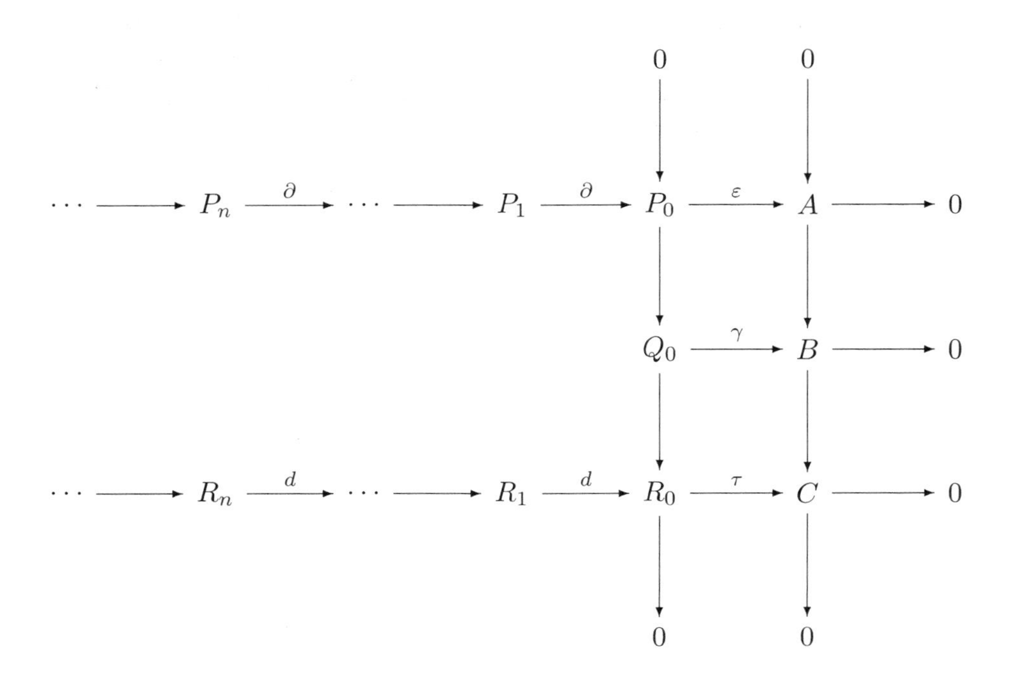

Step $n+1$. Suppose inductively we have constructed the following commutative diagram with exact rows and columns.

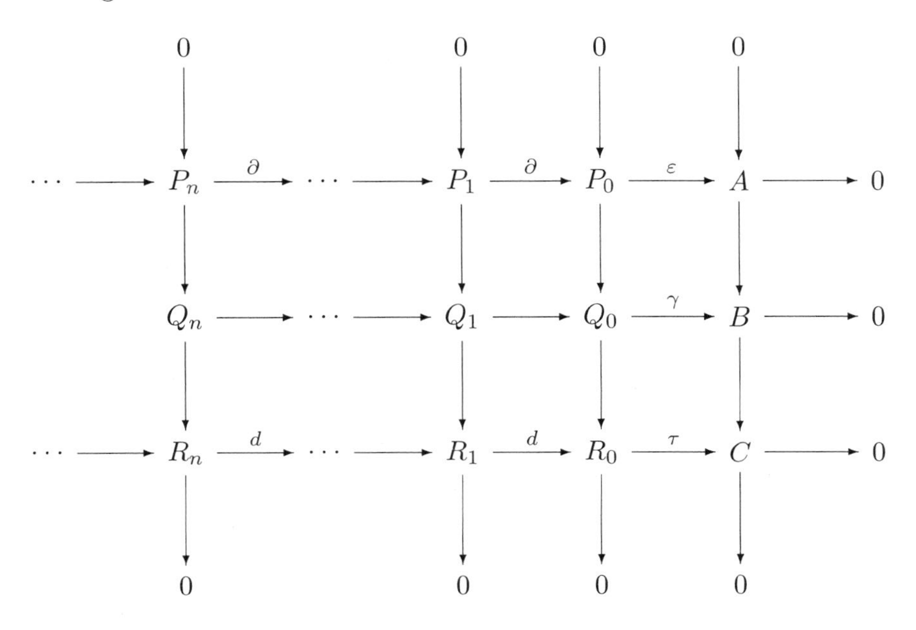

Let $K_n = \ker(P_n \to P_{n-1})$, $L_n = \ker(Q_n \to Q_{n-1})$ and $M_n = \ker(R_n \to R_{n-1})$. We then have the diagram

$$\begin{array}{ccccccccc}
 & & & & & & 0 & & \\
 & & & & & & \downarrow & & \\
\cdots & \longrightarrow & P_{n+2} & \xrightarrow{\partial} & P_{n+1} & \xrightarrow{\partial} & K_n & \longrightarrow & 0 \\
 & & & & & & \downarrow & & \\
 & & & & & & L_n & & \\
 & & & & & & \downarrow & & \\
\cdots & \longrightarrow & R_{n+2} & \xrightarrow{d} & R_{n+1} & \xrightarrow{d} & M_n & \longrightarrow & 0 \\
 & & & & & & \downarrow & & \\
 & & & & & & 0 & &
\end{array}$$

By Step 0 we can fill in the next column and horizontal arrow. We then splice this diagram with the previous one to obtain the inductive step. $\square$

It is important to notice that the short exact sequence $0 \to \mathbf{P}_A \to \mathbf{P}_B \to \mathbf{P}_C \to 0$ is not (necessarily) a split short exact sequence of chain complexes, even though each chain module is projective. (What might a projective object in the category of chain complexes be?)

Corollary 2.27. *Let $0 \to A \to B \to C \to 0$ be a short exact sequence of R-modules and let $0 \to \mathbf{P}_A \to \mathbf{P}_B \to \mathbf{P}_C \to 0$ be a short exact sequence of deleted projective resolutions provided by the horseshoe lemma. Let N be an R-module. Then there are long exact sequences:*

$$\cdots \to H_{n+1}(\mathbf{P}_C \otimes_R N) \to H_n(\mathbf{P}_A \otimes_R N) \to H_n(\mathbf{P}_B \otimes_R N) \to H_n(\mathbf{P}_C \otimes_R N) \to \cdots$$

and

$$\begin{aligned} \cdots \to H^n(\mathrm{Hom}_R(\mathbf{P}_C, N)) \to H^n(\mathrm{Hom}_R(\mathbf{P}_B, N)) \to \\ H^n(\mathrm{Hom}_R(\mathbf{P}_A, N)) \to H^{n+1}(\mathrm{Hom}_R(\mathbf{P}_C, N)) \to \cdots \end{aligned}$$

Proof. Since we have a short exact sequence of deleted projective resolutions, in degree n, the short exact sequence

$$0 \to (\mathbf{P}_A)_n \to (\mathbf{P}_B)_n \to (\mathbf{P}_C)_n \to 0$$

is split; hence

$$0 \to (\mathbf{P}_A \otimes_R N)_n \to (\mathbf{P}_B \otimes_R N)_n \to (\mathbf{P}_C \otimes_R N)_n \to 0$$

is split and hence exact. Thus

$$0 \to \mathbf{P}_A \otimes_R N \to \mathbf{P}_B \otimes_R N \to \mathbf{P}_C \otimes_R N \to 0$$

is a short exact sequence of chain complexes; the zig-zag lemma gives the long exact sequence in homology above. We leave the cohomology proof as an exercise. □

We can finally safely make the following definition.

Definition 2.28.

1. $\mathrm{Tor}_n^R(M, N) = H_n(\mathbf{P}_M \otimes_R N)$.
2. $\mathrm{Ext}_R^n(M, N) = H^n(\mathrm{Hom}_R(\mathbf{P}_M, N))$.

With these definitions, the existence theorem, Theorem 2.2, and the primed version, Theorem 2.6, follow from Corollaries 2.25 and 2.27.

We have not proven that $\mathrm{Ext}_R^n(M, N) = H^n(\mathrm{Hom}_R(M, \mathbf{I}_N))$. This follows by using injective versions of the fundamental lemma and the horseshoe lemma to show that the axioms are also satisfied here. For these facts, see any book on homological algebra, or, better, prove it yourself. Once we have defined tensor products and Hom for chain complexes, one can show $\mathrm{Tor}_n^R(M, N) = H_n(\mathbf{P}_M \otimes_R \mathbf{P}_N)$ and $\mathrm{Ext}_R^n(M, N) = \mathrm{Hom}_R(\mathbf{P}_M, \mathbf{I}_N))$; this is an intermediate way between resolving on the left and the right.

Earlier in this chapter you were asked in an exercise to compute $\mathrm{Tor}(A,B)$ and $\mathrm{Ext}(A,B)$ for finitely generated abelian groups. Lest you learn all the theory without any examples, we give a way of stating the result. Let $\mathrm{torsion}(A)$ denote the subgroup of A consisting of elements of finite order. Then

$$\mathrm{Tor}(A,B) \cong \mathrm{torsion}(A) \otimes_{\mathbf{Z}} \mathrm{torsion}(B)$$

and

$$\mathrm{Ext}(A,B) \cong \mathrm{torsion}(A) \otimes_{\mathbf{Z}} B.$$

Exercise 28. For any commutative ring R show that

$$\mathrm{Ext}^q_R(A \oplus B, M) \cong \mathrm{Ext}^q_R(A,M) \oplus \mathrm{Ext}^q_R(B,M)$$

and

$$\mathrm{Tor}^R_q(A \oplus B, M) \cong \mathrm{Tor}^R_q(A,M) \oplus \mathrm{Tor}^R_q(B,M).$$

We quote the famous exercise from Lang's *Algebra*, Chapter IV ([**21**]):

> Take any book on homological algebra, and prove all the theorems without looking at the proofs given in that book.
>
> Homological algebra was invented by Eilenberg-MacLane. General category theory (i.e. the theory of arrow-theoretic results) is generally known as *abstract nonsense* (the terminology is due to Steenrod).

2.6. Universal coefficient theorems

Let (C_*, ∂) be a chain complex over a ring R. Then there is an evaluation map

$$\begin{aligned} \mathrm{Hom}_R(C_q, M) \times C_q &\to M \\ (f,z) &\mapsto f(z). \end{aligned}$$

You have already come across this pairing in Exercise 11 and have shown that this pairing passes to the *Kronecker* pairing

$$\langle\ ,\ \rangle : H^q(C_*; M) \times H_q(C_*) \to M$$

of cohomology with homology. This pairing is bilinear, and its adjoint is a homomorphism

$$H^q(C_*, M) \to \mathrm{Hom}(H_q(C_*); M).$$

The example following Exercise 11 shows that this adjoint need not be an isomorphism. To understand the kernel and cokernel of this map is a subtle question. Universal coefficient theorems among other things provide a measure of how this adjoint fails to be an isomorphism in terms of the derived functors Ext^q and Tor_q. The answer can be quite difficult for general commutative rings and arbitrary chain complexes.

We will answer the question completely when R is a PID and C_* is a free chain complex. In this case $H^q(C_*, M) \to \operatorname{Hom}(H_q(C_*); M)$ is surjective with kernel $\operatorname{Ext}(H_{q-1}(C_*), M)$. This will cover the topological situation in the most important cases of coefficients in the integers or in a field, since the singular and cellular complexes of a space are free.

Theorem 2.29 (universal coefficient theorem for cohomology). *Let R be a principal ideal domain. Suppose that M is a module over R, and (C_*, ∂) is a free chain complex over R (i.e. each C_q is a free R-module).*

Then the sequence

$$0 \to \operatorname{Ext}_R(H_{q-1}(C_*), M) \to H^q(C_*; M) \to \operatorname{Hom}(H_q(C_*), M) \to 0$$

is exact and natural with respect to chain maps of free chain complexes. Moreover, the sequence splits, but not naturally.

We will give a proof of this based on the concept of an exact triangle.

Definition 2.30. An *exact triangle of R-modules* is a diagram of R-modules

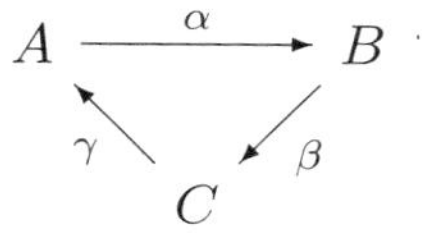

$$\begin{array}{ccc} A & \xrightarrow{\alpha} & B \\ {\scriptstyle\gamma}\nwarrow & & \swarrow{\scriptstyle\beta} \\ & C & \end{array}$$

satisfying $\ker(\beta) = \operatorname{im}(\alpha)$, $\ker(\gamma) = \operatorname{im}(\beta)$, and $\ker(\alpha) = \operatorname{im}(\gamma)$.

Similarly one defines an exact triangle of *graded* R-modules A_*, B_*, C_* (see Definition 3.1). In this case we require the homomorphisms α, β, and γ each to have a *degree*; so for example if α has degree 2, then $\alpha(A_q) \subset A_{q+2}$.

The basic example of an exact triangle of graded R-modules is the long exact sequence in homology

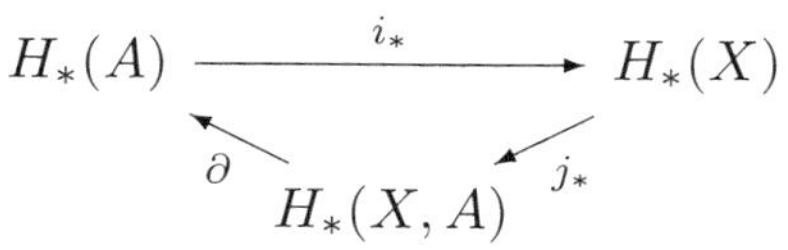

$$\begin{array}{ccc} H_*(A) & \xrightarrow{i_*} & H_*(X) \\ {\scriptstyle\partial}\nwarrow & & \swarrow{\scriptstyle j_*} \\ & H_*(X, A) & \end{array}$$

For this exact triangle i_* and j_* have degree 0, and ∂ has degree -1.

Exercise 29. Suppose that

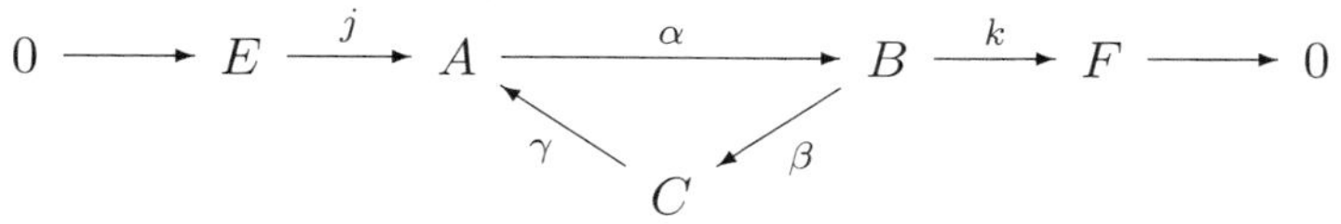

$$\begin{array}{ccccccccc} 0 & \longrightarrow & E & \xrightarrow{j} & A & \xrightarrow{\alpha} & B & \xrightarrow{k} & F & \longrightarrow & 0 \\ & & & & {\scriptstyle\gamma}\nwarrow & & \swarrow{\scriptstyle\beta} & & & & \\ & & & & & C & & & & & \end{array}$$

is a diagram with the top row exact and the triangle exact. Prove that there is a short exact sequence

$$0 \longrightarrow F \xrightarrow{\beta \circ k^{-1}} C \xrightarrow{j^{-1} \circ \gamma} E \longrightarrow 0.$$

State and prove the graded version of this exercise.

Proof of Theorem 2.29. There is a short exact sequence of graded, free R-modules

$$0 \to Z_* \xrightarrow{i} C_* \xrightarrow{\partial} B_* \to 0 \tag{2.1}$$

where Z_q denotes the q-cycles and B_q denotes the q-boundaries. The homomorphism i has degree 0 and ∂ has degree -1. This sequence is in fact a short exact sequence of chain complexes where Z_* and B_* are given the *zero* differential.

Since the sequence (2.1) is an exact sequence of free chain complexes, applying the functor $\mathrm{Hom}(-, M)$ gives another short exact sequence of chain complexes

$$0 \to \mathrm{Hom}(B_*, M) \xrightarrow{\partial^*} \mathrm{Hom}(C_*, M) \xrightarrow{i^*} \mathrm{Hom}(Z_*, M) \to 0.$$

Applying the zig-zag lemma we obtain a long exact sequence (i.e. exact triangle) in homology, which, since the differentials for the complexes $\mathrm{Hom}(B_*, M)$ and $\mathrm{Hom}(Z_*, M)$ are zero, reduces to the exact triangle

$$\begin{array}{ccc} \mathrm{Hom}(Z_*, M) & \xrightarrow{\quad\delta\quad} & \mathrm{Hom}(B_*, M) \\ & {\scriptstyle i^*}\nwarrow \quad \swarrow{\scriptstyle \partial^*} & \\ & H^*(C_*; M) & \end{array} \tag{2.2}$$

There is also a short exact sequence of graded R-modules

$$0 \to B_* \xrightarrow{j} Z_* \to H_* \to 0 \tag{2.3}$$

coming from the definition of homology, that is

$$Z_* = \ker \partial : C_* \to C_*,$$

$$B_* = \mathrm{im}\, \partial : C_* \to C_*,$$

and

$$H_* = H_*(C_*) = Z_*/B_*.$$

Notice that in the sequence (2.3), B_* and Z_* are free, since R is a PID and these are submodules of the free module C_*. Thus using axiom E2) of Theorem 2.2 and using the fact that $\mathrm{Ext}(Z_*, M) = 0$, we obtain an exact sequence

(2.4)

$$0 \to \mathrm{Hom}(H_*, M) \to \mathrm{Hom}(Z_*, M) \xrightarrow{j^*} \mathrm{Hom}(B_*, M) \to \mathrm{Ext}(H_*, M) \to 0.$$

Exercise 30. Complete the proof of the universal coefficient theorem as follows.

1. Show, taking special care with the grading, that the homomorphism δ of the exact triangle (2.2) coincides with the homomorphism j^* of (2.4). Thus there is a commutative diagram

$$\begin{array}{ccccccccc} 0 \to \operatorname{Hom}(H_*, M) \to & \operatorname{Hom}(Z_*, M) & \xrightarrow{j^*} & \operatorname{Hom}(B_*, M) & \to \operatorname{Ext}(H_*, M) \to 0 \\ & {\scriptstyle i^*}\nwarrow & & \swarrow{\scriptstyle \partial^*} & \\ & & H^*(C_*, M) & & \end{array}$$

obtained by putting together (2.2) and (2.4).

2. Apply Exercise 29 to obtain a short exact sequence of graded R-modules

$$0 \to \operatorname{Ext}(H_*, M) \to H^*(C_*; M) \to \operatorname{Hom}(H_*, M) \to 0.$$

Verify that the map $H^*(C_*; M) \to \operatorname{Hom}(H_*, M)$ is induced by evaluating a cochain on a cycle.

3. By taking the grading into account, finish the proof of Theorem 2.29.

Corollary 2.31. *If R is a field, M is a vector space over R, and C_* is a chain complex over R, then*

$$H^q(C_*; M) \cong \operatorname{Hom}(H_q(C_*), M).$$

Moreover the Kronecker pairing is nondegenerate. □

Applying the universal coefficient theorem to the singular or cellular complexes of a space or a pair of spaces, one obtains the following.

Corollary 2.32. *If (X, A) is a pair of spaces $A \subset X$, R a PID, M a module over R, then for each q the sequence*

$$0 \to \operatorname{Ext}_R(H_{q-1}(X, A; R), M) \to H^q(X, A; M) \to \operatorname{Hom}(H_q(X, A; R), M) \to 0$$

is short exact, natural, and splits (though the splitting is not natural). □

Exercise 31. Let $f : \mathbf{R}P^2 \to S^2$ be the map pinching the 1-skeleton to a point. Compute the induced map on $\mathbf{Z}$ and $\mathbf{Z}/2$ cohomology to show the splitting is not natural.

The most important special case of the universal coefficient theorem for cohomology is its use in the computation of $H^q(X)$ (cohomology with integer coefficients). For an abelian group A, we have denoted the torsion subgroup (i.e. the subgroup of finite order elements) by torsion(A). Let free$(A) = A/\text{torsion}(A)$. Then for space X whose homology is finitely generated in every dimension (e.g. a finite CW-complex), the universal coefficient theorem shows that

$$H^q(X) \cong \text{free}(H_q(X)) \oplus \text{torsion}(H_{q-1}(X)).$$

Another formulation is to define the dual of an abelian group A by $A^* = \operatorname{Hom}(A, \mathbf{Z})$ and the torsion dual $A^\frown \cong \operatorname{Hom}(A, \mathbf{Q}/\mathbf{Z})$. The universal coefficient theorem then says that

$$H^q(X) \cong H_q(X)^* \oplus (\text{torsion}(H_{q-1}(X)))\hat{}.$$

The right hand side is then a contravariant functor in X, but the isomorphism is still not natural.

There are other universal coefficient theorems for more complicated rings and other algebraic situations. Knowing these can speed up computations. A useful example is the following theorem. (See [**36**, pg. 246] for a proof.)

Theorem 2.33. *If R is a PID, M is a finitely generated R-module, and C_* is a free chain complex over R then there is a split short exact sequence*

$$0 \to H^q(C_*) \otimes M \to H^q(C_*; M) \to \mathrm{Tor}_1^R(H^{q+1}(C_*), M) \to 0.$$

□

Notice the extra hypothesis that M be finitely generated in this statement.

There are also universal coefficient theorems for homology, and we turn to these now. The following universal coefficient theorem measures the difference between first tensoring a complex with a module M and then passing to homology versus first passing to homology and then tensoring with M.

Theorem 2.34 (universal coefficient theorem for homology)**.** *Suppose that R is a PID, C_* a free chain complex over R, and M a module over R. Then there is a natural short exact sequence.*

$$0 \to H_q(C_*) \otimes M \to H_q(C_* \otimes M) \to \mathrm{Tor}_1^R(H_{q-1}(C_*), M) \to 0$$

which splits, but not naturally.

Sketch of Proof. The proof is similar to the proof given above of Theorem 2.29. As before, there is a short exact sequence of chain complexes

$$0 \to Z_* \to C_* \to B_* \to 0$$

which remains exact when tensoring with M, since B_* is free.

Applying the zig-zag lemma to the tensored sequence, one obtains the exact triangle

$$\begin{array}{ccc} B_* \otimes M & \longrightarrow & Z_* \otimes M \\ & \nwarrow \quad \swarrow & \\ & H_*(C_*; M) & \end{array} \tag{2.5}$$

The short exact sequence of graded R-modules

$$0 \to B_* \to Z_* \to H_*(C_*) \to 0$$

gives, using axiom T2) of Theorem 2.2, an exact sequence

$$(2.6) \quad 0 \to \operatorname{Tor}(H_*(C_*), M) \to B_* \otimes M \to Z_* \otimes M \to H_*(C_*) \otimes M \to 0.$$

Assembling the triangle (2.5) and the sequence (2.6) as in Exercise 29, one obtains the short exact sequence

$$0 \to H_*(C_*) \otimes M \to H_*(C_*; M) \to \operatorname{Tor}(H_*(C_*), M) \to 0.$$

Taking the grading into account finishes the proof. □

Corollary 2.35. *If (X, A) is a pair of spaces $A \subset X$, R a PID, M a module over R, then for each q the sequence*

$$0 \to H_q(X, A; R) \otimes M \to H_q(X, A; M) \to \operatorname{Tor}_1^R(H_{q-1}(X, A; R), M) \to 0$$

is short exact, natural, and splits, but not naturally. □

There is another universal coefficient theorem for homology (see [**36**, pg. 248] for the proof). It addresses the question of how a different version of the Kronecker pairing fails to pass to a perfect pairing on (co)homology.

In this case, the pairing

$$\operatorname{Hom}_R(C_*, R) \times (C_* \otimes M) \to M$$

is defined by

$$(f, z \otimes m) \mapsto f(z) \cdot m.$$

This pairing passes to a pairing on homology

$$H^q(C_*; R) \times H_q(C_* \otimes M) \to M.$$

Taking the adjoint yields the homomorphism

$$\alpha : H_q(C_* \otimes M) \to \operatorname{Hom}_R(H^q(C_*), M).$$

The following theorem computes the kernel of this homomorphism if R is a PID and C_* has finitely generated homology.

Theorem 2.36. *Let R be a PID, let C_* be a free chain complex over R such that $H_q(C_*)$ is finitely generated for each q, and let M be an R-module. Then the sequence*

$$0 \to \operatorname{Ext}_R^1(H^{q+1}(C_*), M) \to H_q(C_*; M) \xrightarrow{\alpha} \operatorname{Hom}(H^q(C_*), M) \to 0$$

is short exact, natural, and splits. □

As an application of the universal coefficient theorems we can identify the different versions of the Betti numbers of a space. Recall that the q-th Betti number $\beta_q(X)$ of a space X is the rank of $H_q(X; \mathbf{Z})$. Since $\mathbf{Q}$ and $\mathbf{R}$ are flat abelian groups (see Exercise 27) $\operatorname{Tor}(-, \mathbf{Q})$, $\operatorname{Tor}(-, \mathbf{R})$, $\operatorname{Ext}(-, \mathbf{Q})$, and $\operatorname{Ext}(-, \mathbf{R})$ all vanish. This implies the following.

Corollary 2.37. *The following are all equal: the Betti number* $\beta_q(X)$, $\dim_{\mathbf{Q}} H_q(X;\mathbf{Q})$, $\dim_{\mathbf{R}} H_q(X;\mathbf{R})$, $\dim_{\mathbf{Q}} H^q(X;\mathbf{Q})$, *and* $\dim_{\mathbf{R}} H^q(X;\mathbf{R})$. □

In particular if X is a compact smooth manifold, by the above corollary and DeRham cohomology we see the q-th Betti number is the dimension of the real vector space of closed q-forms modulo exact q-forms.

Example. Since $H_q(\mathbf{R}P^2;\mathbf{Z}) = \mathbf{Z}, \mathbf{Z}/2, 0, \ldots$ for $q = 0, 1, 2, \ldots$, by the universal coefficient theorem $H_q(\mathbf{R}P^2;\mathbf{Z}/2) = \mathbf{Z}/2, \mathbf{Z}/2, \mathbf{Z}/2, 0, \ldots$ for $q = 0, 1, 2, 3, \ldots$ and $H^q(\mathbf{R}P^2;\mathbf{Z}) = \mathbf{Z}, 0, \mathbf{Z}/2, 0, \ldots$ for $q = 0, 1, 2, 3, \ldots$ What is the geometric meaning of the torsion? Let α be a cycle representing the generator of $H_1(\mathbf{R}P^2;\mathbf{Z})$, i.e. α is a "half-equator". Then $2\alpha = \partial\beta$. The generators of $H_1(\mathbf{R}P^2;\mathbf{Z}/2)$ and $H_2(\mathbf{R}P^2;\mathbf{Z}/2)$ are represented by $\alpha \otimes 1$ and $\beta \otimes 1$ respectively. A representative of the generator of $H^2(\mathbf{R}P^2;\mathbf{Z})$ is represented by a cocycle ω where $\omega(\beta) = 1$.

2.7. Projects for Chapter 2

2.7.1. The acyclic models theorem and the Eilenberg-Zilber map.

First state the acyclic models theorem very carefully.

Theorem 2.38 (acyclic models theorem)**.** *Suppose that* $(\mathcal{A}, \mathcal{M})$ *is a category with models. Let* $\mathcal{C}$ *be the category of augmented chain complexes over* R. *Let* $F, F' : \mathcal{A} \to \mathcal{C}$ *be functors so that* F *is free and* F' *acyclic. Then there exists a natural transformation* $\Phi : F \to F'$ *unique up to chain homotopy.*

In particular, if both F *and* F' *are free and acyclic, then* F *and* F' *are chain equivalent and any natural transformation between them is a chain equivalence.* □

In this theorem $\mathcal{A}$ is a category, for example the category TOP of spaces or the category TOP^2 consisting of pairs (X, Y) where X and Y are spaces (but Y need not be a subspace of X), and $\mathcal{M}$ is a fixed collection of objects of $\mathcal{A}$. The functor $F : \mathcal{A} \to \mathcal{C}$ is called acyclic if $F(M)$ is an acyclic chain complex for each model $M \in \mathcal{M}$; it is called free if for every $X \in \mathcal{A}$, the chain groups $F_q(X)$ are free with basis contained in the set $\{F_q(u)F_q(M) | M \in \mathcal{M}, u \in \mathrm{Hom}(M, X)\}$.

Then describe the functors

$$(X, Y) \mapsto S_*(X \times Y)$$

and

$$(X, Y) \mapsto S_*(X) \otimes S_*(Y)$$

and show that they are both free and acyclic. Conclude the Eilenberg–Zilber theorem, Theorem 3.4. Prove the acyclic models theorem.

If time is left, prove the homotopy axiom for homology and cohomology: homotopic maps give chain homotopic maps on the singular chains. References include [**14**, pp. 265–270]. Also see [**36**].

Chapter 3

Products

The theory of products in the homology and cohomology of a space comes about by combining two basic constructions. The first is purely algebraic; one forms the tensor product of chain complexes and their dual cochain complexes and studies their relationships. The second construction is topological and applies to the singular complex of a space. It is a natural chain homotopy equivalence between the singular chain complex $S_*(X \times Y)$ and the tensor product of S_*X and S_*Y. This result is called the Eilenberg-Zilber theorem, and it is a consequence of the acyclic models theorem.

3.1. Tensor products of chain complexes and the algebraic Künneth theorem

We begin with a discussion about tensor products and Hom of graded R-modules.

Definition 3.1.

1. A *graded R-module* A_* can be thought of either as a collection of R-modules $\{A_k\}_{k\in\mathbf{Z}}$ or as a module $A = \bigoplus_k A_k$ with a direct sum decomposition.
2. A *homomorphism of graded R-modules* is an element of the product $\prod_k \operatorname{Hom}(A_k, B_k)$.
3. The *tensor product of graded R-modules* A_* and B_* is the graded R-module

$$(A_* \otimes B_*)_n = \bigoplus_{p+q=n} (A_p \otimes B_q).$$

4. Define $\operatorname{Hom}(A_*, B_*)$ to be the graded R-module, with
$$\operatorname{Hom}(A_*, B_*)_n = \prod_k \operatorname{Hom}(A_k, B_{k+n}).$$

The functors $- \otimes B_*$ and $\operatorname{Hom}(B_*, -)$ are adjoint functors from the category of graded R-modules to itself.

5. A *graded ring* is a graded abelian group R_* together with a (degree zero) map
$$R_* \otimes R_* \to R_*$$
which is associative in the sense that $(ab)c = a(bc)$ where we write ab for the image of $a \otimes b$. Alternatively, a graded ring can be thought of as a ring R with a direct sum decomposition $R = \bigoplus R_k$, satisfying $R_k \cdot R_l \subset R_{k+l}$.
6. A graded ring is *commutative* if
$$ab = (-1)^{|a||b|} ba,$$
where $a \in R_{|a|}$ and $b \in R_{|b|}$.
7. A *graded module M_* over a graded ring R_** is a module over R satisfying $R_k \cdot M_l \subset M_{k+l}$.

We apply these constructions to chain complexes (C_*, ∂) and (C'_*, ∂'_*). We allow C_q and C'_q to be nonzero for *any* $q \in \mathbf{Z}$.

Definition 3.2. The *tensor product chain complex* $(C_* \otimes C'_*, d)$ is defined by taking the tensor product of the underlying graded modules, i.e.
$$(C_* \otimes C'_*)_n = \bigoplus_{p+q=n} C_p \otimes C'_q$$
and giving it the differential
$$d(z \otimes w) = \partial z \otimes w + (-1)^p z \otimes \partial' w, \quad \text{if } z \in C_p.$$
(The differential d is sort of a "graded derivative"; it satisfies the product rule by definition.)

One computes:
$$\begin{aligned} d^2(z \otimes w) &= d(\partial z \otimes w + (-1)^p z \otimes \partial' w) \\ &= \partial^2 z \otimes w + (-1)^{p-1} \partial z \otimes \partial' w \\ &\qquad + (-1)^p \partial z \otimes \partial' w + (-1)^{2p} z \otimes \partial'^2 w \\ &= 0. \end{aligned}$$
Thus $(C_* \otimes C'_*, d)$ is indeed a chain complex.

One geometric motivation for this construction is the following. If X and Y are CW-complexes with cells $\{e_i\}$ and $\{f_j\}$ respectively, then $X \times Y$ is a

CW-complex with cells $\{e_i \times f_j\}$. The cellular chain complex $C_*(X \times Y)$ can be identified with (i.e. is isomorphic to) the tensor product $C_*(X) \otimes C_*(Y)$.

The question we wish to understand is: To what extent and how does the homology of C_* and C'_* determine the homology of $C_* \otimes C'_*$? A connection between the two is provided by the algebraic homology cross product.

Exercise 32. If C_*, D_* are chain complexes, there is a natural map

$$\times_{\text{alg}} : H_p C_* \otimes H_q D_* \to H_{p+q}(C_* \otimes D_*)$$

called the *algebraic homology cross product* defined by

$$[z] \otimes [w] \mapsto [z \otimes w].$$

Write $[z] \times_{\text{alg}} [w]$ (or just $[z] \times [w]$) for $[z \otimes w]$.

The following theorem measures the extent to which this map is an isomorphism, at least if the ground ring R is a PID.

Theorem 3.3 (Künneth exact sequence). *Suppose C_*, D_* are chain complexes over a PID R, and suppose C_q is a free R-module for each q. Then there is a natural exact sequence*

$$0 \to \bigoplus_{p+q=n} H_p(C_*) \otimes H_q(D_*) \xrightarrow{\times_{\text{alg}}} H_n(C_* \otimes D_*) \to \bigoplus_{p+q=n} \operatorname{Tor}^R(H_p(C_*), H_{q-1}(D_*)) \to 0$$

which splits (non-naturally).

Proof. The proof is similar to the proof of the universal coefficient theorem (Theorem 2.29), and so we only sketch the argument, leaving details, notably issues about the grading, to the reader.

Setting $Z_q = \ker \partial : C_q \to C_{q-1}$ and $B_q = \operatorname{im} \partial : C_{q+1} \to C_q$, we obtain the short exact sequence

$$0 \to Z_* \to C_* \xrightarrow{\partial} B_* \to 0 \tag{3.1}$$

which we view as a short exact sequence of free chain complexes by giving Z_* and B_* the zero differential (the modules Z_q and B_q are free since they are submodules of the free module C_q and R is a PID).

Since B_q is free, tensoring the short exact sequence (3.1) with D_* yields a new short exact sequence of chain complexes.

$$0 \to Z_* \otimes D_* \to C_* \otimes D_* \to B_* \otimes D_* \to 0. \tag{3.2}$$

Since the differential in the chain complex Z_* is zero, the differential $\partial : Z_* \otimes D_* \to Z_* \otimes D_*$ reduces to

$$z \otimes d \mapsto (-1)^{|z|} z \otimes \partial d,$$

and so passing to homology one gets

$$H_*(Z_* \otimes D_*) = Z_* \otimes H_*(D_*).$$

Similarly

$$H_*(B_* \otimes D_*) = B_* \otimes H_*(D_*).$$

Thus, the long exact sequence in homology obtained by applying the zig-zag lemma to the complex (3.2) reduces to the exact triangle

$$\begin{array}{ccc} B_* \otimes H_*(D_*) & \longrightarrow & Z_* \otimes H_*(D_*) \\ & \nwarrow \quad \swarrow & \\ & H_*(C_* \otimes D_*) & \end{array} \tag{3.3}$$

On the other hand, applying Theorem 2.2 to the tensor product of the short exact sequence

$$0 \to B_* \to Z_* \to H_*(C_*) \to 0,$$

with $H_*(D_*)$ yields an exact sequence

(3.4)

$$0 \to \mathrm{Tor}(H_*(C_*), H_*(D_*)) \to B_* \otimes H_*(D_*) \to Z_* \otimes H_*(D_*) \to H_*(C_*) \otimes H_*(D_*) \to 0.$$

Combining (3.3) and (3.4), applying Exercise 29, taking care with the grading, and chasing down the definitions of the maps induced finishes the proof. □

3.2. The Eilenberg-Zilber maps

Recall the statement of the Eilenberg-Zilber theorem. Until further notice, homology and cohomology with coefficients in a ring R are understood, and we omit writing " ; R ".

Theorem 3.4 (Eilenberg-Zilber theorem). *Let TOP^2 be the category whose objects are pairs of spaces (X, Y) (we do not assume $Y \subset X$) and whose morphisms are pairs $(f : X' \to X, g : Y' \to Y)$ of continuous maps. Then the two functors*

$$F : (X, Y) \mapsto S_*(X \times Y)$$

and

$$F' : (X, Y) \mapsto S_*(X) \otimes S_*(Y)$$

from TOP^2 to the category of chain complexes are naturally equivalent; i.e. there exist natural transformations $A : F \to F'$ and $B : F' \to F$ so that for any pair (X, Y) the composites

$$S_*(X \times Y) \xrightarrow{A} S_*(X) \otimes S_*(Y) \xrightarrow{B} S_*(X \times Y)$$

and

$$S_*(X) \otimes S_*(Y) \xrightarrow{B} S_*(X \times Y) \xrightarrow{A} S_*(X) \otimes S_*(Y)$$

are chain homotopic to the identity. Moreover, any two choices of A (resp. B) are naturally chain homotopic.

In particular, there exist natural isomorphisms

$$H_n(X \times Y) \to H_n(S_*(X) \otimes S_*(Y))$$

for each n. □

The proof of this theorem is an easy application of the acyclic models theorem. See the project on the acyclic models theorem at the end of Chapter 2.

The natural transformations A and B determine chain homotopy equivalences

$$A : S_*(X \times Y) \to S_*(X) \otimes S_*(Y)$$

and

$$B : S_*(X) \otimes S_*(Y) \to S_*(X \times Y)$$

for any pair of spaces X and Y. We will call these maps the *Eilenberg-Zilber maps.*

The confusing, abstract, but important point is that A and B are not *canonical*, but only *natural.* That is, they are obtained by the method of acyclic models, and so constructed step by step by making certain arbitrary choices. However, these choices are made consistently for all spaces.

In what follows, we will show how a choice of A and B determines natural additional structure, namely products, on the singular complex and homology of a space. But you should keep in mind that all the constructions depend at core on the noncanonical choice of the transformations A and B.

An alternative approach to this material is to just give specific formulas for A and B. It is easy to imagine a chain map $B : S_*(X) \otimes S_*(Y) \to S_*(X \times Y)$. Given singular simplices $\sigma : \Delta_p \to X$ and $\tau : \Delta_q \to Y$, there is the product map $\sigma \times \tau : \Delta_p \times \Delta_q \to X \times Y$. Unfortunately the product of simplices is not a simplex, but it can be chopped up into a union of $p+q$-simplices (consider a square chopped into triangles or a prism chopped into tetrahedra). Then one could choose $B(\sigma \otimes \tau)$ to be a sum of singular $p+q$-simplices – the "shuffle product".

It is harder to imagine a natural candidate for the reverse map $A : S_*(X \times Y) \to S_*(X) \otimes S_*(Y)$, but one can can be explicitly defined using projections to X and Y – the "Alexander-Whitney diagonal approximation". However, even if one constructs the maps A and B explicitly, they will only be chain homotopy equivalences, not isomorphisms; $S_*(X \times Y)$ is simply bigger than $S_*(X) \otimes S_*(Y)$.

In any case, invoking a technical formula can obscure the point of a construction (just look at any page of a differential geometry book for evidence of this principle). Thus for what follows, fix natural transformations A and B whose existence is asserted in Theorem 3.4. Each product on chain complexes constructed below depends on the choice of A or B, but this dependence disappears when passing to homology and cohomology.

3.3. Cross and cup products

3.3.1. The homology cross product and the Künneth formula. Exercise 32 implies that the natural map

$$\times_{\text{alg}} : H_pX \otimes H_qY \to H_{p+q}(S_*X \otimes S_*Y)$$

given on the chain level by $[a] \otimes [b] \mapsto [a \otimes b]$ is well defined. Denote by B_* the isomorphism induced by the Eilenberg-Zilber map on homology, so

$$B_* : H_*(S_*(X) \otimes S_*(Y)) \to H_*(S_*(X \times Y)) = H_*(X \times Y).$$

Composing $\times_{\text{alg}}$ with B_*, we obtain

$$\times : H_pX \otimes H_qY \to H_{p+q}(X \times Y).$$

Definition 3.5. If $\alpha \in H_pX$, $\beta \in H_qY$, the image of $\alpha \otimes \beta$ under this map is called the *homology cross product* of α and β and is denoted by $\alpha \times \beta$.

The Eilenberg-Zilber theorem has the following important consequence.

Theorem 3.6 (Künneth formula). *If R is a PID, there exists a split exact sequence*

$$0 \to \bigoplus_{p=0}^{n} H_pX \otimes H_{n-p}Y \to H_n(X \times Y) \to \bigoplus_{p=0}^{n-1} \operatorname{Tor}(H_pX, H_{n-1-p}Y) \to 0.$$

The first map is given by cross products.

Proof. This follows easily by combining the Künneth exact sequence (Theorem 3.3) to the free chain complexes S_*X and S_*Y with the Eilenberg-Zilber theorem. □

Corollary 3.7. *If R is a field, then*

$$H_*(X \times Y) = H_*(X) \otimes H_*(Y).$$

Exercise 33. Compute $H_*(\mathbf{R}P^2 \times \mathbf{R}P^2)$, both with $\mathbf{Z}$ and $\mathbf{Z}/2$-coefficients. Give a geometric interpretation of the class coming from the Tor term in the Künneth formula.

The Künneth formula implies that if R is a PID, $\alpha \times \beta \neq 0$ if $\alpha \neq 0$ and $\beta \neq 0$.

3.3.2. The cohomology cross product. Let C_* and D_* be chain complexes over a ring R and let C^* and D^* be the dual chain complexes $\mathrm{Hom}_R(C_*, R)$ and $\mathrm{Hom}_R(D_*, R)$ respectively.

Exercise 34. If C_*, D_* are chain complexes, there is a natural map

$$\times^{\mathrm{alg}} : H^p C^* \otimes H^q D^* \to H^{p+q}((C_* \otimes D_*)^*)$$

defined by $[\alpha] \otimes [\beta] \mapsto [\sum z_i \otimes w_i \mapsto \sum \alpha(z_i) \cdot \beta(w_i)]$. In this formula if α and z_i are of different degrees, then $\alpha(z_i)$ is zero, and likewise for $\beta(w_i)$. The notation $\alpha(z_i) \cdot \beta(w_i)$ refers to multiplication in the ring R.

This map is called the *algebraic cohomology cross product.*

Applying this product to the singular complexes, we see that for any spaces X and Y we have a map

$$\times^{\mathrm{alg}} : H^p X \otimes H^q Y \to H^{p+q}((S_* X \otimes S_* Y)^*).$$

Using the Eilenberg-Zilber theorem we can further map to $H^{p+q}(X \times Y)$. Explicitly, the dual of the Eilenberg-Zilber map $A : S_*(X \times Y) \to S_* X \otimes S_* Y$ is a chain homotopy equivalence $A^* : (S_* X \otimes S_* Y)^* \to S^*(X \times Y)$. Passing to cohomology one obtains an isomorphism

$$A^* : H^*((S_* X \otimes S_* Y)^*) \to H^*(X \times Y).$$

This map is independent of the choice of Eilenberg-Zilber map A since any two choices for A are naturally chain homotopic. (We will be somewhat casual with notation and denote by A^* the dual of A as well as the induced map on cohomology. This should not cause any confusion and will keep the notation under control.)

Definition 3.8. If $a \in H^p X$, $b \in H^q Y$, the image of $a \otimes b$ under the composite

$$H^p X \otimes H^q Y \xrightarrow{\times^{\mathrm{alg}}} H^{p+q}((S_* X \otimes S_* Y)^*) \xrightarrow{A^*} H^*(X \times Y)$$

map is called the *cohomology cross product* of a and b and is denoted by $a \times b$.

3.3.3. The cup product. Combining the diagonal map $\Delta : X \to X \times X$, $x \mapsto (x, x)$ with the cross product leads to the important cup product:

Definition 3.9. Given $a \in H^p X$ and $b \in H^q X$, then the *cup product* of a and b is defined by $a \cup b = \Delta^*(a \times b) \in H^{p+q} X$.

To review, the cup product gives a homomorphism

$$H^p X \otimes H^q X \to H^{p+q} X, \quad a \otimes b \mapsto a \cup b.$$

It is defined as the composite

$$H^pX \otimes H^qX \xrightarrow{\times^{\mathrm{alg}}} H^{p+q}((S_*X \otimes S_*X)^*) \xrightarrow{A^*} H^{p+q}(X \times X) \xrightarrow{\Delta^*} H^{p+q}X$$

where the first map is given by $\times^{\mathrm{alg}}$, the second by the dual of the Eilenberg-Zilber map A and the third by the diagonal map.

It is usually more intuitive to think of the cup product as a bilinear pairing

$$\cup : H^pX \times H^qX \to H^{p+q}X,$$

by precomposing with the canonical map $H^pX \times H^qX \to H^pX \otimes H^qX$.

The next lemma shows that the cross product determines the cup product and conversely that the cup product determines the cross product. Both are functorial.

Lemma 3.10. *Let $f : X' \to X$ and $g : Y' \to Y$ be continuous maps. Let $a, b \in H^*X$ and $c \in H^*Y$.*

1. $a \cup b = \Delta^*(a \times b)$.
2. $a \times c = p_X^* a \cup p_Y^* c$, *where p_X and p_Y are the projections in $X \times Y$.*
3. $f^*(a \cup b) = f^*a \cup f^*b$.
4. $(f \times g)^*(a \times c) = f^*a \times g^*c$.

Proof. We prove these in reverse order. (4) follows from the naturality of the Eilenberg-Zilber map and the algebraic cohomology cross product with respect to pairs of maps $(f : X' \to X, g : Y' \to Y)$.

(3) follows since

$$\begin{aligned} f^*a \cup f^*b &= \Delta^*(f^*a \times f^*b) \\ &= \Delta^*(f \times f)^*(a \times b) \\ &= ((f \times f) \circ \Delta)^*(a \times b) \\ &= (\Delta \circ f)^*(a \times b) \\ &= f^*(\Delta^*(a \times b)) \\ &= f^*(a \cup b). \end{aligned}$$

(2) follows since

$$\begin{aligned} p_X^*a \cup p_Y^*c &= \Delta_{X\times Y}^*(p_X^*a \times p_Y^*c) \\ &= \Delta_{X\times Y}^*((p_X \times p_Y)^*(a \times c)) \\ &= ((p_X \times p_Y) \circ \Delta_{X\times Y})^*(a \times c) \\ &= \mathrm{Id}_{X\times Y}^*(a \times c). \end{aligned}$$

(1) is the definition of cup product. □

At this point it is convenient to introduce the concept of a diagonal approximation.

Definition 3.11. A *diagonal approximation* τ is a chain map

$$\tau : S_*X \to S_*X \otimes S_*X$$

for every space X, so that

1. $\tau(\sigma) = \sigma \otimes \sigma$ for every 0-simplex σ.
2. τ is natural with respect to continuous maps of spaces.

Now the functor S_*X is free on the models $\{\Delta^n\}$ and $S_*X \otimes S_*X$ is acyclic on these models, so the acyclic models theorem says that there exists a diagonal approximation and any two such are natural chain homotopic. If A is an Eilenberg-Zilber map and $\Delta : X \to X \times X$ is the diagonal map, then $\tau = A \circ \Delta_*$ is a diagonal approximation. Thus we can rephrase the definition of the cup product

$$a \cup b = \tau^*(a \times^{\text{alg}} b).$$

Lemma 3.12. *A choice of Eilenberg-Zilber map* $A : S_*(X \times Y) \to S_*(X) \otimes S_*(Y)$ *determines a diagonal approximation* $\tau : S_*X \to S_*X \otimes S_*X$ *by the formula*

$$\tau = A \circ \Delta_*.$$

Conversely, a diagonal approximation τ *determines an Eilenberg-Zilber map* A *by the formula* $A = (p_X \otimes p_Y) \circ \tau$, *that is, as the composite*

$$S_*(X \times Y) \xrightarrow{\tau} S_*(X \times Y) \otimes S_*(X \times Y) \xrightarrow{p_X \otimes p_Y} S_*(X) \otimes S_*(Y).$$

Proof. Given an Eilenberg-Zilber map A, clearly $A \circ \Delta_*$ is a diagonal approximation.

Conversely, the map $(\mathrm{p}_X \otimes \mathrm{p}_Y) \circ \tau$ is a natural transformation from the functor $F : (X, Y) \mapsto S_*(X \times Y)$ to the functor $F' : (X, Y) \mapsto S_*(X) \otimes S_*(Y)$. By the uniqueness part of the Eilenberg-Zilber theorem, this must be an Eilenberg-Zilber map. $\square$

In light of Lemma 3.12, we see that the cup product and the cohomology cross product could just as well have been defined starting with a diagonal approximation. We will use τ or A depending on which one leads to simpler notation, using the formulas of Lemma 3.12 to pass between the two.

The following theorem says that H^*X is a graded commutative ring with unit. (Remember, coefficients in a commutative ring R are understood.) Part 3 of Lemma 3.10 shows that a map of spaces $f : X \to Y$ induces a map $f^* : H^*Y \to H^*X$ of graded rings.

Theorem 3.13.

1. *Let* $1 \in H^0(X)$ *be represented by the cocycle which takes every singular 0-simplex to* $1 \in R$. *Then* $1 \cup a = a = a \cup 1$.
2. $(a \cup b) \cup c = a \cup (b \cup c)$.
3. $a \cup b = (-1)^{|a||b|} b \cup a$. *Here* $|a|$ *represents the degree of* A, *i.e.* $a \in H^{|a|}X$.

Proof. All three parts involve the uniqueness part of the method of acyclic models.

1. Represent $1 \in H^0 X$ by the cochain $1 \in S^0 X$. The algebraic cohomology cross product is represented on the chain level by the canonical identification

$$\times^{\mathrm{alg}} : C^* \otimes D^* \to (C_* \otimes D_*)^*.$$

Consider the natural transformation of functors

$$C : S^* X \to S^* X$$

defined by

$$C(\alpha) = \tau^*(1 \times^{\mathrm{alg}} \alpha).$$

This induces the identity map on H^0; if σ is a singular 0-simplex, then

$$\begin{aligned} C(\alpha)(\sigma) &= \tau^*(1 \times^{\mathrm{alg}} \alpha)(\sigma) \\ &= (1 \times^{\mathrm{alg}} \alpha)(\tau(\sigma)) \\ &= (1 \times^{\mathrm{alg}} \alpha)(\sigma \otimes \sigma) \\ &= 1(\sigma) \cdot \alpha(\sigma) = \alpha(\sigma). \end{aligned}$$

Since $S^* X$ is free and acyclic on the models $\{\Delta^p\}$, the map C must be naturally chain equivalent to the identity by the acyclic models theorem. Passing to cohomology gives 1.

2. The compositions of Eilenberg-Zilber maps

$$S_* X \otimes S_* Y \otimes S_* Z \to S_*(X \times Y) \otimes S_* Z \to S_*(X \times Y \times Z)$$

$$S_* X \otimes S_* Y \otimes S_* Z \to S_* X \otimes S_*(Y \times Z) \to S_*(X \times Y \times Z)$$

are natural transformations of functors on TOP^3 (triples of spaces), which are free and acyclic on the models $\{\Delta^p, \Delta^q, \Delta^r\}$. They induce the same map on H_0 so must be naturally chain homotopic. Associativity follows.

3. The key observation here is that for chain complexes C_* and D_*, the interchange map

$$T : C_* \otimes D_* \to D_* \otimes C_*$$

$$z \otimes w \mapsto (-1)^{|z||w|} w \otimes z$$

gives an *isomorphism* of chain complexes. Let $S : X \times Y \to Y \times X$ be the geometric switch map $S(x,y) = (y,x)$. Thus the Eilenberg-Zilber maps

$$T \circ A : S_*(X \times Y) \to S_*Y \otimes S_*X$$

$$A \circ S_* : S_*(X \times Y) \to S_*Y \otimes S_*X$$

are naturally chain homotopic by the uniqueness part of the acyclic models theorem. The result follows.

More explicitly, if S is the switch map on $X \times X$, note that

$$S \circ \Delta = \Delta : X \to X \times X.$$

Thus

$$\begin{aligned} a \cup b &= \Delta^*(a \times b) \\ &= \Delta^* S^*(a \times b) \\ &= \Delta^* S^* A^*(a \times^{\text{alg}} b) \\ &= \Delta^*(A \circ S)^*(a \times^{\text{alg}} b) \\ &= \Delta^*(T \circ A)^*(a \times^{\text{alg}} b) \\ &= (-1)^{|a||b|}\Delta^* A^*(b \times^{\text{alg}} a) \\ &= (-1)^{|a||b|} b \cup a. \end{aligned}$$

□

Exercise 35. Give another proof of the graded commutativity of the cup product as follows. Let $T : S_*X \otimes S_*X \to S_*X \otimes S_*X$ be the algebraic switch map, $x \otimes y \mapsto (-1)^{|x||y|} y \otimes x$. Show that if τ is a diagonal approximation, so is $T \circ \tau$. Use this to show that H^*X is graded commutative.

Sometimes one wishes to use products on homology and cohomology with coefficients in various R-modules. The following exercise shows how to accomplish this. The basic idea is that multiplication in the ring R was used in the definition of cup products (in fact in the definition of $\times^{\text{alg}}$), and so when passing to more general modules an auxiliary multiplication is needed.

Exercise 36. If M, N, P are R modules, and $M \times N \to P$ a bilinear map, show how to construct a cross product

$$\times : H^p(X; M) \times H^q(Y; N) \to H^{p+q}(X \times Y; P)$$

and a cup product

$$\cup : H^p(X; M) \times H^q(X; N) \to H^{p+q}(X; P).$$

3.3.4. The cap product. Recall that the Kronecker pairing is a natural bilinear evaluation map (sometimes called "integration" by analogy with the deRham map)

$$\langle\ ,\ \rangle : S^*X \times S_*X \to R$$

defined for $a \in S^qX, z \in S_pX$ by

$$\langle a, z\rangle = \begin{cases} a(z) & \text{if } p = q, \\ 0 & \text{otherwise.} \end{cases}$$

This pairing can be extended to a "partial evaluation" or "partial integration" map

$$E : S^*X \otimes S_*X \otimes S_*X \to S_*X$$

by "evaluating the first factor on the last factor", i.e.

$$E(a \otimes z \otimes w) = a(w) \cdot z.$$

We will define the cap product on the chain level first.

Definition 3.14. The *cap product*

$$S^q(X) \times S_{p+q}(X) \to S_q(X)$$

is defined for $a \in S^q(X),\ z \in S_{p+q}(X)$ by

$$a \cap z = E(a \otimes A \circ \Delta_*(z)).$$

The definition can be given in terms of a diagonal approximation τ instead of the Eilenberg-Zilber map A:

$$a \cap z = E(a \otimes \tau(z)).$$

Lemma 3.15. *For $\alpha \in S^qX,\ z \in S_{p+q}X$,*

$$\partial(\alpha \cap z) = (-1)^p \delta\alpha \cap z + \alpha \cap \partial z.$$

Proof. Suppose $\tau(z) = \sum x_i \otimes y_i$ so that $|x_i| + |y_i| = p + q$. Then since α only evaluates nontrivially on chains in degree q, we have

$$\begin{aligned} \partial(\alpha \cap z) &= \partial E(\alpha \otimes \tau(z)) \\ &= \partial \sum_{|y_i|=q} \alpha(y_i) \cdot x_i \\ &= \sum_{|y_i|=q} \alpha(y_i) \cdot \partial x_i \end{aligned}$$

and

$$\begin{aligned} \delta\alpha \cap z &= E(\delta\alpha \otimes \tau(z)) \\ &= \sum \delta\alpha(y_i) \cdot x_i \\ &= \sum_{|y_i|=q+1} \alpha(\partial y_i) \cdot x_i. \end{aligned}$$

Moreover,

$$\begin{aligned} \alpha \cap \partial z &= E(\alpha \otimes \tau(\partial z)) \\ &= E(\alpha \otimes \partial\tau(z)) \\ &= E(\alpha \otimes (\sum \partial x_i \otimes y_i + \sum (-1)^{|x_i|} x_i \otimes \partial y_i)) \\ &= \sum_{|y_i|=q} \alpha(y_i) \cdot \partial x_i + \sum_{|y_i|=q+1} (-1)^{p-1} \alpha(\partial y_i) \cdot x_i \\ &= \partial(\alpha \cap z) + (-1)^{p-1} (\delta\alpha) \cap z. \end{aligned}$$

□

Lemma 3.15 immediately implies:

Corollary 3.16. *The cap product descends to a well defined product*

$$\cap : H^q X \times H_{p+q} X \to H_p X$$
$$([\alpha], [z]) \mapsto [\alpha \cap z]$$

after passing to (co)homology. □

Exercise 37. Let $a, b \in H^* X$ and $z \in H_* X$. Show that

1. $\langle a, b \cap z \rangle = \langle a \cup b, z \rangle$.
2. $a \cap (b \cap z) = (a \cup b) \cap z$.

Thus the cap product makes the homology $H_*(X)$ a module over the ring $H^*(X)$.

3.3.5. The slant product. We next introduce the slant product which bears the same relation to the cross product as the cap product does to the cup product (this could be on an SAT test). Again we give the definition on the chain level first.

Definition 3.17. The *slant product*

$$\backslash : S^q(Y) \times S_{p+q}(X \times Y) \to S_p(X)$$

is defined for $a \in S^q Y$, and $z \in S_{p+q}(X \times Y)$ by

$$a \backslash z = E(a \otimes A(z)) \in S_p X$$

where $A : S_*(X \times Y) \to S_*(X) \otimes S_*(Y)$ is the Eilenberg-Zilber map.

Similar arguments to those given for the other products given above show that

$$\langle a, b\backslash z\rangle = \langle a \times b, z\rangle$$

for all $a \in S^qY$ and that passing to (co)homology one obtains a well defined bilinear map

$$\backslash : H^qY \times H_{p+q}(X \times Y) \to H_pX.$$

If M, N, P are R modules and $M \times N \to P$ a bilinear map, one can define cap products

$$\cap : H^q(X; M) \times H_{p+q}(X; N) \to H_p(X; P)$$

and slant products

$$\backslash : H^q(Y; M) \times H_{p+q}(X \times Y; N) \to H_p(X; P).$$

(See Exercise 36.)

There are even more products (the book by Dold [**9**] is a good reference). For example, there is another slant product

$$/ : H^{p+q}(X \times Y) \times H_qY \to H^pX.$$

Often one distinguishes between *internal products* which are defined in terms of one space X (such as the cup and cap products) and *external products* which involve the product of two spaces $X \times Y$. Of course, one can go back and forth between the two by thinking of $X \times Y$ as a single space and using the two projections p_X and p_Y and the diagonal map $\Delta : X \to X \times X$.

3.4. The Alexander-Whitney diagonal approximation

When considered on the chain level, the various products we have defined above *do depend* on the choice of Eilenberg-Zilber map $A : S_*(X \times Y) \to S_*(X) \otimes S_*(Y)$. Only by passing to (co)homology does the choice of A disappear. It is nevertheless often useful to work on the chain level, since there is subtle homotopy-theoretic information contained in the singular complex which leads to extra structure, such as Steenrod operations and Massey products.

We will give an explicit formula due to Alexander and Whitney for a specific choice of A. This enables one to write down formulas for the products on the chain level, and in particular gives the singular cochain complex of a space an explicit natural *associative* ring structure.

Definition 3.18. Given a singular n-simplex $\sigma : \Delta^n \to X$, and integers $0 \leq p, q \leq n$, define the *front p-face of σ* to be the singular p-simplex ${}_p\sigma : \Delta^p \to X$

$${}_p\sigma(t_0, \ldots, t_p) = \sigma(t_0, \ldots, t_p, 0, \ldots, 0)$$

and the *back q-face of* σ to be the singular q-simplex $\sigma_q : \Delta^q \to X$

$$\sigma_q(t_0, \dots , t_q) = \sigma(0, \dots , 0, t_0, \dots , t_q).$$

Let $p_X : X \times Y \to X$ and $p_Y : X \times Y \to Y$ denote the two projections.

Definition 3.19. The *Alexander-Whitney map*

$$A : S_*(X \times Y) \to S_*(X) \otimes S_*(Y)$$

is the natural transformation given by the formula

$$A(\sigma) = \sum_{p+q=n} {}_p(p_X \circ \sigma) \otimes (p_Y \circ \sigma)_q.$$

Thus

$$A : S_n(X \times Y) \to (S_*X \otimes S_*Y)_n = \bigoplus_{p+q=n} S_pX \otimes S_qY.$$

The Alexander-Whitney map A is a natural chain map since it is given by a specific formula involving geometric simplices which is independent of the choice of X and Y.

Exercise 38. Show directly that A induces an isomorphism on H_0.

From the uniqueness part of the Eilenberg-Zilber theorem, it follows that A is a chain homotopy equivalence and can be used to define cross products and cup products. (This illustrates the power of the acyclic models theorem; the naturality of the Alexander-Whitney map and the map on H_0 suffice to conclude that A is a chain homotopy equivalence.)

To an Eilenberg-Zilber map A one can associate the corresponding diagonal approximation $\tau = A \circ \Delta_*$. Taking A to be the Alexander-Whitney map, one gets the following.

Definition 3.20. The *Alexander-Whitney diagonal approximation* is the map

$$\tau(\sigma) = \sum_{p+q=n} {}_p\sigma \otimes \sigma_q.$$

This allows one to define a specific product structure on S^*X: for a cochain $\alpha \in S^pX$ and $\beta \in S^qX$, define $\alpha \cup \beta \in S^{p+q}X$ by

$$(\alpha \cup \beta)(\sigma) = \alpha({}_p\sigma) \cdot \beta(\sigma_q).$$

Exercise 39. By tracing through this definition of the cup product, show that $[\alpha] \cup [\beta] = [\alpha \cup \beta]$.

Exercise 40. Using the Alexander-Whitney diagonal approximation,

1. prove that $S^*(X; R)$ is an *associative ring* with unit 1 represented by the cochain $c \in S^0(X; R) = \mathrm{Hom}(S_0(X); R)$ which takes the value 1 on *every* singular 0-simplex in X.

2. Compute cap products: show that if $\alpha \in S^q X$ and σ is a singular $(p+q)$-simplex, then
$$\alpha \cap \sigma = \alpha(\sigma_q) \cdot {}_p\sigma.$$
3. Show that $(\alpha \cup \beta) \cap z = \alpha \cap (\beta \cap z)$, and so the cap product makes $S_*(X)$ into a $S^*(X)$ module.

We have already seen that cohomology is an associative and graded commutative ring with unit in Theorem 3.13. However, the methods used there cannot be used to show that $S^*(X)$ is an associative ring; in fact it is not for a random choice of Eilenberg-Zilber map A.

The Alexander-Whitney map is a particularly nice choice of Eilenberg-Zilber map because it does give an associative ring structure on $S^*(X)$. This ring structure, alas, is not (graded) commutative (Steenrod squares give obstructions to its being commutative), while the ring structure on H^*X is commutative by Theorem 3.13.

Notice that the deRham cochain complex of differential forms on a smooth manifold is graded commutative, since differential forms satisfy $a \wedge b = \pm b \wedge a$. It is possible to give a natural construction of a commutative chain complex over the rationals which gives the rational homology of a space; this was done using rational differential forms on a simplicial complex by Sullivan. This fact is exploited in the subject of rational homotopy theory [**15**]. On the other hand it is impossible to construct a functor from spaces to commutative, associative chain complexes over $\mathbf{Z}$ which gives the integral homology of a space.

Exercise 41. Give an example of two singular 1-cochains α_1 and α_2 such that $\alpha_1 \cup \alpha_2 \neq -\alpha_2 \cup \alpha_1$ using the Alexander-Whitney diagonal approximation to define the cup product.

See Vick's book [**41**] for a nice example of computing the cohomology ring of the torus directly using the Alexander-Whitney diagonal approximation.

If X is a CW-complex, then the diagonal $\Delta : X \to X \times X$ is not cellular (consider X $= [0,1]$). However the cellular approximation theorem says that Δ is homotopic to a cellular map Δ'. If $\Delta'_*(z) = \sum x_i \otimes y_i$, then the cup product on cellular cohomology can be defined by $(\alpha \cup \beta)(z) = \sum \alpha(x_i) \cdot \beta(y_i)$. The geometric root of the Alexander-Whitney diagonal approximation is finding a simplicial map (i.e. takes simplices to simplices and is affine on the simplices) homotopic to the diagonal map $\Delta^n \to \Delta^n \times \Delta^n$.

Let $\epsilon : S_0X \to R$ be the *augmentation map* $\epsilon(\sum r_i\sigma_i) = \sum r_i$. This passes to homology $\epsilon_* : H_0X \to R$ and is an isomorphism if X is path connected.

The cap product of a q-dimensional cocycle with a q-dimensional cycle generalizes the Kronecker pairing in the following sense.

Proposition 3.21. *For $\alpha \in H^q X$ and $z \in H_q X$,*

$$\langle \alpha, z \rangle = \epsilon_*(\alpha \cap z).$$

Proof. We show that for any cochain $\alpha \in S^q X$ and for any chain $z \in S_q X$, using the Alexander-Whitney definition,

$$\alpha(z) = \epsilon(\alpha \cap z).$$

By linearity it suffices to check this for $z = \sigma$.

$$\epsilon(\alpha \cap \sigma) = \epsilon(\alpha(\sigma_q) \cdot {}_0\sigma)) = \alpha(\sigma).$$

□

Notice that the argument shows that the equation $\langle \alpha, z \rangle = \epsilon_*(\alpha \cap z)$ holds even on the chain level. Since ϵ_* is a canonical isomorphism for a path–connected space we will usually just write

$$\langle \alpha, z \rangle = \alpha \cap z.$$

Can you prove Proposition 3.21 for an arbitrary choice of diagonal approximation using the acyclic models theorem?

3.5. Relative cup and cap products

The constructions of cup and cap products carry over without any difficulty to the singular chains and singular (co)homology of a pair (X, A). Naturality then implies that there is a cup product

$$H^*(X, A) \times H^*(X, B) \to H^*(X, A \cap B). \tag{3.5}$$

However, it turns out that by applying a construction that comes about in proving the excision theorem via acyclic models, one can obtain a very useful form of cup and cap products. For example there is a well defined natural cup product

$$H^*(X, A) \times H^*(X, B) \to H^*(X, A \cup B), \tag{3.6}$$

provided A and B are open. (Explain to yourself why (3.6) is better than (3.5).)

That the pairing (3.6) exists is not so surprising if you think in terms of the Alexander-Whitney definition of cup product. Recall

$$(a \cup b)\sigma = \sum a({}_p\sigma) \cdot b(\sigma_q).$$

If the image of σ is contained in either A or B, then the sum will be zero, since a is zero on simplices in A and b is zero on simplices in B. However, if

A and B are open, then one can subdivide σ so that each piece is contained in A or B. The existence of this cup product follows since subdivision disappears when passing to cohomology. We now give a formal argument.

We begin with some algebraic observations. Suppose (X, A) and (Y, B) are two pairs of spaces. Then

$$\frac{S_*X}{S_*A} \otimes \frac{S_*Y}{S_*B} \cong \frac{S_*X \otimes S_*Y}{S_*X \otimes S_*B + S_*A \otimes S_*Y}. \tag{3.7}$$

This is a natural isomorphism, induced by the surjection

$$S_*X \otimes S_*Y \to \frac{S_*X}{S_*A} \otimes \frac{S_*Y}{S_*B}.$$

Exercise 42. Prove that (3.7) is a natural isomorphism.

Now assume $X = Y$; i.e. let A and B be subsets of X. The diagonal approximation τ satisfies $\tau(S_*A) \subset S_*A \otimes S_*A$ and $\tau(S_*B) \subset S_*B \otimes S_*B$. Thus τ induces a map

$$\tau : \frac{S_*X}{S_*A + S_*B} \to \frac{S_*X \otimes S_*X}{S_*X \otimes S_*A + S_*B \otimes S_*X}.$$

The composite

$$\operatorname{Hom}\left(\frac{S_*X}{S_*A}, R\right) \otimes \operatorname{Hom}\left(\frac{S_*X}{S_*B}, R\right) \xrightarrow{\tau^* \circ \times^{\text{alg}}} \operatorname{Hom}\left(\frac{S_*X}{S_*A + S_*B}, R\right)$$

induces a cup product

$$H^p(X, A) \times H^q(X, B) \to H^{p+q}\left(\operatorname{Hom}\left(\frac{S_*X}{S_*A + S_*B}, R\right)\right). \tag{3.8}$$

Definition 3.22. If A, B are subspaces of a topological space X, we say $\{A, B\}$ is an *excisive pair* if the inclusion map

$$S_*(A) + S_*(B) \subset S_*(A \cup B)$$

is a chain homotopy equivalence.

Since a chain map between free chain complexes inducing an isomorphism on homology is a chain homotopy equivalence, this is equivalent to requiring that the inclusion map induces an isomorphism on homology.

If $\{A, B\}$ is an excisive pair, the induced map on cochain complexes

$$S^*(A \cup B) \to \operatorname{Hom}(S_*(A) + S_*(B), R)$$

is also a chain homotopy equivalence and hence induces an isomorphism on cohomology.

The excision theorem implies that if

$$A \cup B = \operatorname{Int}_{A \cup B} A \cup \operatorname{Int}_{A \cup B} B,$$

then $\{A, B\}$ is an excisive pair. Standard arguments show that if A, B are subcomplexes of a CW-complex X, then $\{A, B\}$ is excisive (replace A, B by homotopy equivalent neighborhoods).

Exercise 43. Let A be a point on a circle in $X = \mathbf{R}^2$ and let B be the complement of A in this circle. Show that $\{A, B\}$ is not an excisive pair.

Exercise 44. Show that for any pair A, B of subspaces of a space X, the sequence of chain complexes

$$0 \to S_*(A \cap B) \to S_*(A) \oplus S_*(B) \to S_*(A) + S_*(B) \to 0$$

is exact. Conclude, using the zig-zag lemma, that if $\{A, B\}$ is an excisive pair, then the *Mayer-Vietoris* sequence

$$\cdots \to H_q(A \cap B) \to H_q(A) \oplus H_q(B) \to H_q(A \cup B) \to H_{q-1}(A \cap B) \to \cdots$$

is exact.

Suppose $\{A, B\}$ is an excisive pair. Then consider the two short exact sequences of chain complexes

$$\begin{array}{ccccccccc}
0 & \longrightarrow & S_*A + S_*B & \longrightarrow & S_*X & \longrightarrow & \dfrac{S_*X}{S_*A + S_*B} & \longrightarrow & 0 \\
 & & \downarrow & & \downarrow & & \downarrow & & \\
0 & \longrightarrow & S_*(A \cup B) & \longrightarrow & S_*X & \longrightarrow & \dfrac{S_*X}{S_*(A \cup B)} & \longrightarrow & 0
\end{array}$$

The zig-zag lemma gives a ladder of long exact sequences on cohomology where two-thirds of the vertical arrows are isomorphisms. The five lemma shows that the rest are isomorphisms; in particular, we conclude that if $\{A, B\}$ is an excisive pair, the natural map

$$H^n(X, A \cup B) \to H^n\left(\operatorname{Hom}\left(\frac{S_*X}{S_*A + S_*B}, R\right)\right)$$

is an isomorphism for all n. Combining this fact with the cup product of Equation (3.8) gives a proof of the following theorem.

Theorem 3.23. *If $\{A, B\}$ is an excisive pair, there is a well defined cup product*

$$\cup : H^p(X, A) \times H^q(X, B) \to H^{p+q}(X, A \cup B).$$

□

Here is a particularly interesting application of Theorem 3.23.

Exercise 45. Show that if X is covered by open, contractible sets U_i, $i = 1, \cdots, n$, then

$$a_1 \cup \cdots \cup a_n = 0$$

for any collection of $a_i \in H^{q_i}(X)$ with $q_i > 0$.

As an example, the torus cannot be covered by two charts, since the cup product of the two 1-dimensional generators of cohomology is nontrivial (by Exercise 49).

Notice that the pair $\{A, A\}$ is always excisive. Thus $H^*(X, A)$ is a ring. Also, $\{A, \phi\}$ is always excisive. This implies the following.

Corollary 3.24. *There is a well defined natural cup product*

$$\cup : H^p(X, A) \times H^q X \to H^{p+q}(X, A).$$

□

Similar arguments apply to cap products. The final result is:

Theorem 3.25. *If $\{A, B\}$ is an excisive pair, then there is a well defined cap product*

$$\cap : H^q(X, A) \times H_{p+q}(X, A \cup B) \to H_p(X, B).$$

Proof. (*Special case when $A = \emptyset$, using the Alexander-Whitney map.*)

If $A = \emptyset$, let $a \in S^q X$ be a cocycle, and let $c \in S_{p+q} X$ so that its image in $S_{p+q}(X, B)$ is a cycle, i.e. $\partial c \in S_{p+q-1} B$. Then $a \cap c \in S_q X$. Since $\partial(a \cap c) = \delta a \cap c + (-1)^q a \cap \partial c$, and $\delta a = 0$, it follows that $\partial(a \cap c) = a \cap \partial c$.

Because $\partial c \in S_* B$, $a \cap \partial c \in S_* B$ also. Indeed, if $\partial c = \sum r_i \sigma_i$, $\sigma_i : \Delta^{p+q-1} \to B$, $a \cap \partial c = \sum r_i a({}_q\sigma_i) \cdot \sigma_{i_{p-1}}$, but $\sigma_{i_{p-1}} : \Delta^{p-1} \to B \in S_* B$.

Thus $\partial(a \cap c) \in S_* B$; i.e. $a \cap c$ is a cycle in $S_*(X, B)$.

It is easy to check that replacing a by $a + \delta x$ and c by $c + \partial y$, $y \in S_{p+q+1}(X, B)$ does not change $a \cap c$ in $H_p(X, B)$. □

Exercise 46. Prove Theorem 3.25 when $B = \emptyset$ and $A \neq \emptyset$.

3.6. Projects for Chapter 3

3.6.1. Algebraic limits and the Poincaré duality theorem. Define both the colimit and limit of modules over a directed system (these are also called direct and inverse limit, respectively). Define an n-dimensional manifold. Define the local orientation and the fundamental class of a manifold. Define the compactly supported cohomology of a manifold; then state and prove the Poincaré duality theorem. State the Poincaré-Lefschetz duality for a manifold with boundary. If time permits, state the Alexander duality theorem. A good reference is Milnor and Stasheff's *Characteristic Classes*, [**30**, pg. 276]. Also see [**14**, pg. 217]. For the definition of limits see [**33**]; see also Section 5.5.2.

Let M be a connected manifold of dimension n.

1. If M is noncompact, then $H_nM = 0$. (Just prove the orientable case if the nonorientable case seems too involved.)
2. If M is a closed (i.e. compact without boundary), then $H_n(M;\mathbf{Z})$ is $\mathbf{Z}$ or 0. It is $\mathbf{Z}$ if and only if M is orientable.

From these facts you can define the Poincaré duality maps. The following theorem forms the cornerstone of the subject of geometric topology.

Theorem 3.26.

1. (Poincaré duality) *Let M be a closed oriented n-dimensional manifold. Then the orientation determines a preferred generator $[M] \in H_n(M;\mathbf{Z}) \cong \mathbf{Z}$. Taking cap products with this generator induces isomorphisms*

$$\cap[M] : H^p(M;\mathbf{Z}) \to H_{n-p}(M;\mathbf{Z}).$$

2. (Poincaré-Lefschetz duality) *Let M be a compact oriented n-manifold with nonempty boundary ∂M. Then the orientation determines a preferred generator $[M, \partial M] \in H_n(M, \partial M; \mathbf{Z})$. The manifold without boundary ∂M is orientable. Let $[\partial M] = \partial([M, \partial M])$ where $\partial : H_n(M, \partial M) \to H_{n-1}(\partial M)$. Then the diagram*

$$\begin{array}{ccccccccc} \cdots \to & H^{p-1}(M) & \to & H^{p-1}(\partial M) & \to & H^p(M,\partial M) & \to & H^p(M) \to \cdots \\ & \downarrow{\scriptstyle \cap[M,\partial M]} & & \downarrow{\scriptstyle \cap[\partial M]} & & \downarrow{\scriptstyle \cap[M,\partial M]} & & \downarrow{\scriptstyle \cap[M,\partial M]} \\ \cdots \to & H_{n-p+1}(M,\partial M) & \to & H_{n-p}(\partial M) & \to & H_{n-p}(M) & \to & H_{n-p}(M,\partial M) \to \cdots \end{array}$$

commutes up to sign, where the horizontal rows are the long exact sequences in cohomology and homology for the pairs, and every vertical map is an isomorphism.

3. (Alexander duality) *Let M be a closed orientable n-manifold, and let $A \subset M$ be a finite subcomplex. Then $H^p(A)$ is isomorphic to $H_{n-p}(M, M - A)$.*

□

3.6.2. Exercises on intersection forms. Let M be a compact, closed, oriented n-dimensional manifold. For each p, define a bilinear form

$$H^p(M;\mathbf{Z}) \times H^{n-p}(M;\mathbf{Z}) \to \mathbf{Z}$$

by $a \cdot b = \langle a \cup b, [M] \rangle$.

Exercise 47. $a \cdot b = (-1)^{p(n-p)} b \cdot a$.

Given a finitely generated abelian group A, let $T = T(A) \subset A$ denote the torsion subgroup. Thus A/T is a free abelian group.

Exercise 48. Show that the pairing $(a, b) \mapsto a \cdot b$ passes to a well defined pairing

$$H^p(M;\mathbf{Z})/T \times H^{n-p}(M;\mathbf{Z})/T \to \mathbf{Z}. \tag{3.9}$$

Show that this pairing is *perfect*; i.e. the adjoint

$$H^p(M;\mathbf{Z})/T \to \mathrm{Hom}(H^{n-p}(M;\mathbf{Z})/T, \mathbf{Z})$$

is an isomorphism of free abelian groups. (Hint: Use the universal coefficient theorem and Poincaré duality.)

The pairing (3.9) is called the *intersection pairing on M*. In Section 10.7 we will see that the pairing can be described by the intersection of submanifolds of M.

Exercise 49. Compute the cohomology rings $H^*(\mathbf{R}P^n;\mathbf{Z}/2)$, $H^*(\mathbf{C}P^n;\mathbf{Z})$, and $H^*(T^n;\mathbf{Z})$ using Poincaré duality and induction on n. (The first two are truncated polynomial rings; the last one is an exterior algebra.)

If dim $M = 2k$, then

$$H^k(M;\mathbf{Z})/T \times H^k(M;\mathbf{Z})/T \to \mathbf{Z}$$

is called the *intersection form of M*. It is well defined and unimodular over $\mathbf{Z}$, i.e. has determinant equal to ± 1.

Let $V = H^k(M,\mathbf{Z})/T$. So $(V, \cdot)$ is an inner product space over $\mathbf{Z}$. This inner product space can have two kinds of symmetry.

Case 1. k is odd. Thus dim $M = 4\ell + 2$. Then $v \cdot w = -w \cdot v$ for $v, w \in V$, so $(V, \cdot)$ is a *skew-symmetric* and unimodular inner product space over $\mathbf{Z}$.

Exercise 50. Prove that there exists a basis $v_1, w_1, v_2, w_2, \cdots, v_r, w_r$ so that $v_i \cdot v_j = 0$ for all i, j; $w_i \cdot w_j = 0$ for all i, j; and $v_i \cdot w_j = \delta_{ij}$. So $(V, \cdot)$ has matrix

$$\begin{bmatrix} 0 & 1 & & & \\ -1 & 0 & & & \\ & & 0 & 1 & \\ & & -1 & 0 & \\ & & & & \ddots \end{bmatrix}$$

(all other entries zero) in this basis. Such a basis is called a *symplectic basis.* The closed surface of genus r is an example; describe a symplectic basis geometrically.

Hence unimodular skew-symmetric pairings over $\mathbf{Z}$ are classified by their rank. In other words, the integer intersection form of a $4k-2$-dimensional manifold M contains no more information than the dimension of $H^{2k+1}(M)$.

Case 2. k is even. Thus dim $M = 4\ell$. Then $v \cdot w = w \cdot v$, so $(V, \cdot\,)$ is a *symmetric* and unimodular inner product space over $\mathbf{Z}$.

There are 3 invariants of such unimodular symmetric forms:

1. The *rank* of $(V, \cdot\,)$ is the rank of V as a free abelian group.
2. The *signature* of $(V, \cdot\,)$ is the difference of the number of positive eigenvalues and the number of negative eigenvalues in any matrix representation of $(V, \cdot)$. (The eigenvalues of a symmetric real matrix are all real.)

Notice that in any basis $\{v_i\}$ for V, the form $\cdot$ defines a matrix Q with $Q_{ij} = v_i \cdot v_j$. Since Q is symmetric, there exists a basis *over the real numbers* so that in this basis Q is diagonal (with real eigenvalues).

Exercise 51. Show that although the eigenvalues of Q are not well defined, their signs are well defined, so that the signature is well defined. (This is often called Sylvester's Theorem of Inertia.)

3. The *type* (odd or even) of $(V, \cdot)$ is defined to be *even* if and only if $v \cdot v$ is even for all $v \in V$. Otherwise the type is said to be *odd.*

The form $(V, \cdot\,)$ is called *definite* if the absolute value of its signature equals its rank; i.e. the eigenvalues of Q are either all positive or all negative.

The main result about unimodular integral forms is the following. For a proof see e.g. [**29**].

Theorem 3.27. *If $(V, \cdot\,), (W, \cdot\,)$ are two unimodular, symmetric, indefinite forms over $\mathbf{Z}$, then V and W are isometric (i.e. there exists an isomorphism $V \to W$ preserving the inner product) if and only if they have the same rank, signature, and type.*

In fact, any odd indefinite form is equivalent to $\bigoplus_\ell (1) \bigoplus_m (-1)$, and any even indefinite form is equivalent to

$$\bigoplus_\ell \begin{bmatrix} 0 & 1 \\ 1 & 0 \end{bmatrix} \bigoplus_m E8$$

where

$$E8 = \begin{bmatrix} 2 & 1 & & & & & & \\ 1 & 2 & 1 & & & & & \\ & 1 & 2 & 1 & & & & \\ & & 1 & 2 & 1 & & & \\ & & & 1 & 2 & 1 & 0 & 1 \\ & & & & 1 & 2 & 1 & 0 \\ & & & & 0 & 1 & 2 & 0 \\ & & & & 1 & 0 & 0 & 2 \end{bmatrix}$$

(all other entries zero). □

Exercise 52. Prove $E8$ is unimodular and has signature equal to 8.

The classification of definite forms is not known. It is known that:

1. For each rank, there are finitely many isomorphism types.
2. If $(V,\cdot)$ is definite and *even*, then $\text{sign}(V,\cdot) \equiv 0 \bmod 8$.
3. There are

1	even, positive definite	rank 8	forms
2	"	rank 16	"
24	"	rank 24	"
710^{51}	"	rank 40	"

Definition 3.28. The *signature*, sign M, of a compact, oriented $4k$-manifold without boundary M is the signature of its intersection form

$$H^{2k}(M;\mathbf{Z})/T \times H^{2k}(M;\mathbf{Z})/T \to \mathbf{Z}.$$

The following sequence of exercises introduces the important technique of *bordism* in geometric topology. The topic will be revisited from the perspective of algebraic topology in Chapter 8.

Exercise 53.

1. Let M be a closed odd–dimensional manifold. Show that the Euler characteristic $\chi(M) = 0$. Prove it for nonorientable manifolds, too.
2. Let M be a closed, orientable manifold of dimension $4k+2$. Show that $\chi(M)$ is even.
3. Let M be a closed, oriented manifold of dimension $4k$. Show that the signature sign M is congruent to $\chi(M)$ mod 2.
4. Let M be the boundary of a compact manifold W. Show $\chi(M)$ is even.
5. Let M be the boundary of an compact, oriented manifold W and suppose the dimension of M is $4k$. Show sign $M = 0$.

6. Give examples of manifolds which are and manifolds which are not boundaries.

We have seen that even–dimensional manifolds admit intersection forms on the free part of their middle dimensional cohomology. For odd-dimensional manifolds one can construct the *linking form* on the torsion part of the middle dimensional cohomology as well. The construction is a bit more involved. We will outline one approach. Underlying this construction is the following exercise.

Exercise 54. If M is a compact, closed, oriented manifold of dimension n, show that the torsion subgroups of $H^p(M)$ and $H^{n-p+1}(M)$ are isomorphic. (Note: you will use the fact that $H_*(M)$ is finitely generated if M is a compact manifold.)

Consider the short exact sequence of abelian groups

$$0 \to \mathbf{Z} \to \mathbf{Q} \to \mathbf{Q}/\mathbf{Z} \to 0.$$

For any space X, one can dualize this sequence with the (integer) singular complex to obtain a short exact sequence of cochain complexes

$$0 \to \mathrm{Hom}_{\mathbf{Z}}(S_*(X), \mathbf{Z}) \to \mathrm{Hom}_{\mathbf{Z}}(S_*(X), \mathbf{Q}) \to \mathrm{Hom}_{\mathbf{Z}}(S_*(X), \mathbf{Q}/\mathbf{Z}) \to 0.$$

The zig-zag lemma gives a long exact sequence in cohomology

$$\cdots \to H^{q-1}(X; \mathbf{Q}/\mathbf{Z}) \xrightarrow{\delta} H^q(X; \mathbf{Z}) \xrightarrow{i} H^q(X; \mathbf{Q}) \to \cdots .$$

Exercise 55. Prove that the map $\delta : H^{q-1}(X; \mathbf{Q}/\mathbf{Z}) \to H^q(X; \mathbf{Z})$ maps onto the torsion subgroup T of $H^q(X; \mathbf{Z})$.

(The map δ is a Bockstein homomorphism; see Section 10.4.)

The bilinear map

$$\mathbf{Q}/\mathbf{Z} \times \mathbf{Z} \to \mathbf{Q}/\mathbf{Z}, \quad (a, b) \mapsto ab$$

is nondegenerate, in fact induces an isomorphism $\mathbf{Q}/\mathbf{Z} \otimes \mathbf{Z} \to \mathbf{Q}/\mathbf{Z}$. This bilinear map can be used to define a cup product

$$H^{q-1}(X; \mathbf{Q}/\mathbf{Z}) \times H^q(X; \mathbf{Z}) \to H^{2q-1}(X; \mathbf{Q}/\mathbf{Z}) \tag{3.10}$$

as in Exercise 36.

Now suppose that M is a closed manifold of dimension $2k-1$. Let $T \subset H^k(M; \mathbf{Z})$ denote the torsion subgroup.

Exercise 56. Prove that the *linking pairing of* M

$$T \times T \to \mathbf{Q}/\mathbf{Z}$$

defined by

$$(a, b) \mapsto \langle \delta^{-1}(a) \cup b, [M] \rangle$$

is well defined. Here $\delta^{-1}(a)$ means any element z in $H^{k-1}(M;\mathbf{Q}/\mathbf{Z})$ with $\delta(z)=a$.

Show that this pairing is skew symmetric if $\dim(M)=4\ell+1$ and symmetric if $\dim(M)=4\ell-1$.

It is a little bit harder to show that this pairing is nonsingular (the proof uses Exercise 54 in the same way that the corresponding fact for the free part of cohomology is used to show that the intersection pairing is nonsingular).

Chapter 4

Fiber Bundles

Fiber bundles form a nice class of maps in topology, and many naturally occurring maps are fiber bundles. A theorem of Hurewicz says that fiber bundles are fibrations, and fibrations are a natural class of maps to study in algebraic topology, as we will soon see. There are several alternate notions of fiber bundles, and their relationships with one another are somewhat technical. The standard reference is Steenrod's book [**37**].

A fiber bundle is also called a Hurewicz fiber bundle or a locally trivial fiber bundle. The word "fiber" is often spelled "fibre", even by people who live in English speaking countries in the Western hemisphere.

4.1. Group actions

Let G be a *topological group.* This means that G is a topological space and also a group so that the multiplication map $\mu : G \times G \to G, \mu(g,h) = gh$ and the inversion map $\iota : G \to G, \iota(g) = g^{-1}$ are continuous.

Definition 4.1. A *topological group G acts on a space X* if there is a group homomorphism $G \to \mathrm{Homeo}(X)$ such that the adjoint

$$G \times X \to X \qquad (g,x) \mapsto g(x)$$

is continuous. We will usually write $g \cdot x$ instead of $g(x)$.

The *orbit* of a point $x \in X$ is the set $Gx = \{g \cdot x | g \in G\}$.

The *orbit space* or *quotient space* X/G is the quotient space $X/\sim$, with the equivalence relation $x \sim g \cdot x$.

The *fixed set* is $X^G = \{x \in X | g \cdot x = x \text{ for all } g \in G\}$.

An action is called *free* if $g(x) \neq x$ for all $x \in X$ and for all $g \neq e$.

An action is called *effective* if the homomorphism $G \to \text{Homeo}(X)$ is injective.

A variant of this definition requires the homomorphism $G \to \text{Homeo}(X)$ to be continuous with respect to the compact-open topology on $\text{Homeo}(X)$, or some other topology, depending on what X is (for example, one could take the C^∞ topology on $\text{Diff}(X)$ if X is a smooth manifold). Also note that we have defined a *left* action of G on X. There is a corresponding notion of right G-action $(x, g) \mapsto x \cdot g$. For example, one usually takes the action of $\pi_1 X$ by covering transformations on the universal cover of X to be a right action.

4.2. Fiber bundles

We can now give the definition of a fiber bundle.

Definition 4.2. Let G be a topological group acting effectively on a space F. A *fiber bundle E over B with fiber F and structure group G* is a map $p : E \to B$ together with a collection of homeomorphisms $\{\varphi : U \times F \to p^{-1}(U)\}$ for open sets U in B (φ is called a *chart over* U) such that:

1. The diagram

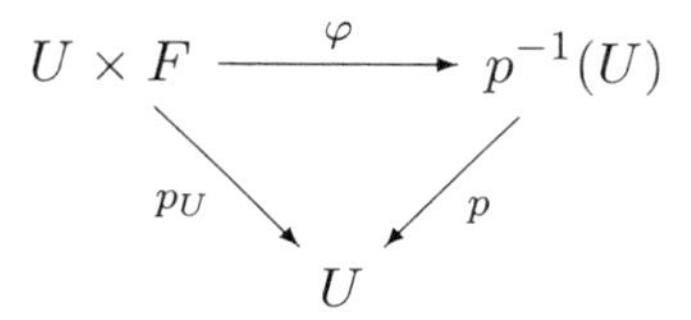

   commutes for each chart φ over U.
2. Each point of B has a neighborhood over which there is a chart.
3. If φ is a chart over U and $V \subset U$ is open, then the restriction of φ to V is a chart over V.
4. For any charts φ, φ' over U, there is a continuous map $\theta_{\varphi,\varphi'} : U \to G$ so that
$$\varphi'(u, f) = \varphi(u, \theta_{\varphi,\varphi'}(u) \cdot f)$$
   for all $u \in U$ and all $f \in F$. The map $\theta_{\varphi,\varphi'}$ is called the *transition function* for φ, φ'.
5. The collection of charts is maximal among collections satisfying the previous conditions.

The standard terminology is to call B the *base*, F the *fiber*, and E the *total space*. For shorthand one often abbreviates (p, E, B, F, G) by E.

This definition of fiber bundle is slick, and some discussion about the various requirements helps one to understand the concept.

A map $p : E \to B$ is called a *locally trivial bundle* if the first 3 requirements of Definition 4.2 are met. There is no need to assume that any group G is acting since this does not enter into the first three axioms. Local triviality is the important distinction between a fiber bundle and an arbitrary map.

The fourth condition invokes the structure group G. To understand the difference between a locally trivial bundle and a fiber bundle, notice that in a locally trivial bundle, if

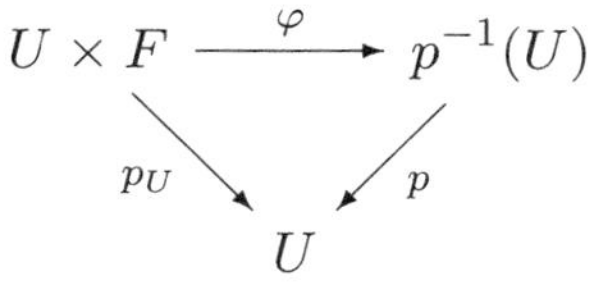

$$\begin{array}{ccc} U \times F & \xrightarrow{\varphi} & p^{-1}(U) \\ & {\scriptstyle p_U}\searrow \quad \swarrow{\scriptstyle p} & \\ & U & \end{array}$$

and

$$\begin{array}{ccc} U \times F & \xrightarrow{\varphi'} & p^{-1}(U) \\ & {\scriptstyle p_U}\searrow \quad \swarrow{\scriptstyle p} & \\ & U & \end{array}$$

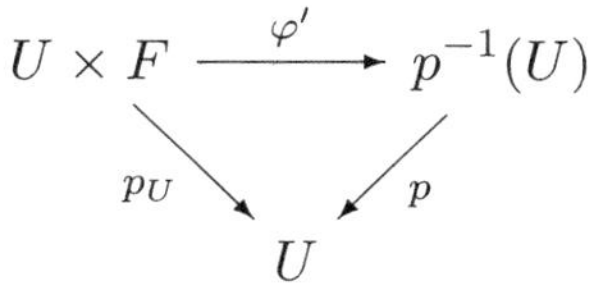

are two local trivializations, then commutativity of the diagram

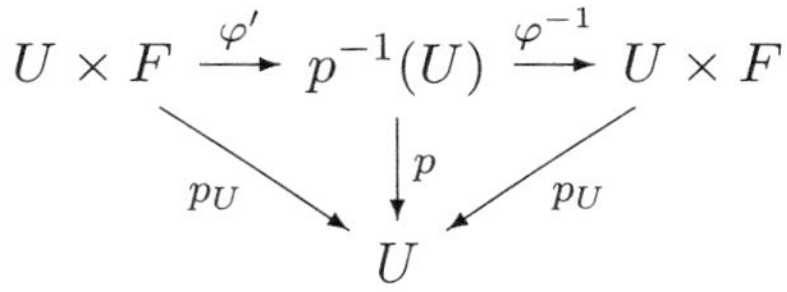

$$\begin{array}{ccccc} U \times F & \xrightarrow{\varphi'} & p^{-1}(U) & \xrightarrow{\varphi^{-1}} & U \times F \\ & {\scriptstyle p_U}\searrow & \downarrow{\scriptstyle p} & \swarrow{\scriptstyle p_U} & \\ & & U & & \end{array}$$

implies that there is a map $\psi_{\varphi,\varphi'} : U \times F \to F$ so that the composite $\varphi^{-1} \circ \varphi' : U \times F \to U \times F$ has the formula

$$(u, f) \mapsto (u, \psi_{\varphi,\varphi'}(u, f)).$$

For each $u \in U$ the map $\psi_{\varphi,\varphi'}(u, -) : F \to F$ is a homeomorphism.

In a fiber bundle, the map $\psi_{\varphi,\varphi'}$ must have a very special form, namely

1. The homeomorphism $\psi_{\varphi,\varphi'}(u, -) : F \to F$ is not arbitrary, but is given by the action of an element of G, i.e. $\psi_{\varphi,\varphi'}(u, f) = g \cdot f$ for some $g \in G$ independent of f. The element g is denoted by $\theta_{\varphi,\varphi'}(u)$.
2. The topology of G is integrated into the structure by requiring that $\theta_{\varphi,\varphi'} : U \to G$ be continuous.

The requirement that G act effectively on F implies that the functions $\theta_{\varphi,\varphi'} : U \to G$ are unique. Although we have included the requirement that G acts effectively on F in the definition of a fiber bundle, there are certain circumstances when we will want to relax this condition, particularly when studying liftings of the structure group, for example, when studying local coefficients.

It is not hard to see that a locally trivial bundle is the same thing as a fiber bundle with structure group Homeo(F). One subtlety about the topology is that the requirement that G be a topological group acting effectively on F says only that the homomorphism $G \to \text{Homeo}(F)$ is injective, but the inclusion $G \to \text{Homeo}(F)$ need not be an embedding, nor even continuous.

Exercise 57. Show that the transition functions determine the bundle. That is, suppose that spaces B and F are given, and an action of a topological group G on F is specified.

Suppose also that a collection of pairs $\mathcal{T} = \{(U_\alpha, \theta_\alpha)\}$ with each U_α an open subset of B and $\theta_\alpha : U_\alpha \to G$ a continuous map is given satisfying:

1. The U_α cover B.
2. If $(U_\alpha, \theta_\alpha) \in \mathcal{T}$ and $W \subset U_\alpha$, then the restriction $(W, \theta_{\alpha|W})$ is in the collection $\mathcal{T}$.
3. If (U, θ_1) and (U, θ_2) are in $\mathcal{T}$, then $(U, \theta_1 \cdot \theta_2)$ is in $\mathcal{T}$, where $\theta_1 \cdot \theta_2$ means the pointwise multiplication of functions to G.
4. The collection $\mathcal{T}$ is maximal with respect to the first three conditions.

Then there exists a fiber bundle $p : E \to B$ with structure group G, fiber F, and transition functions θ_α.

The third condition in Exercise 57 is a hidden form of the famous *cocycle condition.* Briefly what this means is the following. In an alternative definition of a fiber bundle one starts with a fixed open cover $\{U_i\}$ and a single function $\varphi_i : U_i \times F \to p^{-1}(U_i)$ of each open set U_i of the cover. Then to each pair of open sets U_i, U_j in the cover one requires that there exists a function $\theta_{i,j} : U_i \cap U_j \to G$ so that (on $U_i \cap U_j$)

$$\varphi_i^{-1} \circ \varphi_j(u, f) = (u, \theta_{i,j}(u) \cdot f).$$

A *G-valued Čech 1-cochain for the cover* $\{U_i\}$ is just a collection of maps $\theta_{i,j} : U_i \cap U_j \to G$, and so a fiber bundle with structure group G determines a Čech 1-cochain.

From this point of view the third condition of the exercise translates into the requirement that for each triple U_i, U_j and U_k the restrictions of the various θ satisfy

$$\theta_{i,j} \cdot \theta_{j,k} = \theta_{i,k} : U_i \cap U_j \cap U_k \to G.$$

In the Čech complex this condition is just the requirement that the Čech 1-cochain defined by the $\theta_{i,j}$ is in fact a cocycle.

This is a useful method of understanding bundles since it relates them to (Čech) cohomology. Cohomologous cochains define isomorphic bundles, and so equivalence classes of bundles over B with structure group G can be identified with $H^1(B;G)$ (this is one starting point for the theory of characteristic classes; we will take a different point of view in a later chapter). One must be extremely cautious when working this out carefully. For example, G need not be abelian (and so what does $H^1(B;G)$ mean?) Also, one must consider *continuous cocycles* since the $\theta_{i,j}$ should be continuous functions. We will not pursue this line of exposition any further in this book.

We will frequently use the notation $F \hookrightarrow E \xrightarrow{p} B$ or

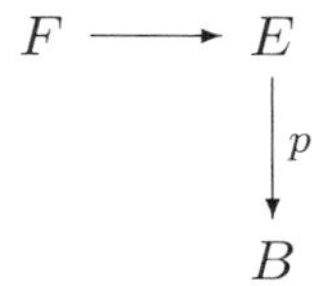

to indicate a fiber bundle $p : E \to B$ with fiber F.

4.3. Examples of fiber bundles

The following are some examples of locally trivial bundles. We will revisit these and many more examples in greater detail in Section 6.14.

1. The *trivial* bundle is the projection $p_B : B \times F \to B$.
2. If F has the discrete topology, any locally trivial bundle over B with fiber F is a covering space; conversely if $p : E \to B$ is a covering space with B path connected, then p is a locally trivial bundle with discrete fiber.
3. The Möbius strip mapping onto its core circle is a locally trivial bundle with fiber $[0,1]$.
4. The tangent bundle of a smooth manifold is a locally trivial bundle.

Exercise 58. Show that a fiber bundle with trivial structure group is (isomorphic to) a trivial bundle.

4.3.1. Vector bundles.

Exercise 59. Let $F = \mathbf{R}^n$, and let $G = GL_n\mathbf{R} \subset \text{Homeo}(\mathbf{R}^n)$. A fiber bundle over B with fiber $\mathbf{R}^n$ and structure group $GL_n(\mathbf{R})$ is called a *vector bundle* of *dimension n over B*. Show that each fiber $p^{-1}\{b\}$ can be given a well defined vector space structure.

(Similarly, one can take $F = \mathbf{C}^n$, $G = GL_n(\mathbf{C})$ to get *complex vector bundles.*)

In particular, if M is a differentiable n-manifold and TM is the set of all tangent vectors to M, then $p : TM \to M$ is a vector bundle of dimension n.

4.3.2. Bundles over S^2. For every integer $n \geq 0$, we can construct an S^1 bundle over S^2 with structure group $SO(2)$; n is called the *Euler number* of the bundle. For $n = 0$, we have the product bundle $p : S^2 \times S^1 \to S^2$. For $n \geq 1$, define a 3-dimensional lens space $L_n^3 = S^3/\mathbf{Z}_n$, where the action is given by letting the generator of $\mathbf{Z}_n$ act on $S^3 \subset \mathbf{C}^2$ by $(z_1, z_2) \mapsto (\zeta_n z_1, \zeta_n z_2)$ (here $\zeta_n = \exp^{2\pi i/n}$ is a primitive n-th root of unity). For $n = 2$, the lens space is just real projective space $\mathbf{R}P^3$. Define the S^1-bundle with Euler number $n \geq 1$ by $p : L_n^3 \to S^2 = \mathbf{C} \cup \infty$ by $[z_1, z_2] \to z_1/z_2$.

When $n = 1$ we obtain the famous Hopf bundle $S^1 \hookrightarrow S^3 \to S^2$. For $n > 1$ the Hopf map $S^3 \to S^2$ factors through the quotient map $S^3 \to L_n^3$, and the fibers of the bundle with Euler number n are $S^1/\mathbf{Z}_n$ which is again homeomorphic to S^1.

Exercise 60. Let $S(TS^2)$ be the sphere bundle of the tangent bundle of the 2-sphere, i.e. the tangent vectors of unit length, specifically

$$S(TS^2) = \{(P, v) \in \mathbf{R}^3 \times \mathbf{R}^3 | P, v \in S^2 \text{ and } P \cdot v = 0\}.$$

Let $SO(3)$ be the 3-by-3 orthogonal matrices of determinant one (the group of orientation preserving rigid motions of $\mathbf{R}^3$ preserving the origin). This is a topological group. Show that the spaces $S(TS^2)$, $SO(3)$, and $\mathbf{R}P^3$ are all homeomorphic.

(Hints:

1. Given two perpendicular vectors in $\mathbf{R}^3$, a third one can be obtained by the cross product.
2. On one hand, every element of $SO(3)$ is rotation about an axis. On the other hand $\mathbf{R}P^3$ is $D^3/\sim$, where you identify antipodal points on the boundary sphere.)

This gives three incarnations of the S^1-bundle over S^2 with Euler number equal to 2:

1. $p : S(TS^2) \to S^2$, $(P, v) \mapsto P$.
2. $p : SO(3) \to S^2$, $A \mapsto A \begin{bmatrix} 1 \\ 0 \\ 0 \end{bmatrix}$.
3. $p : \mathbf{R}P^3 \to S^2$, the lens space bundle above.

4.3.3. Clutching. Suppose a topological group G acts on a space F. Let X be a space and let ΣX be the unreduced suspension of X,

$$\Sigma X = \frac{X \times I}{(x, 0) \sim (x', 0), (x, 1) \sim (x', 1)}.$$

Then given a map $\beta : X \to G$, define

$$E = \frac{(X \times [0,1/2] \times F) \amalg (X \times [1/2,1] \times F)}{\sim}$$

where the equivalence relation is given by identifying $(x, 0, f) \sim (x', 0, f)$, $(x, 1, f) \sim (x', 1, f)$, and $(x, 1/2, f) \sim (x, 1/2, \beta(x)f)$; the last relation glues the summands of the disjoint union. Projecting to the first two factors defines a map $E \to \Sigma X$ called the *bundle over* ΣX *with clutching function* $\beta : X \to G \subset \mathrm{Aut}(F)$.

Exercise 61. Show that projection onto the first two coordinates gives a fiber bundle $p : E \to \Sigma X$ with fiber F and structure group G. Give some examples with $X = S^0$ and $X = S^1$. In particular, show that the S^1-bundle over $S^2 = \Sigma S^1$ with Euler number equal to n is obtained by clutching using a degree n map $S^1 \to S^1$.

Clutching provides a good way to describe fiber bundles over spheres. For X a CW-complex, all bundles over ΣX arise by this clutching construction. This follows from the fact that any fiber bundle over a contractible CW-complex is trivial (this can be proven using obstruction theory). Since ΣX is the union of two contractible spaces, $X \times [0, \frac{1}{2}]/\sim$ and $X \times [\frac{1}{2}, 1]/\sim$, any bundle over ΣX is obtained by clutching two trivial bundles over X.

4.3.4. Local coefficients and other structures. An important type of fiber bundle is the following. Let A be a group and G a subgroup of the automorphism group $\mathrm{Aut}(A)$. Then any fiber bundle E over B with fiber A and structure group G has the property that each *fiber* $p^{-1}\{b\}$ has a group structure. This group is isomorphic to A, but the isomorphism is not canonical in general.

We have already run across an important case of this, namely vector bundles, where $A = \mathbf{R}^n$ and $G = GL_n(\mathbf{R})$.

In particular, if A is an abelian group with the discrete topology, then $p : E \to B$ is a covering space and is called a *system of local coefficients on* B. The terminology will be explained later.

Exercise 62. Define local coefficient systems for R-modules, R a commutative ring, generalizing the case of $\mathbf{Z}$-modules above.

The basic principle at play here is *if the structure group preserves a certain structure on* F*, then every fiber* $p^{-1}\{b\}$ *has this structure.* For example, a local coefficient system corresponds to the case when the structure group is a subgroup of the group of automorphisms of the fiber, a discrete abelian group. A vector bundle corresponds to the case when the structure group corresponds to the group of linear transformations of a vector space.

Other examples of fibers with a structure include the following.

1. F is a real vector space with an inner product; $G = O(F, \langle \ , \ \rangle) \subset GL(F)$ consists of those linear isomorphisms which preserve the inner product. The resulting fiber bundle is called a *vector bundle with an orthogonal structure.*
2. Similarly one can define a *complex vector bundle with hermitian structure* by taking F to be a complex vector space with a hermitian inner product.
3. Taking this further, let F be a riemannian manifold and suppose that G acts isometrically on F. Then each fiber in a fiber bundle with structure group G and fiber F will be (non-canonically) isometric to F.
4. Take F to be a smooth manifold and G a subgroup of the diffeomorphism group of F (with the C^∞ strong topology, say). Then each fiber in a fiber bundle with structure group G will be diffeomorphic to F.

Exercise 63. Invent your own examples of fibers with structure and the corresponding fiber bundles.

4.4. Principal bundles and associated bundles

Principal bundles are special cases of fiber bundles, but nevertheless can be used to construct any fiber bundle. Conversely any fiber bundle determines a principal bundle. A principal bundle is technically simpler, since the fiber is just $F = G$ with a canonical action.

Let G be a topological group. It acts on itself by *left translation*.

$$G \to \mathrm{Homeo}(G), \quad g \mapsto (x \mapsto gx).$$

Definition 4.3. A *principal G-bundle over B* is a fiber bundle $p : P \to B$ with fiber $F = G$ and structure group G acting by left translations.

Proposition 4.4. *If $p : P \to B$ is a principal G-bundle, then G acts freely on P on the right with orbit space B.*

Proof. Notice first that G acts on the local trivializations *on the right*:

$$(U \times G) \times G \to U \times G$$

$$(u, g) \cdot g' = (u, gg').$$

This commutes with the action of G on itself by left translation (i.e. $(g''g)g' = g''(gg')$), so one gets a well defined right action of G on E using the identification provided by a chart

$$U \times G \xrightarrow{\varphi} p^{-1}(U).$$

More explicitly, define $\varphi(u,g)\cdot g' = \varphi(u,gg')$. If φ' is another chart over U, then

$$\varphi(u,g) = \varphi'(u, \theta_{\varphi,\varphi'}(u)g),$$

and $\varphi(u,gg') = \varphi'(u,\theta_{\varphi,\varphi'}(u)(gg')) = \varphi'(u,(\theta_{\varphi,\varphi'}(u)g)g')$, so the action is independent of the choice of chart. The action is free, since the local action $(U \times G) \times G \to U \times G$ is free, and since $U \times G/G = U$ it follows that $E/G = B$. □

As a familiar example, any regular covering space $p : E \to B$ is a principal G-bundle with $G = \pi_1 B / p_* \pi_1 E$. Here G is given the discrete topology. In particular, the universal covering $\tilde{B} \to B$ of a space is a principal $\pi_1 B$-bundle. A non-regular covering space is not a principal G-bundle.

Exercise 64. Any free (right) action of a finite group G on a (Hausdorff) space E gives a regular cover and hence a principal G-bundle $E \to E/G$.

The converse to Proposition 4.4 holds in some important cases. We state the following fundamental theorem without proof, referring you to [**5**, Theorem II.5.8].

Theorem 4.5. *Suppose that X is a compact Hausdorff space, and G is a compact Lie group acting freely on X. Then the orbit map*

$$X \to X/G$$

is a principal G-bundle. □

4.4.1. Construction of fiber bundles from principal bundles. Exercise 57 shows that the transition functions $\theta_\alpha : U_\alpha \to G$ and the action of G on F determine a fiber bundle over B with fiber F and structure group G.

As an application note that if a topological group G acts effectively on spaces F and F', and if $p : E \to B$ is a fiber bundle with fiber F and structure group G, then one can use the transition functions from p to define a fiber bundle $p' : E' \to B$ with fiber F' and structure group G with *exactly the same transition functions.*

This is called *changing the fiber from F to F'.* This can be useful because the topology of E and E' may change. For example, take $G = GL_2(\mathbf{R})$, $F = \mathbf{R}^2$, $F' = \mathbf{R}^2 - \{0\}$ and the tangent bundle of the 2-sphere:

$$\begin{array}{ccc} \mathbf{R}^2 & \longrightarrow & TS^2 \\ & & \downarrow p \\ & & S^2 \end{array}$$

After changing the fiber from $\mathbf{R}^2$ to $\mathbf{R}^2 - \{0\}$, we obtain

$$\begin{array}{ccc} \mathbf{R}^2 - \{0\} & \longrightarrow & TS^2 - z(S^2) \\ & & \downarrow p \\ & & S^2 \end{array}$$

where $z : S^2 \to TS^2$ denotes the zero section.

With the second incarnation of the bundle the twisting becomes revealed in the homotopy type, because the total space of the first bundle has the homotopy type of S^2, while the total space of the second has the homotopy type of the sphere bundle $S(TS^2)$ and hence of $\mathbf{R}P^3$ according to Exercise 60.

A fundamental case of changing fibers occurs when one lets the fiber F' be the group G itself, with the left translation action. Then the transition functions for the fiber bundle

$$\begin{array}{ccc} F & \longrightarrow & E \\ & & \downarrow p \\ & & B \end{array}$$

determine, via the construction of Exercise 57, a principal G-bundle

$$\begin{array}{ccc} G & \longrightarrow & P(E) \\ & & \downarrow p \\ & & B. \end{array}$$

We call this principal G-bundle the *principal G-bundle underlying the fiber bundle $p : E \to B$ with structure group G.*

Conversely, to a principal G-bundle and an action of G on a space F one can associate a fiber bundle, again using Exercise 57. An alternative construction is given in the following definition.

Definition 4.6. Let $p : P \to B$ be a principal G-bundle. Suppose G acts on the left on a space F; i.e. an action $G \times F \to F$ is given. Define the *Borel construction*

$$P \times_G F$$

to be the quotient space $P \times F/\sim$ where

$$(x, f) \sim (xg, g^{-1}f).$$

(We are continuing to assume that G acts on F on the left and by Proposition 4.4 it acts freely on the principal bundle P on the right.)

Let $[x, f] \in P \times_G F$ denote the equivalence classes of (x, f). Define a map

$$q : P \times_G F \to B$$

by the formula $[x, f] \mapsto p(x)$.

The following important exercise shows that the two ways of going from a principal G-bundle to a fiber bundle with fiber F and structure group G are the same.

Exercise 65. If $p : P \to B$ is a principal G-bundle and G acts on F, then

$$\begin{array}{ccc} F & \longrightarrow & P \times_G F \\ & & \downarrow{\scriptstyle q} \\ & & B \end{array}$$

(where $q[x, f] = p(x)$) is a fiber bundle over B with fiber F and structure group G which has the same transition functions as $p : P \to B$.

We say $q : E \times_G F \to B$ is *the fiber bundle associated to the principal bundle $p : E \to B$ via the action of G on F.*

Thus principal bundles are more basic than fiber bundles, in the sense that the fiber and its G-action are explicit, namely G acting on itself by left translation. Moreover, any fiber bundle with structure group G is associated to a principal G-bundle by specifying an action of G on a space F. Many properties of bundles become more visible when stated in the context of principal bundles.

The following exercise gives a different method of constructing the principal bundle underlying a vector bundle, without using transition functions.

Exercise 66. Let $p : E \to B$ be a vector bundle with fiber $\mathbf{R}^n$ and structure group $GL(n, \mathbf{R})$. Define a space $F(E)$ to be the space of *frames* in E, so that a point in $F(E)$ is a pair $(b, \mathbf{f})$ where $b \in B$ and $\mathbf{f} = (f_1, \cdots, f_n)$ is a basis for the vector space $p^{-1}(b)$. There is an obvious map $q : F(E) \to B$.

Prove that $q : F(E) \to B$ is a principal $GL(n, \mathbf{R})$-bundle, and that

$$E = F(E) \times_{GL(n,\mathbf{R})} \mathbf{R}^n$$

where $GL(n, \mathbf{R})$ acts on $\mathbf{R}^n$ in the usual way.

For example, given a representation of $GL(n, \mathbf{R})$, that is, a homomorphism $\rho : GL(n, \mathbf{R}) \to GL(k, \mathbf{R})$, one can form a new vector bundle

$$F(E) \times_\rho \mathbf{R}^k$$

over B.

An important set of examples comes from this construction by starting with the tangent bundle of a smooth manifold M. The principal bundle

$F(TM)$ is called the *frame bundle* of M. Any representation of $GL(n, \mathbf{R})$ on a vector space V gives a vector bundle with fiber isomorphic to V. Important representations include the *alternating representations* $GL(n, \mathbf{R}) \to GL\big(\Lambda^p(\text{Hom}(\mathbf{R}^n, \mathbf{R}))\big)$ from which one obtains the vector bundles of differential p-forms over M.

We next give one application of the Borel construction. Recall that a local coefficient system is a fiber bundle over B with fiber A and structure group G where A is a (discrete) abelian group and G acts via a homomorphism $G \to \text{Aut}(A)$.

Lemma 4.7. *Every local coefficient system over a path-connected (and semi-locally simply connected) space B is of the form*

$$\begin{array}{ccc} A & \longrightarrow & \tilde{B} \times_{\pi_1 B} A \\ & & \big\downarrow q \\ & & B \end{array}$$

i.e. is associated to the principal $\pi_1 B$-bundle given by the universal cover $\tilde{B}$ of B where the action is given by a homomorphism $\pi_1 B \to \text{Aut}(A)$.

In other words the group $G \subset \text{Aut}(A)$ can be replaced by the discrete group $\pi_1 B$. Notice that in general one cannot assume that the homomorphism $\pi_1 B \to \text{Aut}(A)$ is injective, and so this is a point where we would wish to relax the requirement that the structure group acts effectively on the fiber. Alternatively, one can take the structure group to be $\pi_1(B)/\ker(\phi)$ where $\phi : \pi_1(B) \to \text{Aut}(A)$ is the corresponding representation.

Sketch of proof. It is easy to check that $q : \tilde{B} \times_{\pi_1 B} A \to B$ is a local coefficient system, i.e. a fiber bundle with fiber on abelian group A and structure group mapping to $\text{Aut}(A)$.

Suppose that $p : E \to B$ is any local coefficient system. Any loop $\gamma : (I, \partial I) \to (B, *)$ has a unique lift to E starting at a given point in $p^{-1}(*)$, since A is discrete so that $E \to B$ is a covering space. Fix an identification of $p^{-1}(*)$ with A, given by a chart. Then the various lifts of γ starting at points of A define, by taking the end point, a function $A \to A$.

The fact that $p : E \to B$ has structure group $\text{Aut}(A)$ easily implies that this function is an automorphism. Since E is a covering space of B, the function only depends on the homotopy class, and so we get a map $\pi_1(B, *) \to \text{Aut}(A)$. This is clearly a homomorphism since if $\tilde{\gamma_1}, \tilde{\gamma_2}$ are lifts starting at a, b, then $\tilde{\gamma_1} + \tilde{\gamma_2}$ (addition in A) is the lift of $\gamma_1 \gamma_2$ (multiplication in π_1) and starts at $a + b$. A standard covering space argument implies that $E = \tilde{B} \times_{\pi_1 B} A$. □

4.5. Reducing the structure group

In some circumstances, given a subgroup H of G and a fiber bundle $p : E \to B$ with structure group G, one can view the bundle as a fiber bundle with structure group H. When this is possible, we say *the structure group can be reduced to H.*

Proposition 4.8. *Let H be a topological subgroup of the topological group G. Let H act on G by left translation. Let $q : Q \to B$ be a principal H-bundle. Then*

$$\begin{array}{ccc} G & \longrightarrow & Q \times_H G \\ & & \downarrow{\scriptstyle q} \\ & & B \end{array}$$

is a principal G-bundle. □

The proof is easy; one approach is to consider the transition functions $\theta : U \to H$ as functions to G using the inclusion $H \subset G$. To satisfy maximality of the charts it may be necessary to add extra charts whose transition functions into G map outside of H.

Exercise 67. Prove Proposition 4.8.

Definition 4.9. Given a principal G-bundle $p : E \to B$ we say *the structure group G can be reduced to H* for some subgroup $H \subset G$ if there exists a principal H-bundle $Q \to B$ and a commutative diagram

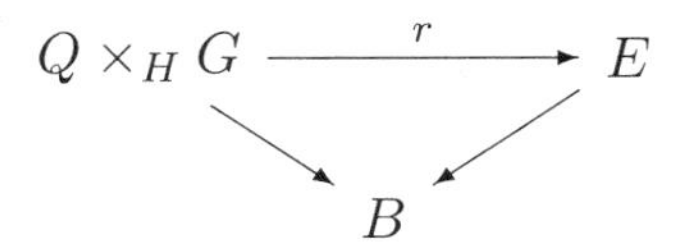

so that the map r is G-equivariant. For a fiber bundle, we say the structure group reduces if the structure group of the underlying principal bundle reduces.

If we are willing to relax the requirement that the structure group acts effectively, then we can just assume that we are given a homomorphism $H \to G$ rather than an inclusion of a subgroup. Proposition 4.8 holds without change. In this more general context, for example, Lemma 4.7 states that any fiber bundle over B with discrete fiber can have its structure group reduced to $\pi_1 B$.

Exercise 68. Show that every real vector bundle (i.e. fiber bundle with structure group $GL(n, \mathbf{R})$ acting on $\mathbf{R}^n$ in the usual way) over a paracompact base can have its structure group reduced to the orthogonal group $O(n)$. (Hint: Use a partition of unity.)

Another subtle point is that there may be several "inequivalent" reductions. An example concerns orientability and orientation of vector bundles.

Definition 4.10. A real vector bundle is called *orientable* if its structure group can be reduced to the subgroup $GL_+(n,\mathbf{R})$ of matrices with positive determinant.

For example, a smooth manifold is orientable if and only if its tangent bundle is orientable. A more detailed discussion of orientability for manifolds and vector bundles can be found in Section 10.7.

For the following exercise it may help to read the definition of a map between fiber bundles in the next section.

Exercise 69. Prove that an orientable vector bundle can be oriented in two incompatible ways; that is, the structure group can be reduced from $GL(n,\mathbf{R})$ to $GL_+(n,\mathbf{R})$ (or, using Exercise 68, from $O(n)$ to $SO(n)$) in two ways so that the identity map Id: $E \to E$ is a not a map of fiber bundles with structure group $GL_+(n,\mathbf{R})$ (or $SO(n)$).

4.6. Maps of bundles and pullbacks

The concept of morphisms of fiber bundles is subtle, especially when there are different fibers and structure groups. Rather than try to work in the greatest generality, we will just define one of many possible notions of morphism.

Definition 4.11. A *morphism of fiber bundles with structure group G and fiber F* from $E \to B$ to $E' \to B'$ is a pair of continuous maps $\tilde{f} : E \to E'$ and $f : B \to B'$ so that the diagram

$$\begin{array}{ccc} E & \xrightarrow{\tilde{f}} & E' \\ \downarrow & & \downarrow \\ B & \xrightarrow{f} & B' \end{array}$$

commutes and so that for each chart $\phi : U \times F \to p^{-1}(U)$ with $b \in U$ and chart $\phi' : U' \times F \to p^{-1}(U')$ and each $b \in U$ with $f(b) \in U'$ the composite

$$\{b\} \times F \xrightarrow{\phi} p^{-1}(b) \xrightarrow{\tilde{f}} (p')^{-1}(f(b)) \xrightarrow{(\phi')^{-1}} \{f(b)\} \times F$$

is a homeomorphism given by the action of an element $\psi_{\phi,\phi'}(b) \in G$. Moreover, $b \mapsto \psi_{\phi,\phi'}(b)$ should define a continuous map from $U \cap f^{-1}(U')$ to G.

As you can see, this is a technical definition. Notice that the fibers are mapped homeomorphically by a map of fiber bundles of this type. In particular, an *isomorphism* of fiber bundles is a map of fiber bundles $(\tilde{f}, f)$ which admits a map $(\tilde{g}, g)$ in the reverse direction so that both composites are the identity.

One important type of fiber bundle map is a *gauge transformation*. This is a bundle map from a bundle to itself which covers the identity map of the base; i.e. the following diagram commutes:

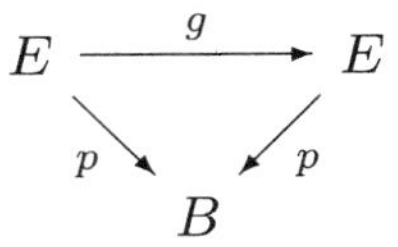

$$\begin{array}{ccc} E & \xrightarrow{g} & E \\ & {\scriptstyle p}\searrow \quad \swarrow{\scriptstyle p} & \\ & B & \end{array}$$

By definition g restricts to an isomorphism given by the action of an element of the structure group on each fiber. The set of all gauge transformations forms a group.

One way in which morphisms of fiber bundles arise is from a pullback construction.

Definition 4.12. Suppose that a fiber bundle $p : E \to B$ with fiber F and structure group G is given, and that $f : B' \to B$ is some continuous function. Define the *pullback of* $p : E \to B$ *by* f to be the space

$$f^*(E) = \{(b', e) \in B' \times E \mid p(e) = f(b')\}.$$

Let $q : f^*(E) \to B$ be the restriction of the projection $E \times B \to B$ to $f^*(E)$. Notice that there is a commutative diagram

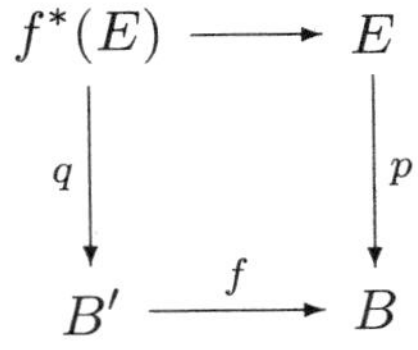

$$\begin{array}{ccc} f^*(E) & \longrightarrow & E \\ {\scriptstyle q}\downarrow & & \downarrow{\scriptstyle p} \\ B' & \xrightarrow{f} & B \end{array}$$

Theorem 4.13. *The map $q : f^*(E) \to B'$ is a fiber bundle with fiber F and structure group G. The map $f^*(E) \to E$ is a map of fiber bundles.*

Proof. This is not hard. The important observation is that if φ is a chart over $U \subset B$, then $f^{-1}(U)$ is open in B' and φ induces a homeomorphism $f^{-1}(U) \times F \to f^*(E)_{|f^{-1}(U)}$. We leave the details as an exercise. □

The following exercise shows that any map of fiber bundles is given by a pullback.

Exercise 70. Let

$$\begin{array}{ccc} E' & \xrightarrow{\tilde{f}} & E \\ {\scriptstyle p'}\downarrow & & \downarrow{\scriptstyle p} \\ B' & \xrightarrow{f} & B \end{array}$$

be a map of bundles with fiber F in the sense of Definition 4.11. Show that there is a factorization

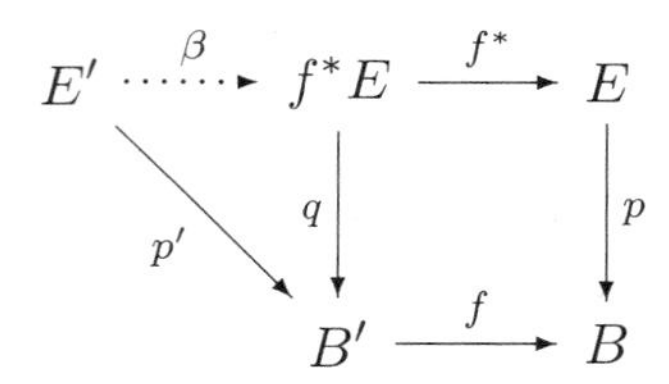

so that $f^* \circ \beta = \tilde{f}$, with (β, Id) a map of bundles over B'.

We have given a rather narrow and rigid definition of fiber bundle morphisms. More general definitions can be given depending on the structure group, fiber, etc.

Exercise 71. Define a morphism between two fiber bundles with structure group G but with different fibers by requiring the map on fibers to be equivariant. Use this to define a morphism of vector bundles.

4.7. Projects for Chapter 4

4.7.1. Fiber bundles over paracompact bases are fibrations. State and prove the theorem of Hurewicz (Theorem 6.8) which says that a map $f : E \to B$ with B paracompact is a fibration (see Definition 6.7) provided that B has an open cover $\{U_i\}$ so that $f : f^{-1}(U_i) \to U_i$ is a fibration for each i. In particular, any locally trivial bundle over a paracompact space is a fibration.

A reference for the proof is [**11**, Chapter XX, §3-4] or [**36**].

4.7.2. Classifying spaces. For any topological group G there is a space BG and a principal G-bundle $EG \to BG$ so that given any paracompact space B, the pullback construction induces a bijection between the set $[B, BG]$ of homotopy classes of maps from B to BG and isomorphism classes of principal G-bundles over B. Explain the construction of the bundle $EG \to BG$ and prove this theorem. Show that the assignment $G \mapsto BG$ is functorial with respect to continuous homomorphisms of topological groups.

Show that a principal G-bundle P is of the form $Q \times_H G$ (as in Proposition 4.8) if and only if the classifying map $f : B \to BG$ lifts to BH

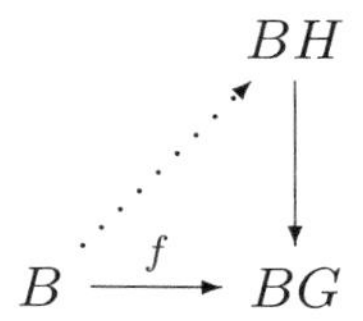

Show that given any action of G on F, any fiber bundle $E \to B$ with structure group G and fiber F is isomorphic to the pullback

$$f^*(EG \times_G F)$$

where $f : B \to BG$ classifies the principal G-bundle underlying $E \to B$. Use this theorem to define characteristic classes for principal bundles.

See Theorem 8.22 and Corollary 6.50 for more on this important topic.

A reference for this material is [**18**]. We will use these basic facts about classifying spaces throughout this book, notably when we study bordism.

Chapter 5

Homology with Local Coefficients

When studying the homotopy theory of non-simply connected spaces, one is often led to consider an action of the fundamental group on some abelian group. Local coefficient systems are a tool to organize this information. The theory becomes more complicated by the fact that one must consider non-commutative rings. It is possible to learn a good deal of homotopy theory by restricting only to simply connected spaces, but fundamental group issues are ubiquitous in geometric topology.

There are two approaches to constructing the complexes giving the homology and cohomology of a space with local coefficients. The first is more algebraic and takes the point of view that the fundamental chain complex associated to a space X is the singular (or cellular) complex of the universal cover $\tilde{X}$, viewed as a chain complex over the group ring $\mathbf{Z}[\pi_1 X]$. From this point of view local coefficients are nothing more than modules over the group ring $\mathbf{Z}[\pi_1 X]$.

The second approach is more topological; one takes a local coefficient system over X (i.e. a fiber bundle over X whose fibers are abelian groups and whose transition functions take values in the automorphisms of the group) and defines a chain complex by taking the chains to be formal sums of singular simplices (or cells) such that the coefficient of a simplex is an element in the fiber over that simplex (hence the terminology *local coefficients*). Each of these two points of view has its strengths; Lemma 4.7 is the basic result which identifies the two.

In this chapter we will work with $\mathbf{Z}$-modules, (i.e. abelian groups) and modules over integral group rings $\mathbf{Z}\pi$. Everything generalizes appropriately for R-modules and $R\pi$-modules for any commutative ring R.

5.1. Definition of homology with twisted coefficients

We begin with the definition of a group ring.

Definition 5.1. The *group ring* $\mathbf{Z}\pi$ is a ring associated to a group π. Additively it is the free abelian group on π; i.e. elements are (finite) linear combinations of the group elements

$$m_1 g_1 + \cdots + m_k g_k \qquad m_i \in \mathbf{Z}, \quad g_i \in \pi.$$

Multiplication is given by the distributive law and multiplication in π:

$$\Big(\sum_i m_i g_i\Big)\Big(\sum_j n_j h_j\Big) = \sum_{i,j} (m_i n_j)(g_i h_j).$$

In working with group rings the group π is always written multiplicatively, and if e is the identity of the group, e is written as 1, since this element forms the unit in the ring $\mathbf{Z}\pi$. To avoid confusing notation we will sometimes write $\mathbf{Z}[\pi]$ instead of $\mathbf{Z}\pi$.

Two examples of group rings (with their standard notation) are

$$\mathbf{Z}[\mathbf{Z}] = \mathbf{Z}[t, t^{-1}] = \{a_{-j}t^{-j} + \cdots + a_0 + \cdots + a_k t^k \mid a_n \in \mathbf{Z}\}$$

(this ring is called the ring of *Laurent polynomials*) and

$$\mathbf{Z}[\mathbf{Z}/2] = \mathbf{Z}[t]/(t^2 - 1) = \{a + bt \mid a, b \in \mathbf{Z}\}.$$

We will work with modules over $\mathbf{Z}\pi$. If π is a non-abelian group, the ring $\mathbf{Z}\pi$ is not commutative, and so one must distinguish between left and right modules.

Let A be an abelian group and

$$\rho : \pi \longrightarrow \mathrm{Aut}_{\mathbf{Z}}(A)$$

be a homomorphism. (The standard terminology is to call either ρ or A a *representation* of π.) The representation ρ endows A with the structure of a *left* $\mathbf{Z}\pi$-module by taking the action

$$\Big(\sum_{g\in\pi} m_g g\Big) \cdot a = \sum_{g\in\pi} m_g \, \rho(g)(a).$$

Conversely if A is a left module over a group ring $\mathbf{Z}\pi$, there is a homomorphism

$$\rho : \pi \longrightarrow \mathrm{Aut}_{\mathbf{Z}}(A)$$

given by $(g \mapsto (a \mapsto ga))$ where ga is multiplying $a \in A$ by $g \in \mathbf{Z}\pi$. Thus *a representation of a group π on an abelian group is the same thing as a $\mathbf{Z}\pi$-module.*

Exercise 72. Let A be a finitely generated (left) module over $\mathbf{Z}[\mathbf{Z}/2]$ so that, as an abelian group, A is finitely generated and torsion free. Show that A is a direct sum of modules of the form $\mathbf{Z}_+$, $\mathbf{Z}_-$, and $\mathbf{Z}[\mathbf{Z}/2]$. Here $\mathbf{Z}_+$ is the trivial $\mathbf{Z}[\mathbf{Z}/2]$-module corresponding to the trivial homomorphism $\rho : \mathbf{Z}/2 \to \mathrm{Aut}(\mathbf{Z})$, and $\mathbf{Z}_-$ corresponds to the non-trivial homomorphism.

We briefly outline the definition of the tensor product in the non-commutative case.

Definition 5.2. If R is a ring (possibly non-commutative), M is a right R-module, and N is a left R-module (sometimes one writes M_R and ${}_RN$), then the *tensor product* $M \otimes_R N$ is an abelian group satisfying the adjoint property

$$\mathrm{Hom}_{\mathbf{Z}}(M \otimes_R N, A) \cong \mathrm{Hom}_R(M, \mathrm{Hom}_{\mathbf{Z}}(N, A))$$

for any abelian group A. The corresponding universal property is that there is a $\mathbf{Z}$-bilinear map $\phi : M \times N \to M \otimes_R N$, so that $\phi(mr, n) = \phi(m, rn)$, and this map is initial in the category of $\mathbf{Z}$-bilinear maps $\overline{\phi} : M \times N \to A$, satisfying $\overline{\phi}(mr, n) = \overline{\phi}(m, rn)$.

The tensor product is constructed by taking the free *abelian* group on $M \times N$ and modding out by the expected relations. Elements of $M \otimes_R N$ are denoted by

$$\sum m_i \otimes n_i.$$

The relation $mr \otimes n = m \otimes rn$ holds. (This is why we take a right module tensored with a left module.)

Exercise 73. Compute the abelian group $\mathbf{Z}_+ \otimes_{\mathbf{Z}[\mathbf{Z}/2]} \mathbf{Z}_-$ (see Exercise 72).

The starting point in the algebraic construction of homology with local coefficients is the observation that *the singular chain complex of the universal cover of a space is a right $\mathbf{Z}\pi$-module.*

To proceed, fix a path connected and locally path-connected space X with a base point which admits a universal cover. For notational ease set $\pi = \pi_1 X$. Let $\widetilde{X} \to X$ be the universal cover of X, with its usual *right* $\pi_1 X$-action obtained by identifying π with the group of covering transformations. Then the singular complex $S_*(\widetilde{X})$ of the universal cover (with integer coefficients) is a right $\mathbf{Z}\pi$-module; the action of $g \in \pi$ on a singular simplex $\sigma : \Delta^k \to \widetilde{X}$ is the singular simplex $\sigma \cdot g$ defined as the composite of σ and the covering transformation $g : \widetilde{X} \to \widetilde{X}$. This is extended from π to $\mathbf{Z}\pi$ by linearity.

We can now give the algebraic definition of homology with local coefficients.

Definition 5.3. Given a $\mathbf{Z}\pi$-module A, form the tensor product

$$S_*(X; A) = S_*(\widetilde{X}) \otimes_{\mathbf{Z}\pi} A.$$

This is a chain complex whose homology is called the *homology of X with local coefficients in A* and is denoted by $H_*(X; A)$.

Notice that since the ring $\mathbf{Z}\pi$ is non-commutative (except if π is abelian), the tensored chain complex only has the structure of a chain complex over $\mathbf{Z}$, not $\mathbf{Z}\pi$. Thus the homology group $H_*(X; A)$ is only a $\mathbf{Z}$-module.

If the $\mathbf{Z}\pi$-module is specified by a representation $\rho : \pi_1 X \to \mathrm{Aut}(A)$ for some abelian group A, and we wish to emphasize the representation, we will sometimes embellish A with the subscript ρ and write $H_*(X; A_\rho)$ for the homology with coefficients in A. It is also common to call $H_*(X; A_\rho)$ the *homology of X twisted by $\rho : \pi_1 X \to \mathrm{Aut}(A)$.*

Before we look at examples, we will give the corresponding definition of cohomology. A new wrinkle which appears is that since the functor $\mathrm{Hom}_{\mathbf{Z}\pi}(-,-)$ is defined on the category of pairs of right R-modules or of pairs of left R-modules, we need to either change $S_*(\widetilde{X})$ to a left $\mathbf{Z}\pi$-module or consider coefficients in right $\mathbf{Z}\pi$-modules. We opt for the former.

Thus transform $S_*(\widetilde{X})$ into a *left* $\mathbf{Z}\pi$-module by the (standard) procedure:

$$g \cdot z \underset{\text{def}}{:=} z \cdot g^{-1}.$$

Definition 5.4. Given a left $\mathbf{Z}\pi$-module A, form the cochain complex

$$S^*(X; A) = \mathrm{Hom}_{\mathbf{Z}\pi}(S_*(\widetilde{X}), A).$$

(This means the set of group homomorphisms $f : S_*(\widetilde{X}) \to A$ which satisfy $f(rz) = rf(z)$ for all $r \in \mathbf{Z}\pi$ and $z \in S_*(\widetilde{X})$.)

The cohomology of this complex is called the *cohomology of X with local coefficients in A* and is denoted by

$$H^*(X; A).$$

If the module A is defined by a representation $\rho : \pi_1 X \to \mathrm{Aut}(A)$ for an abelian group A, the cohomology with local coefficients may be denoted by $H^*(X; A_\rho)$ and is often called the *cohomology of X twisted by ρ.*

5.2. Examples and basic properties

The (ordinary) homology and cohomology groups are just special cases of the homology and cohomology with local coefficients corresponding to twisting by the *trivial* representations ρ as we now show.

If $\rho : \pi_1 X \to \text{Aut}(A)$ is the trivial homomorphism, then the definition of tensor product gives a chain map

$$S_*(\widetilde{X}) \otimes_{\mathbf{Z}\pi} A_\rho \ \to \ S_* X \otimes_{\mathbf{Z}} A$$

which we will see is an isomorphism. (In the chain complex on the left A is considered only as an abelian group.) This follows since both $S_*(\widetilde{X})$ and $S_*(X)$ are chain complexes of *free* modules, so it is easy to compute tensor products. The complex $S_*(\widetilde{X})$ is a free $\mathbf{Z}\pi$-chain complex since π acts freely on $\widetilde{X}$, and hence on the set of all singular simplices in $\widetilde{X}$. We obtain a $\mathbf{Z}\pi$ basis by choosing a representative simplex for each orbit. Better yet, for each singular simplex $\sigma : \Delta^n \to X$, choose a single lift $\tilde{\sigma} : \Delta^n \to \widetilde{X}$. Then the set $\{\tilde{\sigma}\}$ gives a basis for $S_*(\widetilde{X})$ over $\mathbf{Z}\pi$, and it follows that $S_*(\widetilde{X}) \otimes_{\mathbf{Z}\pi} A_\rho \ \to \ S_* X \otimes_{\mathbf{Z}} A$ is an isomorphism of graded abelian groups; from this description it is not hard to check that this isomorphism is a chain map, and so $H_k(X; A_\rho) \ = \ H_k(X; A)$, the usual homology with coefficients in (the underlying $\mathbf{Z}$-module) A.

Similarly,

$$\text{Hom}_{\mathbf{Z}\pi}(S_* \widetilde{X}, A) \ \cong \ \text{Hom}_{\mathbf{Z}}(S_* X, A)$$

so $H^k(X; A_\rho) \ \cong \ H^k(X; A)$, the usual cohomology with coefficients in A.

Exercise 74. Show that the natural map

$$\text{Hom}_{\mathbf{Z}}(S_* X, A) \ \to \ \text{Hom}_{\mathbf{Z}\pi}(S_* \widetilde{X}, A)$$

is a chain isomorphism.

At the other extreme we consider what happens if A is a (finitely generated) free $\mathbf{Z}\pi$-module. Since the tensor product and Hom functors respect direct sums, it suffices to consider the case when $A = \mathbf{Z}\pi$.

Then,

$$S_* \widetilde{X} \otimes_{\mathbf{Z}\pi} \mathbf{Z}\pi \ = \ S_* \widetilde{X},$$

and therefore

$$H_k(X; \mathbf{Z}\pi) \ = \ H_k(\widetilde{X}; \mathbf{Z}),$$

the (untwisted) integral homology of the universal cover.

In other words, the homology with local coefficients given by the *tautological representation* $\rho : \pi \to \text{Aut}(\mathbf{Z}\pi)$

$$\rho(g) = (\sum m_h h \mapsto \sum m_h gh)$$

equals the homology of $\widetilde{X}$ with (untwisted) $\mathbf{Z}$ coefficients.

Exercise 75. Let M be an abelian group with the trivial left π action. Let $A = \mathbf{Z}\pi \otimes_{\mathbf{Z}} M$; notice that A has a left $\mathbf{Z}\pi$-module structure defined by $g \cdot (x \otimes m) \ = \ (gx) \otimes M$. Show that the homology $H_*(X; A)$ is just the (ordinary) homology of $\widetilde{X}$ with coefficients in M.

Exercise 76. (Shapiro's Lemma) Show that if $H \subset \pi$ is a subgroup and $A = \mathbf{Z}[\pi/H]$, viewed as a left π-module, then the corresponding homology is isomorphic to the homology of the H-cover of X. Generalize this as in the previous exercise to include other coefficients. (Hint: Try the case when H is normal first.)

These examples and the two exercises show that the (untwisted) homology of any cover of X with any coefficients can be obtained as a special case of the homology of X with appropriate local coefficients.

One might ask whether the same facts hold for cohomology. They do not without some modification. If $A = \mathbf{Z}\pi$, then the cochain complex $\mathrm{Hom}_{\mathbf{Z}\pi}(S_*\widetilde{X}, A)$ is not in general isomorphic to $\mathrm{Hom}_{\mathbf{Z}}(S_*X, \mathbf{Z})$ and so $H^k(X; \mathbf{Z}\pi)$ is not equal to $H^k(\widetilde{X}; \mathbf{Z})$. It turns out that if X is compact, $H^k(X; \mathbf{Z}\pi) \cong H^k_c(\widetilde{X}; \mathbf{Z})$, the *compactly supported* cohomology of $\widetilde{X}$.

5.2.1. Cellular methods. If X is a (connected) CW-complex, then homology and cohomology with local coefficients can be defined using the cellular chain complex; this is much better for computations. If $p : \widetilde{X} \to X$ is the universal cover, then $\widetilde{X}$ inherits a CW-structure from X – the cells of $\widetilde{X}$ are the path components of the inverse images of cells of X. The action of $\pi = \pi_1 X$ on $\widetilde{X}$ gives $C_*(\widetilde{X})$ the structure of a $\mathbf{Z}\pi$-chain complex. For each cell e of X, choose a cell $\widetilde{e}$ above $\widetilde{X}$; this gives a $\mathbf{Z}\pi$-basis for $C_*(\widetilde{X})$.

For example, let $X = \mathbf{R}P^n$ with $n > 1$. Then $X = e^0 \cup e^1 \cup \cdots \cup e^n$. Then $\widetilde{X} = S^n$ and the corresponding cell decomposition is

$$S^n = e^0_+ \cup e^0_- \cup e^1_+ \cup e^1_- \cup \cdots \cup e^n_+ \cup e^n_-$$

with $e^i_\pm$ being the upper and lower hemispheres of the i-sphere. A basis for the free (rank 1) $\mathbf{Z}\pi$-module $C_i(\widetilde{X})$ is e^i_+. With this choice of basis the $\mathbf{Z}\pi$-chain complex $C_*(\widetilde{X})$ is isomorphic to

$$\mathbf{Z}[\mathbf{Z}/2] \to \cdots \xrightarrow{1-t} \mathbf{Z}[\mathbf{Z}/2] \xrightarrow{1+t} \mathbf{Z}[\mathbf{Z}/2] \xrightarrow{1-t} \mathbf{Z}[\mathbf{Z}/2] \to 0.$$

Writing down this complex is the main step in the standard computation of $C_*(\mathbf{R}P^n)$ as in [**41**]: first use the homology of S^n and induction on n to compute $C_*(\widetilde{X})$ as a $\mathbf{Z}[\mathbf{Z}/2]$-chain complex, then compute $C_*(\mathbf{R}P^n) = C_*(\widetilde{X}) \otimes_{\mathbf{Z}\pi} \mathbf{Z}$.

The following exercises are important in gaining insight into what information homology with local coefficients captures.

Exercise 77. Compute the cellular chain complex $C_*(\widetilde{S^1})$ as a $\mathbf{Z}[t, t^{-1}]$-module. Compute $H_k(S^1; A_\rho)$ and $H^k(S^1; A_\rho)$ for any abelian group A and any homomorphism $\rho : \pi_1 S^1 = \mathbf{Z} \to \mathrm{Aut}(A)$.

Exercise 78. Let $\rho : \pi_1(\mathbf{R}P^n) \xrightarrow{\cong} \mathbf{Z}/2 = \mathrm{Aut}(\mathbf{Z})$. Compute $H_k(\mathbf{R}P^n; \mathbf{Z}_\rho)$ and $H^k(\mathbf{R}P^n; \mathbf{Z}_\rho)$ and compare to the untwisted homology and cohomology.

Exercise 79. Let p and q be a relatively prime pair of integers and denote by $L(p,q)$ the 3-dimensional Lens space $L(p,q) = S^3/(\mathbf{Z}/p)$, where $\mathbf{Z}/p = \langle t \rangle$ acts on $S^3 \subset \mathbf{C}^2$ via

$$t(Z,W) = (\zeta Z, \zeta^q W)$$

($\zeta = e^{2\pi i/p}$). Let $\rho : \mathbf{Z}/p \to \mathrm{Aut}(\mathbf{Z}/n) = \mathbf{Z}/(n-1)$ for n *prime.* Compute $H_k(L(p,q); (\mathbf{Z}/n)_\rho)$ and $H^k(L(p,q); (\mathbf{Z}/n)_\rho)$.

Exercise 80. Let K be the Klein bottle. Compute $H_n(K; \mathbf{Z}_\rho)$ for all twistings ρ of $\mathbf{Z}$ (i.e. all $\rho : \pi_1 K \to \mathbf{Z}/2 = \mathrm{Aut}(\mathbf{Z})$).

5.2.2. The orientation double cover and Poincaré duality. An important application of local coefficients is their use in studying the algebraic topology of non-orientable manifolds.

Theorem 5.5. *Any n-dimensional manifold M has a double cover*

$$p : M_O \to M$$

where M_O is an oriented manifold. Moreover, for any point $x \in M$, if $p^{-1}\{x\} = \{x_1, x_2\}$, then the orientations $\mu_{x_1} \in H_n(M_O, M_O - \{x_1\})$ and $\mu_{x_2} \in H_n(M_O, M_O - \{x_2\})$ map (by the induced homomorphism p_*) to the two generators of $H_n(M, M - \{x\})$.

Proof. As a set $M_O = \{a \in H_n(M, M - \{x\}) \mid a \text{ is a generator and } x \in M\}$. As for the topology, let V be an open set in X and $z \in Z_n(M, M - V)$ a relative cycle. Then let

$$V_z = \{\mathrm{im}[z] \in H_n(M, M - \{x\}) \mid x \in V \text{and } \mathbf{Z} \cdot \mathrm{im}[z] = H_n(M, M - \{x\})\}.$$

Then $\{V_z\}$ is a basis for the topology on M_O. For more details see [**23**]. □

For example, consider $\mathbf{R}P^n$ for n even. The orientation double cover is S^n; the deck transformation reverses orientation. For $\mathbf{R}P^n$ for n odd, the orientation double cover is a disjoint union of two copies of $\mathbf{R}P^n$, oriented with the opposite orientations.

If M is a connected manifold, define the *orientation character* or *the first Stiefel–Whitney class*

$$w : \pi_1 M \to \{\pm 1\}$$

by setting $w[\gamma] = 1$ if γ lifts to a loop in the orientation double cover and setting $w[\gamma] = -1$ if γ lifts to a path which is not a loop. Intuitively, $w[\gamma] = -1$ if going around the loop γ reverses the orientation. M is orientable if and only if w is trivial. Clearly w is a homomorphism.

Corollary 5.6. *Any manifold with $H^1(M;\mathbf{Z}/2)=0$ is orientable.*

Proof. This is because

$$H^1(M;\mathbf{Z}/2)\cong \operatorname{Hom}(H_1M;\mathbf{Z}/2)\cong \operatorname{Hom}(\pi_1M;\mathbf{Z}/2),$$

where the first isomorphism follows from the universal coefficient theorem and the second from the Hurewicz theorem

$$H_1M\cong \pi_1M/[\pi_1M,\pi_1M].$$

□

Notice that $\operatorname{Aut}(\mathbf{Z})=\{\pm 1\}$ and so the orientation character defines a representation $w:\pi_1X\to \operatorname{Aut}(\mathbf{Z})$. The corresponding homology and cohomology $H_k(X;\mathbf{Z}_w)$, $H^k(X;\mathbf{Z}_w)$ are called the homology and cohomology of X *twisted by the orientation character* w, or with *local coefficients in the orientation sheaf.*

The Poincaré duality theorem (Theorem 3.26) has an extension to the non-orientable situation.

Theorem 5.7 (Poincaré duality theorem). *If X is an n-dimensional manifold, connected, compact and without boundary, then*

$$H_n(X;\mathbf{Z}_w)\cong \mathbf{Z},$$

and if $[X]$ denotes a generator, then

$$\cap[X]:H^k(X;\mathbf{Z}_w)\to H_{n-k}(X;\mathbf{Z})$$

and

$$\cap[X]:H^k(X;\mathbf{Z})\to H_{n-k}(X;\mathbf{Z}_w)$$

are isomorphisms. (This statement of Poincaré duality applies to non-orientable manifolds as well as orientable manifolds.) □

The Poincaré–Lefschetz duality theorem also holds in this more general context.

The cap products in Theorem 5.7 are induced by the bilinear maps on coefficients $\mathbf{Z}\times\mathbf{Z}_w\to\mathbf{Z}_w$ and $\mathbf{Z}_w\times\mathbf{Z}_w\to\mathbf{Z}$ as in Exercise 36.

Exercise 81. Check that this works for $\mathbf{R}P^n$, n even.

More generally, for a manifold X and any right $\mathbf{Z}\pi$-module A given by a representation $\rho:\pi\to \operatorname{Aut} A$, let A_w be the module given by the representation $\rho_w:\pi\to\operatorname{Aut}(A)$, $g\mapsto w(g)\rho(g^{-1})$. Then a stronger form of Poincaré duality says

$$\cap[X]:H^k(X;A)\to H_{n-k}(X;A_w)$$

is an isomorphism.

5.3. Definition of homology with a local coefficient system

The previous (algebraic) definition of homology and cohomology with local coefficients may appear to depend on base points, via the representation

$$\rho : \pi_1(X, *) \to \mathrm{Aut}(A),$$

and the identification of $\pi_1 X$ with the covering translations of $\widetilde{X}$. In fact, it does not. We now give an alternative definition, which takes as input only the local coefficient system itself, i.e. the fiber bundle with discrete abelian group fibers. This definition is more elegant in that it does not depend on the arbitrary choice of a base point, but it is harder to compute with.

Let $p : E \to X$ be a system of local coefficients with fiber a discrete abelian group A and structure group $G \subset \mathrm{Aut}(A)$. Denote the fibers $p^{-1}(x)$ by E_x; for each x this is an abelian group non-canonically isomorphic to A.

We construct a chain complex as follows. Let $S_k(X; E)$ denote the set of formal sums

$$\sum_{i=1}^{m} a_i \sigma_i$$

where:

1. $\sigma_i : \Delta^k \to X$ is a singular k-simplex, and
2. a_i is an element of the group $E_{\sigma_i(e_0)}$ where $e_0 \in \Delta^k$ is the *base point* $(1, 0, 0, \cdots, 0)$ of Δ^k. More precisely, $\sigma_i(e_0) \in X$, and we require $a_i \in E_{\sigma_i(e_0)} = p^{-1}(\sigma_i(e_0))$.

The obvious way to add elements of $S_k(X; E)$ makes sense and is well defined. Thus $S_k(X; E)$ is an abelian group. This is somewhat confusing since the coefficients lie in different groups depending on the singular simplex. One way to lessen the confusion is to view $S_k(X; E)$ as a subgroup of the direct sum over every point $x \in X$ of $S_k(X; E_x)$.

Think of $S_*(X; E)$ as a graded abelian group. We next describe the differential. The formula would be the usual one were it not for the fact that given any k-simplex, one of its faces does not contain the base point. We will use the local coefficient system to identify fibers over different points of the simplex to resolve this problem.

Recall there are *face maps* $f_m^k : \Delta^{k-1} \to \Delta^k$ defined by

$$f_m^k(t_0, t_1, \cdots, t_{k-1}) = (t_0, \cdots, t_{m-1}, 0, t_m, \cdots, t_{k-1}).$$

Note that $f_m^k(e_0) = e_0$ if $m > 0$, but

$$f_0^k(e_0) = f_0^k(1, 0, \cdots, 0) = (0, 1, 0, \cdots, 0).$$

This will make the formulas for the differential a little bit more complicated than usual, since this one face map does not preserve base points.

Given a singular simplex $\sigma : \Delta^k \to X$, let $\gamma_\sigma : [0,1] \to X$ be the path $\sigma(t, 1-t, 0, 0, \cdots, 0)$. Then because $p : E \to X$ is a covering space (the fiber is discrete), the path γ_σ defines an isomorphism of groups $\gamma_\sigma : E_{\sigma(0,1,\cdots,0)} \to E_{\sigma(1,0,0,\cdots,0)}$ via path lifting.

Thus, define the differential $\partial : S_k(X; E) \to S_{k-1}(X; E)$ by the formula

$$a\sigma \mapsto \gamma_\sigma(a)(\sigma \circ f_0^k) + \sum_{m=1}^{k} (-1)^m \, a \, (\sigma \circ f_m^k).$$

Theorem 5.8. *This is a differential, i.e. $\partial^2 = 0$. Moreover the homology $H_k(S_*(X; E), \partial)$ equals $H_k(X; A_\rho)$, where $\rho : \pi_1 X \to \mathrm{Aut}(A)$ is the homomorphism determined by the local coefficient system $p : E \to X$ as in Lemma 4.7.* □

Exercise 82. Prove Theorem 5.8.

The homology of the chain complex $(S_k(X; E), \partial)$ is called the *homology with local coefficients in E*. Theorem 5.8 says that this is isomorphic to the homology with coefficients twisted by ρ. Notice that the definition of homology with local coefficients does not involve a choice of base point for X. It follows from Theorem 5.8 that the homology twisted by a representation ρ also does not depend on the choice of base point.

Similar constructions apply to cohomology, as we now indicate. Let $S^k(X; E)$ be the set of all functions, c, which assign to a singular simplex $\sigma : \Delta^k \to X$ an element $c(\sigma) \in E_{\sigma(e_0)}$. Then $S^k(X; E)$ is an abelian group and has coboundary operator $\delta : S^k(X; E) \to S^{k+1}(X; E)$ defined by

$$(\delta c)\ (\sigma) = (-1)^k \left(\gamma_\sigma^{-1}(c\ (\partial_0 \sigma)) + \sum_{i=1}^{k+1} (-1)^i \ c\ (\partial_i \sigma) \right).$$

Then $\delta^2 = 0$ and,

Theorem 5.9. *The cohomology of the chain complex $(S^*(X; E), \delta)$ equals the cohomology $H^*(X; A_\rho)$, where $\rho : \pi_1 X \to \mathrm{Aut}(A)$ is the homomorphism determined by the local coefficient system $p : E \to X$.* □

For the proof see [**43**].

Here is the example involving orientability of manifolds, presented in terms of local coefficients instead of the orientation representation. Let M be an n-dimensional manifold. Define a local coefficient system $E \to M$ by setting

$$E = \bigcup_{x \in M} H_n(M, M - \{x\}).$$

A basis for the topology of E is given by

$$V^z = \{\mathrm{im}[z] \in H_n(M, M - \{x\}) \mid x \in V\}$$

where V is open in X and $z \in Z_n(M, M - V)$ is a relative cycle. Then $E \to X$ is a local coefficient system with fibers $H_n(M, M - \{x\}) \cong \mathbf{Z}$, called the *orientation sheaf of* M. (Note the orientation double cover M_O is the subset of E corresponding to the subset $\pm 1 \in \mathbf{Z}$.) Then $H_*(M; E)$ can be identified with $H_*(M; \mathbf{Z}_w)$.

5.4. Functoriality

The functorial properties of homology and cohomology with local coefficients depend on more than just the spaces involved; they also depend on the coefficient systems.

Definition 5.10. A *morphism* $(E \to X) \to (E' \to X)$ *of local coefficients over* X is a commutative diagram

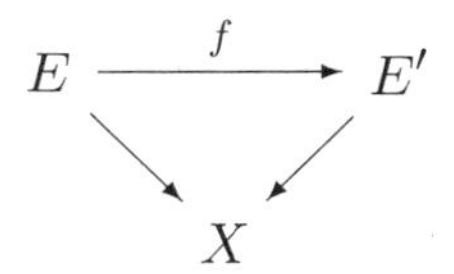

so that for each point $x \in X$, the restriction of $f : E \to E'$ to the fibers $f_{|E_x} : E_x \to E'_x$ is a group homomorphism.

Notice that we do not require the maps on fibers to be isomorphisms, and so this is more general than the concept of bundle map we introduced in Section 4.6.

It follows immediately from the definition of pullbacks that a commutative diagram

$$\begin{array}{ccc} E & \xrightarrow{\tilde{f}} & E' \\ {\scriptstyle p}\downarrow & & \downarrow{\scriptstyle p'} \\ X & \xrightarrow{f} & X' \end{array}$$

with $\tilde{f}$ inducing homomorphisms on fibers induces a morphism of local coefficients $(E \to X) \to (f^*(E') \to X)$ over X.

Theorem 5.11. *Homology with local coefficients is a covariant functor from a category* $\mathcal{L}$ *of pairs of spaces* (X, A) *with the following extra structure.*

1. *The objects of $\mathcal{L}$ are pairs (X, A) (allowing A empty) with a system of local coefficients $p : E \to X$.*
2. *The morphisms of $\mathcal{L}$ are the continuous maps $f : (X, A) \to (X', A')$ together with a morphism of local coefficients*

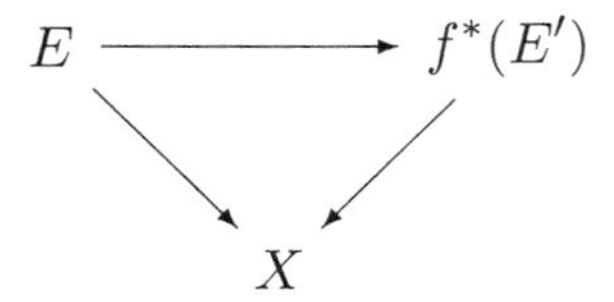

where $f^*(E')$ denotes the pullback of E' via f.

Sketch of proof. The basic idea comes from looking at the definition of the chain complex. Given a formal sum $\sum_i a_i\sigma_i$ with $a_i \in E_{\sigma_i(e_0)}$, the simplices σ_i push forward to simplices $f \circ \sigma$ in X'. Thus one needs a way to assign to a_i an element b_i' in $E'_{f(\sigma_i(e_0))}$. This is exactly what the morphism of local coefficients does. □

Cohomology with local coefficients is a functor on a slightly different category, owing to the variance of cohomology with respect to coefficients.

Theorem 5.12. *Cohomology with local coefficients is a contravariant functor on the category $\mathcal{L}^*$, where:*

1. *The objects of $\mathcal{L}^*$ are the same as the objects of $\mathcal{L}$, i.e. pairs (X, A) with a local coefficient system $p : E \to X$.*
2. *A morphism in $\mathcal{L}^*$ from $(p : E \to X)$ to $(p' : E' \to X')$ is a continuous map $f : (X, A) \to (X', A')$ together with a morphism of local coefficients*

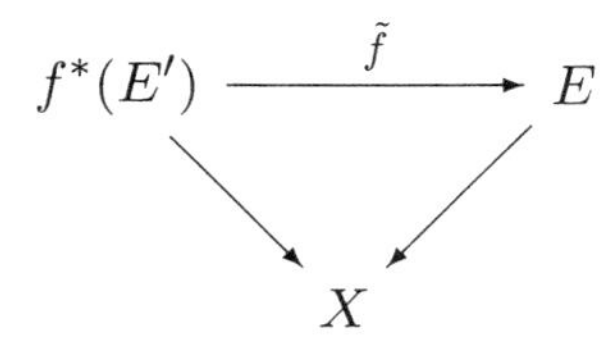

In other words, $\tilde{f}$ induces a group homomorphism from $E'_{f(x)}$ to E_x for all $x \in X$.

Sketch of proof. This is similar to the previous argument. A cochain c in $S^k(X'; E')$ is a function that assigns to each singular simplex $\sigma : \Delta^k \to X'$ an element $c(\sigma)$ in $E'_{\sigma(e_0)}$.

We need to construct $f^*(c) \in S^k(X; E)$. Given a simplex $\tau : \Delta^k \to X$, compose with f to get $f \circ \tau : \Delta^k \to X'$. Next apply c to get an element

$c(f \circ \tau(e_0)) \in E'_{f(\tau(e_0))}$. Finally apply $\tilde{f}$ to get

$$f^*(c)(\tau) = \tilde{f}(c(f \circ \tau(e_0))) \in E_{\tau(e_0)}.$$

□

Exercise 83. Give an alternative description of these two functoriality properties in terms of representations using the algebraic definition of homology and cohomology with local coefficients. More precisely, if $\rho : \pi_1 X \to \mathrm{Aut}(A)$ is a representation defining the homology of X with coefficients in A_ρ, and similarly $\rho' : \pi_1 X' \to \mathrm{Aut}(A')$ defines the homology of X' with coefficients in $A'_{\rho'}$, construct a commutative diagram which must exist for the homology of X with coefficient in ρ to map to the homology of X' with coefficients in $A'_{\rho'}$. Do the same for cohomology.

A straightforward checking that all the usual constructions continue to hold with local coefficients proves the following theorem.

Theorem 5.13. *Homology with local coefficients forms a homology theory on $\mathcal{L}$. More precisely, for any object in $\mathcal{L}$ there exists a connecting homomorphism and a natural long exact sequence. The excision and homotopy axioms hold.*

Similarly cohomology with local coefficients forms a cohomology theory on the category $\mathcal{L}^*$. □

In particular, there is a Mayer-Vietoris sequence for homology with local coefficients which gives a method for computing. Some care must be taken in using this theorem because local coefficients do not always extend. For example, given a homomorphism $\rho : \pi_1(X - U) \to \mathrm{Aut}(A)$ and an inclusion of pairs $(X - U, B - U) \to (X, B)$, excision holds (i.e. the inclusion of pairs induces isomorphisms in homology with local coefficients) only if ρ extends over $\pi_1(X)$. In particular the morphism of local coefficients must (exist and) be isomorphisms on fibers.

We end this section with a useful proposition which explicitly describes the 0th homology and cohomology with local coefficients.

Proposition 5.14. *Let B be a path connected CW-complex, $\pi = \pi_1(B)$ and V a $\mathbf{Z}\pi$-module. Then*

1. *Let V_π denote the quotient of V by the subgroup generated by the elements $\{v - \gamma \cdot v \mid v \in V,\ \gamma \in \pi_1 B\}$ (the group V_π is called the group of* coinvariants*). Then*

$$H_0(B; V) \cong V_\pi.$$

2. *Let V^π denote the subgroup of V consisting of elements fixed by π, i.e. $V^\pi = \{v \in V \mid \gamma \cdot v = v$ for all $\gamma \in \pi\}$ (the group V^π is called the group of* invariants*). Then*
$$H^0(B;V) \cong V^\pi.$$

Proof. Since B is path connected, up to homotopy we may as well assume that the 0-skeleton of B consists of a single vertex, and so the 0-cells of the universal cover $\tilde{B}$ are all of the form $\tau \cdot \gamma$ where τ is a fixed 0-cell and $\gamma \in \pi_1 B$. Thus, any 1-cell σ in $\tilde{B}$ has boundary
$$\partial(\sigma) = \tau \cdot \gamma_1(\sigma) - \tau \cdot \gamma_2(\sigma)$$
for some $\gamma_i(\sigma) \in \pi_1 B$.

Let
$$\Phi : V \to H_0(B;V) = H_0(C_*(\tilde{B} \otimes_{\mathbf{Z}\pi} V))$$
be defined by
$$\Phi(v) = [\tau \otimes v].$$

The 0-cycles are generated by $(\tau \cdot \gamma) \otimes v$ for $\gamma \in \pi_1 B$ and $v \in V$. Given a 1-cell σ in $\tilde{B}$,
$$\begin{aligned} \partial(\sigma \otimes v) &= \partial(\sigma) \otimes v \\ &= (\tau \cdot \gamma_1(\sigma) - \tau \cdot \gamma_2(\sigma)) \otimes v \\ &= \tau \otimes (\gamma_1(\sigma) - \gamma_2(\sigma)) \cdot v \end{aligned}$$
and so
$$0 = \Phi(\gamma_1(\sigma) \cdot v) - \Phi(\gamma_2(\sigma) \cdot v). \tag{5.1}$$

There exists a collection of 1-cells $\sigma_i, i \in \Lambda$ for $\tilde{B}$ with $\partial(\sigma_i) = \tau \cdot \xi_i - \tau$ for a set of generators $\{\xi_i\}$ of $\pi_1 B$ (just lift each 1-cell of B to a 1-cell of $\tilde{B}$ starting at τ).

Using Equation (5.1) and writing an arbitrary element of $\pi_1 B$ as a word in the σ_i, one finds easily that Φ is onto with kernel generated by $\{v - \gamma \cdot v \mid v \in V,\ \gamma \in \pi_1 B\}$.

The proof of the second assertion is similar and is left as an exercise. □

Exercise 84. Prove the second assertion in Proposition 5.14.

5.5. Projects for Chapter 5

5.5.1. The Hopf degree theorem. This theorem states that the degree of a map $f : S^n \to S^n$ determines its homotopy class. See Theorems 6.67 and 8.5. Prove the theorem using the simplicial approximation theorem. One place to find a proof is [**43**, pp. 13–17].

5.5.2. Colimits and Limits. The categorical point of view involves defining an object in terms of its (arrow theoretic) properties and showing that the properties uniquely define the object up to isomorphism. Colimits and limits are important categorical constructions in algebra and topology. Special cases include the notions of a cartesian product, a disjoint union, a pullback, a pushout, a quotient space X/A, and the topology of a CW-complex.

Define a product and coproduct of two objects in a category, and show that cartesian product and disjoint union give the product and coproduct in the category of topological spaces. Define the colimit of a sequence of topological spaces

$$X_0 \xrightarrow{f_0} X_1 \xrightarrow{f_1} X_2 \xrightarrow{f_2} X_3 \to \cdots,$$

show that it is unique up to homeomorphism, and show existence by taking

$$\operatorname*{colim}_{i\to\infty} X_i = \frac{\coprod X_i}{(x_i \sim f_i(x_i))}.$$

If all the X_i are subsets of a set A and if all the f_i's are inclusions of subspaces, show that the colimit can be taken to $X = \cup X_i$. The topology is given by saying $U \subset X$ is open if and only if $U \cap X_i$ is open for all i. Thus such a colimit can be thought of as some sort of generalization of a union. Define the limit of a sequence of topological spaces

$$\cdots \to X_3 \xrightarrow{f_3} X_2 \xrightarrow{f_2} X_1 \xrightarrow{f_1} X_0,$$

and show existence by taking

$$\varprojlim X_i = \{(x_i) \in \prod X_i : f_i(x_i) = x_{i-1} \quad \text{for all } i > 0\}.$$

Interpret the limit as a generalized form of intersection.

Now let $\mathcal{I}$ be a category and let $\mathcal{T}$ be the category of topological spaces. Let $X : \mathcal{I} \to \mathcal{T}$, $i \mapsto X_i$ be a functor, so you are given a topological space for every object i, and the morphisms of $\mathcal{I}$ give oodles of maps between the X_i satisfying the same composition laws as the morphisms in $\mathcal{I}$ do. Define

$$\operatorname*{colim}_{\mathcal{I}} X_i \qquad \text{and} \qquad \lim_{\mathcal{I}} X_i.$$

Consider the categories $\{\cdot \to \cdot \to \cdot \to \cdot \to \cdots\}$, $\{\cdots \leftarrow \cdot \leftarrow \cdot \leftarrow \cdot \leftarrow\}$, $\{\cdot \to \cdot \leftarrow \cdot\}$, $\{\cdot \leftarrow \cdot \to \cdot\}$, $\{\cdot \quad \cdot\}$, and discuss how colimits and limits over these categories give the above colimit, the above limit, the pullback, the pushout, the cartesian product and the disjoint union.

A definition of a CW-complex can be given in terms of colimits. A CW-complex is a space X together with an increasing sequence of subspaces

$X^0 \subset X^1 \subset X^2 \subset \cdots$, where X^{-1} is the empty set, each X^i is the pushout of

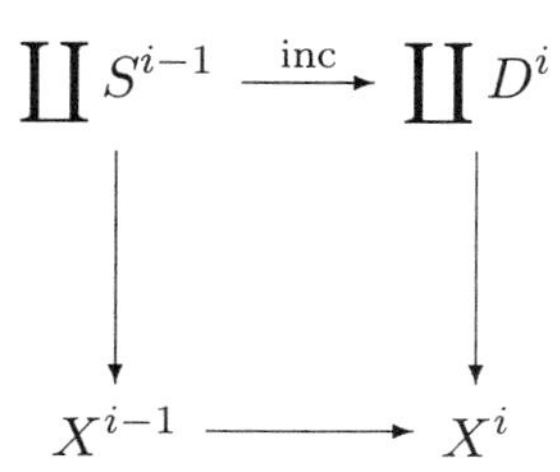

and $X = \operatorname{colim}_{i\to\infty} X^i$. This definition incorporates all the properties of the topology of a CW-complex that you use in practice. Show that this definition is equivalent to your favorite definition of a CW-complex.

Finally suppose that Y is a CW-complex and $Y_0 \subset Y_1 \subset Y_2 \subset Y_3 \subset \cdots$ is an increasing union of subcomplexes whose union is Y. Show that

$$H_n Y = \operatorname*{colim}_{i\to\infty} H_n(Y_i).$$

Define Milnor's $\lim^1$ (see [**25**] or [**42**]) and show that there is an exact sequence

$$0 \to \varprojlim{}^1 H^{n-1}(Y_i) \to H^n Y \to \varprojlim H^n(Y_i) \to 0.$$

We are using the more modern notation of colimit. Other authors use the terms "direct limit" or "inductive limit", and restrict the categories they consider. Other authors use the words "inverse limit" or "projective limit", while we just use the term "limit".

For an old-fashioned approach to limits in the special case of a directed system see [**33**], and for the more modern approach see [**42**].

Chapter 6

Fibrations, Cofibrations and Homotopy Groups

The material in this chapter forms the topological foundation for algebraic topology.

6.1. Compactly generated spaces

Given a map $f : X \times Y \to Z$, we would like to topologize the set of continuous functions $C(Y, Z)$ in such a way that f is continuous if and only if the adjoint

$$\tilde{f} : X \to C(Y, Z), \qquad \tilde{f}(x)(y) = f(x, y)$$

is continuous. Here are three examples:

1. We would like an action of a topological group $G \times Z \to Z$ to correspond to a continuous function $G \to \text{Homeo}(Z)$, where $\text{Homeo}(Z)$ is given the subspace topology inherited from $C(Z, Z)$.
2. We would like a homotopy $f : I \times Y \to Z$ to correspond to a path $\tilde{f} : I \to C(Y, Z)$ of functions.
3. The evaluation map

   $$C(Y, Z) \times Y \to Z, \qquad (f, y) \mapsto f(y)$$

   should be continuous. (Is the evaluation map an adjoint?)

Unfortunately, such a topology on $C(Y, Z)$ is not possible, even for Hausdorff topological spaces, unless you bend your point of view. Although many of the constructions we will give are set-theoretically simple, the issue of how to appropriately topologize these sets can become a nuisance. The category of compactly generated spaces is a framework which permits one to make such constructions without worrying about these technical issues. The reference for the material in this section is Steenrod's paper "A convenient category of topological spaces" [**38**].

Definition 6.1. A topological space X is said to be *compactly generated* if X is Hausdorff and a subset $A \subset X$ is closed if and only if $A \cap C$ is closed for every compact $C \subset X$.

Examples of compactly generated spaces include:

1. locally compact Hausdorff spaces (e.g. manifolds),
2. metric spaces, and
3. CW-complexes with finitely many cells in each dimension.

We will use the notation $\mathcal{K}$ for the category of compactly generated spaces. (This is taken as a *full* subcategory of the category of all topological spaces; i.e. every continuous function between compactly generated spaces is a morphism in $\mathcal{K}$.)

Any Hausdorff space can be turned into a compactly generated space by the following trick.

Definition 6.2. If X is Hausdorff, let $k(X)$ be the set X with the new topology defined by declaring a subset $A \subset X$ to be closed in $k(X)$ if and only if $A \cap C$ is closed in X for all $C \subset X$ compact.

Exercise 85. Show that $k(X)$ is compactly generated.

Thus $k(X)$ is the underlying set of X topologized with more (closed and hence more) open sets than X. This construction defines a *functor*

$$k : \mathcal{T}_2 \to \mathcal{K}$$

from the category $\mathcal{T}_2$ of Hausdorff spaces to the category $\mathcal{K}$ of compactly generated spaces.

Exercise 86. Show that k is a right adjoint for the inclusion functor $i : \mathcal{K} \to \mathcal{T}_2$. You will end up having to verify several of the facts below.

6.1.1. Basic facts about compactly generated spaces.

1. If $X \in \mathcal{K}$, then $k(X) = X$.
2. If $f : X \to Y$ is a function, then $k(f) : k(X) \to k(Y)$ is continuous if and only if $f|_C : C \to Y$ is continuous for each compact $C \subset X$.
3. Let $C(X, Y)$ denote the set of continuous functions from X to Y. Then $k_* : C(X, k(Y)) \to C(X, Y)$ is a bijection if X is in $\mathcal{K}$.

4. The singular chain complexes of a Hausdorff space Y and the space $k(Y)$ are the same.
5. The homotopy groups (see Definition 6.43) of Y and $k(Y)$ are the same.
6. Suppose that $X_0 \subset X_1 \subset \cdots \subset X_n \subset \cdots$ is an expanding sequence of compactly generated spaces so that X_n is closed in X_{n+1}. Topologize the union $X = \cup_n X_n$ by defining a subset $C \subset X$ to be closed if $C \cap X_n$ is closed for each n. Then if X is Hausdorff, it is compactly generated. In this case every compact subset of X is contained in some X_n.

6.1.2. Products in $\mathcal{K}$.

Unfortunately, the product of compactly generated spaces need not be compactly generated. However, this causes little concern, as we now see.

Definition 6.3. Let X, Y be compactly generated spaces. The *categorical product* of X and Y is the space $k(X \times Y)$.

The following useful facts hold about the categorical product.

1. $k(X \times Y)$ is in fact a product in the category $\mathcal{K}$.
2. If X is locally compact and Y is compactly generated, then $X \times Y = k(X \times Y)$. In particular, $I \times Y = k(I \times Y)$. Thus the notion of homotopy is unchanged.

From now on, if X and Y are compactly generated, we will denote $k(X \times Y)$ by $X \times Y$.

6.1.3. Function spaces.

The standard way to topologize the set of functions $C(X, Y)$ is to use the compact-open topology.

Definition 6.4. If X and Y are compactly generated spaces, let $C(X, Y)$ denote the set of continuous functions from X to Y, topologized with the *compact-open topology.* This topology has as a subbasis sets of the form

$$U(K, W) = \{f \in C(X, Y) | f(K) \subset W\}$$

where K is a compact set in X and W an open set in Y.

If Y is a metric space, this is the notion, familiar from complex analysis, of uniform convergence on compact sets. Unfortunately, even for compactly generated spaces X and Y, $C(X, Y)$ need not be compactly generated. We know how to handle this problem: define

$$\text{Map}(X, Y) \;=\; k(C(X, Y)).$$

As a *set*, $\text{Map}(X, Y)$ is the set of continuous maps from X to Y, but its topology is slightly different from the compact open topology.

Theorem 6.5 (adjoint theorem). *For X, Y, and Z compactly generated, $f(x,y) \mapsto \tilde{f}(x)(y)$ gives a homeomorphism*

$$\mathrm{Map}((X \times Y), Z) \to \mathrm{Map}(X, \mathrm{Map}(Y, Z)).$$

Thus $- \times Y$ and $\mathrm{Map}(Y, -)$ are adjoint functors from $\mathcal{K}$ to $\mathcal{K}$. □

The following useful properties of $\mathrm{Map}(X,Y)$ hold.

1. Let $e : \mathrm{Map}(X,Y) \times X \to Y$ be the *evaluation* $e(f,x) = f(x)$. Then if $X, Y \in \mathcal{K}$, e is continuous.
2. If $X, Y, Z \in \mathcal{K}$, then:
   (a) $\mathrm{Map}(X, Y \times Z)$ is homeomorphic to $\mathrm{Map}(X,Y) \times \mathrm{Map}(X,Z)$.
   (b) Composition defines a continuous map
   $$\mathrm{Map}(X,Y) \times \mathrm{Map}(Y,Z) \to \mathrm{Map}(X,Z).$$

We will also use the notation $\mathrm{Map}(X, A; Y, B)$ to denote the subspace of $\mathrm{Map}(X,Y)$ consisting of those functions $f : X \to Y$ which satisfy $f(A) \subset B$. A variant of this notation is $\mathrm{Map}(X, x_0; Y, y_0)$ denoting the subspace of basepoint preserving functions.

6.1.4. Quotient maps. We discuss yet another convenient property of compactly generated spaces. For topological spaces, one can give an example of quotient maps $p : W \to Y$ and $q : X \to Z$ so that $p \times q : W \times X \to Y \times Z$ is not a quotient map. However, one can show the following.

Theorem 6.6.

1. *If $p : W \to Y$ and $q : X \to Z$ are quotient maps, and X and Z are locally compact Hausdorff, then $p \times q$ is a quotient map.*
2. *If $p : W \to Y$ and $q : X \to Z$ are quotient maps and all spaces are compactly generated, then $p \times q$ is a quotient map, provided we use the categorical product.* □

From now on, *we assume all spaces are compactly generated.* If we ever meet a space which is not compactly generated, we immediately apply k. Thus, for example, if X and Y are Hausdorff spaces, then by our convention $X \times Y$ really means $k(k(X) \times k(Y))$. By this convention, we lose no information concerning homology and homotopy, but we gain the adjoint theorem.

6.2. Fibrations

There are two kinds of maps of fundamental importance in algebraic topology: fibrations and cofibrations. Geometrically, fibrations are more complicated than cofibrations. However, your garden variety fibration tends to be a fiber bundle, and fiber bundles over paracompact spaces are always fibrations, so that we have seen many examples so far.

Definition 6.7. A continuous map $p : E \to B$ is a *fibration* if it has the *homotopy lifting property (HLP)*; i.e. the problem

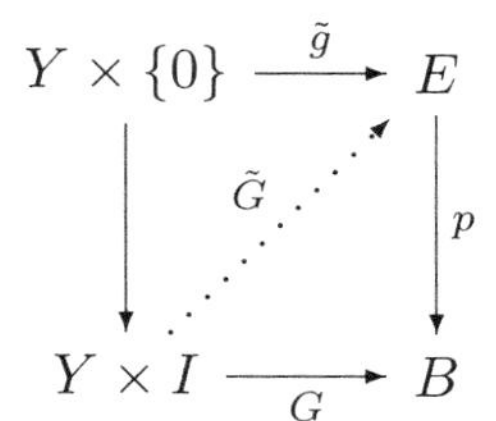

has a solution for every space Y.

In other words, given the continuous maps p, G, $\tilde{g}$, and the inclusion $Y \times \{0\} \to Y \times I$, the problem is to find a continuous map $\tilde{G}$ making the diagram commute.

Remark. Recall that whenever a commutative diagram is given with one dotted arrow, we consider it as a problem whose solution is a map which can be substituted for the dashed arrow to give a commutative diagram.

A covering map is a fibration. In studying covering space theory this fact is called the covering homotopy theorem. For covering maps the lifting is unique, but this is not true for an arbitrary fibration.

Exercise 87. Show that the projection to the first factor $p : B \times F \to B$ is a fibration. Show by example that the liftings need not be unique.

The following theorem of Hurewicz says that if a map is locally a fibration, then it is so globally.

Theorem 6.8. *Let $p : E \to B$ be a continuous map. Suppose that B is paracompact and suppose that there exists an open cover $\{U_\alpha\}$ of B so that $p : p^{-1}(U_\alpha) \to U_\alpha$ is a fibration for each U_α.*

Then $p : E \to B$ is a fibration. □

Proving this theorem is one of the projects for Chapter 4. The corollary of most consequence for us is the following.

Corollary 6.9. *If $p : E \to B$ is a fiber bundle over a paracompact space B, then p is a fibration.*

Proof. Exercise 87 says that the projection $U \times F \to U$ is a fibration. Since fiber bundles have this local product structure, Theorem 6.8 implies that a fiber bundle is a fibration. □

Exercise 88. Give an example of a fibration which is not a fiber bundle.

Maps between fibrations are analogous to (and simpler than) maps of fiber bundles.

Definition 6.10. If $p : E \to B$ and $p' : E' \to B'$ are fibrations, then a *map of fibrations* is a pair of maps $f : B \to B'$, $\tilde{f} : E \to E'$ so that the diagram

$$\begin{array}{ccc} E & \xrightarrow{\tilde{f}} & E' \\ \downarrow & & \downarrow \\ B & \xrightarrow[f]{} & B' \end{array}$$

commutes.

Pullbacks make sense and exist in the world of fibrations.

Definition 6.11. If $p : E \to B$ is a fibration, and $f : X \to B$ a continuous map, define the *pullback of* $p : E \to B$ *by* f to be the map $f^*(E) \to X$ where

$$f^*(E) = \{(x,e) \in X \times E | f(x) = p(e)\} \subset X \times E$$

and the map $f^*(E) \to B$ is the restriction of the projection $X \times E \to X$.

The following exercise is a direct consequence of the universal property of pullbacks.

Exercise 89. Show that $f^*(E) \to X$ is a fibration.

The following notation will be in effect for the rest of the book. If $H : Y \times I \to B$ is a homotopy, then $H_t : Y \to B$ is the homotopy at time t, i.e.

$$H_t(y) = H(y, t).$$

6.3. The fiber of a fibration

A fibration need not be a fiber bundle. Indeed, the definition of a fibration is less rigid than that of a fiber bundle, and it is not hard to alter a fiber bundle slightly to get a fibration which is not locally trivial. Nevertheless, a fibration has a well defined fiber *up to homotopy.* The following theorem asserts this and also states that a fibration has a substitute for the structure group of a fiber bundle, namely the group of homotopy classes of self-homotopy equivalences of the fiber.

It is perhaps at first surprising that the homotopy lifting property in itself is sufficient to endow a map with the structure of a "fiber bundle up to homotopy". But as we will see, the notion of a fibration is central in studying spaces up to homotopy.

Theorem 6.12. *Let $p : E \to B$ be a fibration. Assume B is path connected.*

Then all fibers $E_b = p^{-1}(b)$ are homotopy equivalent. Moreover every path $\alpha : I \to B$ defines a homotopy class α_* of homotopy equivalences $E_{\alpha(0)} \to E_{\alpha(1)}$ which depends only on the homotopy class of α rel end points, in such a way that multiplication of paths *corresponds to* composition of homotopy equivalences.

In particular, there exists a well defined group homomorphism

$$[\alpha] \mapsto (\alpha^{-1})_*$$

$$\pi_1(B, b_0) \to \textit{Homotopy classes of self-homotopy equivalences of } E_{b_0}.$$

Remark. The reason why we use $\alpha \mapsto (\alpha^{-1})_*$ instead of $\alpha \mapsto \alpha_*$ is because by convention, multiplication of paths in B is defined so that $\alpha\beta$ means first follow α, then β. This implies that $(\alpha\beta)_* = \beta_* \circ \alpha_*$, and so we use the inverse to turn this anti-homomorphism into a homomorphism.

Proof. Let $b_0, b_1 \in B$ and let α be a path in B from b_0 to b_1. The inclusion $E_{b_0} \hookrightarrow E$ completes to a diagram

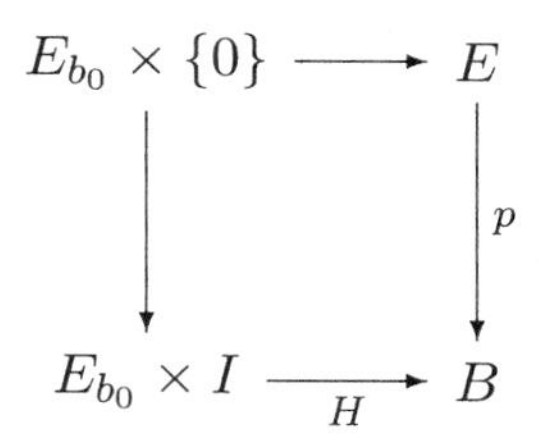

where $H(e, t) = \alpha(t)$. Since $E \to B$ is a fibration, H lifts to E; i.e. there exists a map $\widetilde{H}$ such that

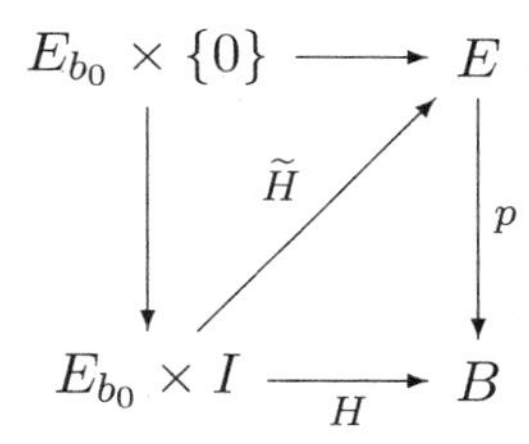

commutes.

Notice that the homotopy at time $t = 0$, $\widetilde{H}_0 : E_{b_0} \to E$, is just the inclusion of the fiber E_{b_0} in E. Furthermore, $p \circ \widetilde{H}_t$ is the constant map at $\alpha(t)$, so the homotopy $\widetilde{H}$ at time $t = 1$ is a map $\widetilde{H}_1 : E_{b_0} \to E_{b_1}$. We will let $\alpha_* = [\widetilde{H}_1]$ denote the homotopy class of this map. Since $\widetilde{H}$ is not unique, we need to show that another choice of lift gives a homotopic map. We will in fact show something more general. Suppose $\alpha' : I \to B$ is another path homotopic to α rel end points. Then as before, we obtain a solution $\widetilde{H}'$ to the problem

$$\begin{array}{ccc} E_{b_0} \times \{0\} & \longrightarrow & E \\ \downarrow & \overset{\widetilde{H}'}{\nearrow} & \downarrow p \\ E_{b_0} \times I & \xrightarrow[H']{} & B \end{array}$$

(where $H' = \alpha' \circ \mathrm{proj}_I$) and hence a map $\widetilde{H}'_1 : E_{b_0} \to E_{b_1}$.

Claim. *$\widetilde{H}_1$ is homotopic to $\widetilde{H}'_1$.*

Proof of Claim. Since α is homotopic rel end points to α', there exists a map $\Lambda : E_{b_0} \times I \times I \to B$ such that

$$\Lambda(e,s,t) = F(s,t)$$

where $F(s,t)$ is a homotopy rel end points of α to α'. (So $F_0 = \alpha$ and $F_1 = \alpha'$.) The solutions $\widetilde{H}$ and $\widetilde{H}'$ constructed above give a diagram

$$\begin{array}{ccc} (E_{b_0} \times I) \times \{0,1\} \cup (E_{b_0} \times \{0\}) \times I & \xrightarrow{\Gamma} & E \\ \downarrow & & \downarrow p \\ (E_{b_0} \times I) \times I & \xrightarrow[\Lambda]{} & B \end{array}$$

where

$$\Gamma(e,s,0) = \widetilde{H}(e,s)$$

$$\Gamma(e,s,1) = \widetilde{H}'(e,s), \text{ and}$$

$$\Gamma(e,0,t) = e.$$

Let $U = I \times \{0,1\} \cup \{0\} \times I \subset I \times I$ There exists a homeomorphism $\varphi : I^2 \to I^2$ taking U to $I \times \{0\}$ as indicated in the following picture.

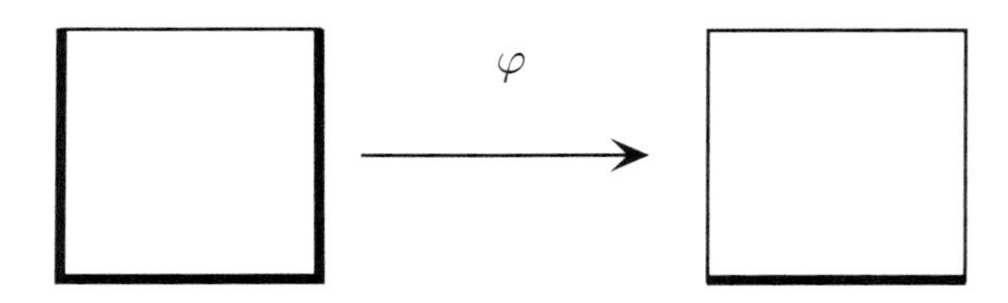

Thus the diagram

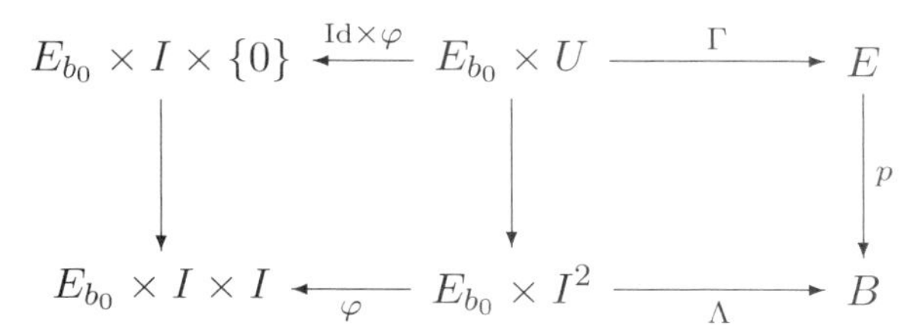

has the left two horizontal maps homeomorphisms. Since the homotopy lifting property applies to the outside square, there exists a lift $\widetilde{\Lambda} : E_{b_0} \times I^2 \to E$ so that

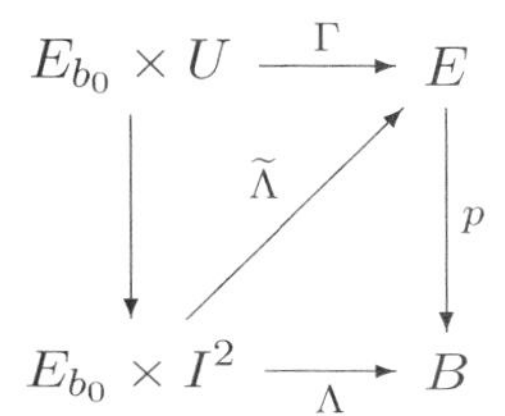

commutes.

But then $\widetilde{\Lambda}$ is a homotopy from $\widetilde{H} : E_{b_0} \times I \to E$ to $\widetilde{H}' : E_{b_0} \times I \to E$. Restricting to $E_{b_0} \times \{1\}$ we obtain a homotopy from $\widetilde{H}_1$ to $\widetilde{H}_1'$. Thus the homotopy class $\alpha_* = [\widetilde{H}_1]$ depends only on the homotopy class of α rel end points, establishing the claim.

Clearly $(\alpha\beta)_* = \beta_* \circ \alpha_*$ if $\beta(0) = \alpha(1)$. In particular, if $\beta = \alpha^{-1}$, then $(\text{const})_* = \beta_* \circ \alpha_*$, where const denotes the constant path at b_0. But clearly

$$(\text{const})_* = [\mathrm{Id}_{E_{b_0}}].$$

Thus β_* is a homotopy inverse of α_*.

This shows that α_* is a homotopy equivalence, and since B is path connected, all fibers are homotopy equivalent.

Applying this construction to $\alpha \in \pi_1(B,b_0)$, we see that α_* defines a homotopy equivalence of E_{b_0}, and products of loops correspond to composites of homotopy equivalences. The following exercise completes the proof. □

Exercise 90. Show that the set of homotopy classes of homotopy equivalences of a space X forms a group under composition. That is, show that multiplication and taking inverses are well defined.

Theorem 6.12 asserts that the fibers $p^{-1}(b) = E_b$ for $b \in B$ are homotopy equivalent. Thus we will abuse terminology slightly and refer to any space in the homotopy equivalence class of the space E_b for any $b \in B$ as *the fiber of the fibration* $p : E \to B$.

Since homotopy equivalences induce isomorphisms in homology or cohomology, a fibration with fiber F gives rise to local coefficients systems whose fiber is the homology or cohomology of F, as the next corollary asserts.

Corollary 6.13. *Let $p : E \to B$ be a fibration and let $F = p^{-1}(b_0)$. Then p gives rise to local coefficient systems over B with fiber $H_n(F;M)$ or $H^n(F;M)$ for any n and any coefficient group M. These local coefficients are obtained from the representations via the composite homomorphism*

$$\pi_1(B,b_0) \to \left\{ \begin{array}{l} \textit{Homotopy classes of self-homotopy} \\ \textit{equivalences } F \to F \end{array} \right\} \to \mathrm{Aut}(A)$$

where $A = H_n(F;M)$ or $A = H^n(F;M)$.

Proof. The maps $f_* : H_n(F;M) \to H_n(F;M)$ and $f^* : H^n(F;M) \to H^n(F;M)$ induced by a homotopy equivalence $f : F \to F$ are isomorphisms which depend only on the homotopy class of f. Thus there is a function from the group of homotopy classes of homotopy equivalences of F to the group of automorphisms of A. This is easily seen to be a homomorphism. The corollary follows. □

We see that a fibration gives rise to many local coefficient systems, by taking homology or cohomology of the fiber. More generally one obtains a local coefficient system given any homotopy functor from spaces to abelian groups (or R-modules), such as the generalized homology theories which we introduce in Chapter 8.

With some extra hypotheses one can also apply this to homotopy functors on the category of based spaces. For example, we will see below that if F is simply connected, or more generally "simple", then taking homotopy groups $\pi_n F$ also gives rise to a local coefficient system. For now, however, observe that the homotopy equivalences constructed by Theorem 6.12 need not preserve base points.

6.4. Path space fibrations

An important type of fibration is the path space fibration. Path space fibrations will be useful in replacing arbitrary maps by fibrations and then in extending a fibration to a fiber sequence.

Definition 6.14. Let (Y, y_0) be a based space. The *path space* $P_{y_0}Y$ is the space of paths in Y starting at y_0, i.e.

$$P_{y_0}Y = \mathrm{Map}(I,0;Y,y_0) \subset \mathrm{Map}(I,Y),$$

topologized as in the previous subsection, i.e. as a compactly generated space. The *loop space* $\Omega_{y_0}Y$ is the space of all loops in Y based at y_0, i.e.

$$\Omega_{y_0}Y = \mathrm{Map}(I,\{0,1\};Y,\{y_0\}).$$

Often the subscript y_0 is omitted in the above notation. Let $Y^I = \mathrm{Map}(I,Y)$. This is called the *free path space.* Let $p : Y^I \to Y$ be the evaluation at the end point of a path: $p(\alpha) = \alpha(1)$.

By our conventions on topologies, $p : Y^I \to Y$ is continuous. The restriction of p to $P_{y_0}Y$ is also continuous.

Exercise 91. Let y_0, y_1 be two points in a path-connected space Y. Prove that $\Omega_{y_0}Y$ and $\Omega_{y_1}Y$ are homotopy equivalent.

Theorem 6.15.

1. *The map $p : Y^I \to Y$, where $p(\alpha) = \alpha(1)$, is a fibration. Its fiber over y_0 is the space of paths which end at y_0, a space homeomorphic to $P_{y_0}Y$.*
2. *The map $p : P_{y_0}Y \to Y$ is a fibration. Its fiber over y_0 is the loop space $\Omega_{y_0}Y$.*
3. *The free path space Y^I is homotopy equivalent to Y. The projection $p : Y^I \to Y$ is a homotopy equivalence.*
4. *The space of paths in Y starting at y_0, $P_{y_0}Y$, is contractible.*

Proof. 1. Let A be a space, and suppose a homotopy lifting problem

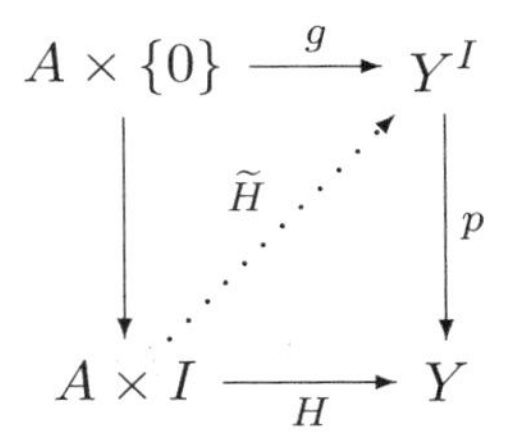

is given. We write $g(a)$ instead of $g(a,0)$. For each $a \in A$, $g(a)$ is a path in Y which ends at $p(g(a)) = H(a,0)$. This point is the start of the path $H(a,-)$.

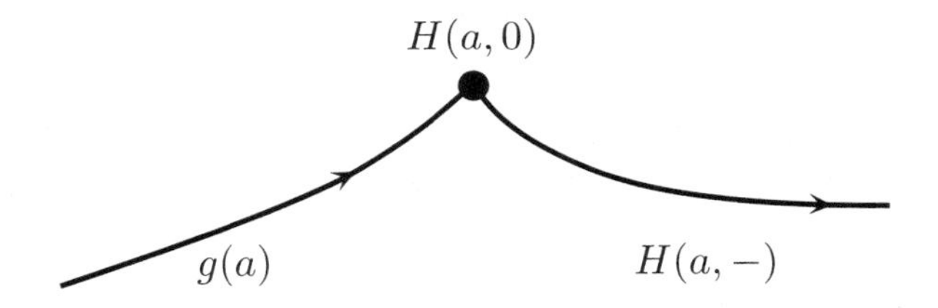

We will define $\widetilde{H}(a,s)(t)$ to be a path running along the path $g(a)$ and then partway along $H(a,-)$, ending at $H(a,s)$.

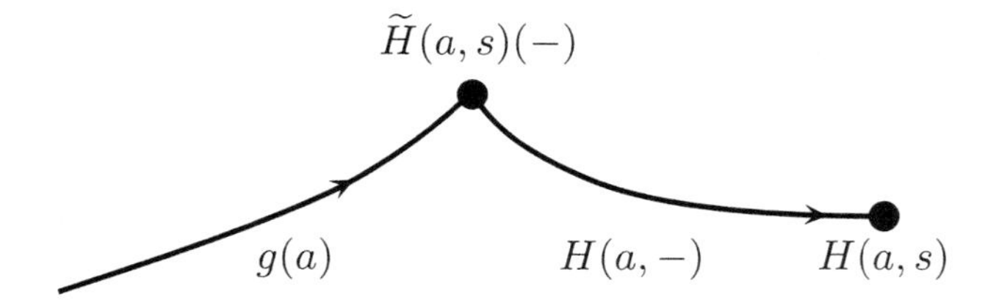

Define

$$\widetilde{H}(a,s)(t) = \begin{cases} g(a)((1+s)t) & \text{if } 0 \leq t \leq 1/(1+s), \\ H(a,((1+s)t-1) & \text{if } 1/(1+s) \leq t \leq 1. \end{cases}$$

Then $\widetilde{H}(a,s)(t)$ is continuous as a function of (a,s,t), so $\widetilde{H}(a,s) \in Y^I$ and by our choice of topologies $\widetilde{H} : A \times I \to Y^I$ is continuous. Also $\widetilde{H}(a,0) = g(a)$ and $p(\widetilde{H}(a,s)) = \widetilde{H}(a,s)(1) = H(a,s)$. Thus the lifting problem is solved and so $p : P_{y_0}Y \to Y$ is a fibration. The fiber $p^{-1}(y_0)$ consists of all paths ending at y_0, and the path space $P_{y_0}Y$ consists of all paths starting at y_0. A homeomorphism is given by

$$\alpha(t) \mapsto \overline{\alpha}(t) = \alpha(1-t).$$

This proves (1).

(2) has the same proof; the fact that $g(a)$ *starts at* y_0 means that $\widetilde{H}(a,s)$ also starts at y_0.

To prove (3), let $i : Y \to Y^I$ be the map taking y to the constant path at y. Then $p \circ i = \mathrm{Id}_Y$. Let $F : Y^I \times I \to Y^I$ be given by

$$F(\alpha,s)(t) = \alpha(s+t-st).$$

Then $F(\alpha,0) = \alpha$ and $F(\alpha,1)$ is the constant path at $\alpha(1)$ which in turn equals $i \circ p(\alpha)$. Thus F shows that the identity is homotopic to $i \circ p$. Hence p and i are homotopy inverses.

(4) has the same proof as (3). □

6.5. Fiber homotopy

Recall a map of fibrations $(p : E \to B)$ to $(p' : E' \to B')$ is a commutative diagram

$$\begin{array}{ccc} E & \xrightarrow{\tilde{f}} & E' \\ {\scriptstyle p}\downarrow & & \downarrow{\scriptstyle p'} \\ B & \xrightarrow[f]{} & B' \end{array}$$

Definition 6.16. A *fiber homotopy* between two morphisms $(\tilde{f}_i, f_i)$ $i = 0, 1$ of fibrations is a commutative diagram

$$\begin{array}{ccc} E \times I & \xrightarrow{\widetilde{H}} & E' \\ {\scriptstyle p\times \mathrm{Id}}\downarrow & & \downarrow{\scriptstyle p'} \\ B \times I & \xrightarrow[H]{} & B' \end{array}$$

with $H_0 = f_0$, $H_1 = f_1$, $\widetilde{H}_0 = \tilde{f}_0$, and $\widetilde{H}_1 = \tilde{f}_1$.

Given two fibrations over B, $p : E \to B$ and $p' : E' \to B$, we say they have the same *fiber homotopy type* if there exists a map $\tilde{f}$ from E to E' covering the identity map of B, and a map $\tilde{g}$ from E' to E covering the identity map of B, such that the composites

$$\begin{array}{ccc} E & \xrightarrow{\tilde{g}\circ\tilde{f}} & E \\ & \searrow \quad \swarrow & \\ & B & \end{array} \qquad \begin{array}{ccc} E' & \xrightarrow{\tilde{f}\circ\tilde{g}} & E' \\ & \searrow \quad \swarrow & \\ & B & \end{array}$$

are each fiber homotopic to the identity via a homotopy which is the identity on B (i.e. there exists $\widetilde{H} : E \times I \to E$ such that $p(\widetilde{H}(e,t)) = p(e)$, $\widetilde{H}_0 = \tilde{g} \circ \tilde{f}$, and $\widetilde{H}_1 = \mathrm{Id}_E$. Similarly for $\tilde{f} \circ \tilde{g}$). One says that $\tilde{f}$ and $\tilde{g}$ are *fiber homotopy equivalences.*

Notice that a fiber homotopy equivalence $\tilde{f} : E \to E'$ induces a homotopy equivalence $E_{b_0} \to E'_{b_0}$ on fibers.

6.6. Replacing a map by a fibration

Let $f : X \to Y$ be a continuous map. We will replace X by a homotopy equivalent space P_f and obtain a map $P_f \to Y$ which is a fibration. In short, every map is equivalent to a fibration. If f is a fibration to begin with, then the construction gives a fiber homotopy equivalent fibration. We assume that Y is path-connected and X is non-empty.

Let $q : Y^I \to Y$ be the path space fibration, with $q(\alpha) = \alpha(0)$; evaluation at the starting point.

Definition 6.17. The pullback $P_f = f^*(Y^I)$ of the path space fibration along f is called the *mapping path space.*

$$\begin{array}{ccc} P_f = f^*(Y^I) & \longrightarrow & Y^I \\ \downarrow & & \downarrow q \\ X & \xrightarrow{f} & Y \end{array} \tag{6.1}$$

An element of P_f is a pair (x, α) where α is a path in Y and x is a point in X which maps via f to the starting point of α.

The *mapping path fibration*

$$p : P_f \to Y$$

is obtained by evaluating at the end point

$$p(x, \alpha) = \alpha(1).$$

Theorem 6.18. *Suppose that $f : X \to Y$ is a continuous map.*

1. *There exists a homotopy equivalence $h : X \to P_f$ so that the diagram*

$$\begin{array}{ccc} X & \xrightarrow{h} & P_f \\ & {}_f\searrow \quad \swarrow_p & \\ & Y & \end{array}$$

   commutes.
2. *The map $p : P_f \to Y$ is a fibration.*
3. *If $f : X \to Y$ is a fibration, then h is a fiber homotopy equivalence.*

Proof. 1. Let $h : X \to P_f$ be the map

$$h(x) = (x, \text{const}_{f(x)})$$

where $\text{const}_{f(x)}$ means the constant path at $f(x)$. Then $f = p \circ h$, so the triangle commutes. The homotopy inverse of h is $p_1 : P_f \to X$, projection on the X-component. Then $p_1 \circ h = \text{Id}_X$. The homotopy from $h \circ p_1$ to Id_{P_f} is given by

$$F((x, \alpha), s) = (x, \alpha_s),$$

where α_s is the path $s \mapsto \alpha(st)$. (We have embedded X in P_f via h and have given a deformation retract of P_f to X by contracting a path to its starting point.)

2. Let the homotopy lifting problem

$$\begin{array}{ccc} A \times \{0\} & \xrightarrow{g} & P_f \\ \Big\downarrow & \overset{\widetilde{H}}{\nearrow} & \Big\downarrow p \\ A \times I & \xrightarrow[H]{} & Y \end{array}$$

be given. For $a \in A$, we write $g(a)$ instead of $g(a, 0)$. Furthermore $g(a)$ has an X-component and a Y^I-component, and we write

$$g(a) = (g_1(a), g_2(a)) \in P_f \subset X \times Y^I.$$

Note that since $g(a)$ is in the pullback, $g_1(a)$ maps via f to the starting point of the path $g_2(a)$ and the square above commutes, so the end point of the path $g_2(a)$ is the starting point of the path $H(a, -)$. Here is a picture of $g(a)$ and $H(a, -)$.

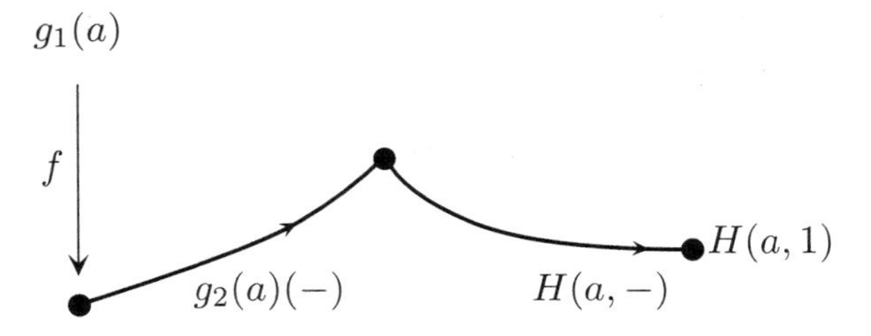

The lift $\widetilde{H}$ will have two components. The X-component will be constant in s,

$$\widetilde{H}_1(a, s) = g_1(a).$$

The Y^I-component of the lift will be a path running along the path $g_2(a)$ and then partway along $H(a, -)$, ending at $H(a, s)$.

Here is a picture of $\widetilde{H}(a, s)$.

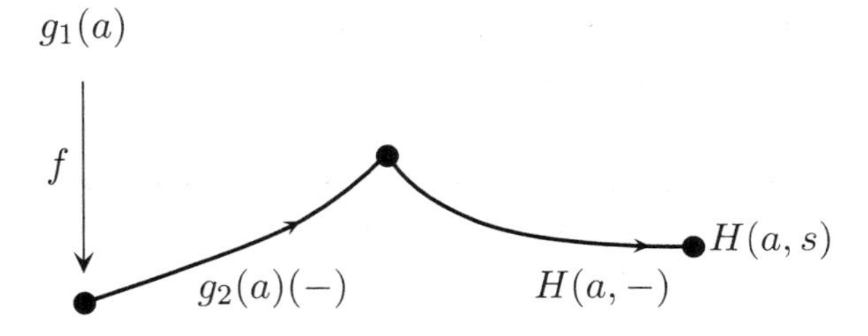

A formula is given by

$$\widetilde{H}(a, s) = (g_1(a), \widetilde{H}_2(a, s)(-)) \in P_f \subset X \times Y^I,$$

where

$$\widetilde{H_2}(a,s)(t) = \begin{cases} g_2(a)((1+s)t) & \text{if } 0 \le t \le 1/(1+s), \\ H(a,(1+s)t-1) & \text{if } 1/(1+s) \le t \le 1. \end{cases}$$

We leave it to the reader to check that $\widetilde{H}$ is continuous and that it is a lift of H extending the map g. Thus we have shown that the mapping path fibration is a fibration.

3. Finally suppose that $f : X \to Y$ is itself a fibration. In the proof of 1 we showed that

$$h : X \to P_f, \qquad h(x) = (x, \mathrm{const}_{f(x)})$$

and

$$p_1 : P_f \to X \qquad p_1(x,\alpha) = x$$

are homotopy inverses. Note h is a map of fibrations (covering the identity), but p_1 is not, since $f \circ p_1(x,\alpha)$ is the starting point of α and $p(x,\alpha)$ is the end point of α.

Let $\gamma : P_f \times I \to Y$ be the map $\gamma(x,\alpha,t) = \alpha(t)$. Since f is a fibration, the homotopy lifting problem

$$\begin{array}{ccc} P_f \times \{0\} & \xrightarrow{p_1} & X \\ \downarrow & \overset{\tilde{\gamma}}{\nearrow} & \downarrow f \\ P_f \times I & \xrightarrow[\gamma]{} & Y \end{array}$$

has a solution. Define $g : P_f \to X$ by $g(x,\alpha) = \tilde{\gamma}(x,\alpha,1)$. Then the diagrams

commute.

Thus h and g are maps of fibrations and in fact homotopy inverses since g is homotopic to p_1. But this is not enough.

To finish the proof, we need to show that $g \circ h$ is homotopic to Id_X by a *vertical homotopy* (i.e. a homotopy over the identity $\mathrm{Id}_Y : Y \to Y$) and that $h \circ g$ is homotopic to Id_{P_f} by a vertical homotopy.

Let $F : X \times I \to X$ be the map

$$F(x,t) = \tilde{\gamma}(x, \mathrm{const}_{f(x)}, t).$$

Then

1. $F(x,0) = \tilde{\gamma}(x,\text{const}_{f(x)},0) = p_1(x,\text{const}_{f(x)}) = x$, and
2. $F(x,1) = \tilde{\gamma}(x,\text{const}_{f(x)},1) = g \circ h(x)$.

Hence F is a homotopy from Id_X to $g \circ h$. Moreover,

$$f(F(x,t)) = f(\tilde{\gamma}(x,\text{const}_{f(x)},t)) = \gamma(x,\text{const}_{f(x)},t) = f(x),$$

so F is a vertical homotopy.

Here is a picture of $\tilde{\gamma}$.

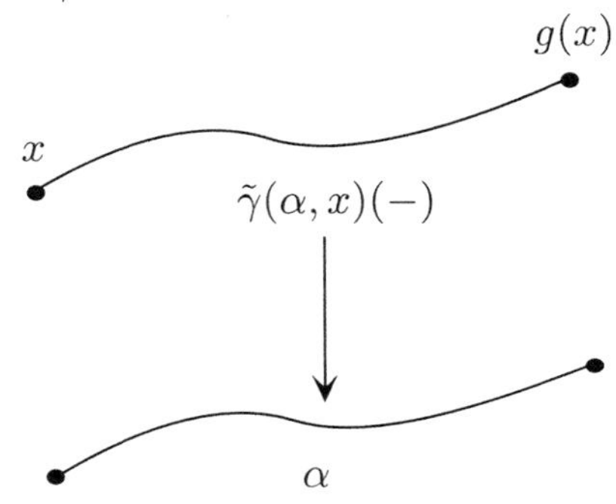

The vertical homotopy from Id_{P_f} to $h \circ g$ is given by contracting along paths to their end points. Explicitly $H : P_f \times I \to P_f$ is

$$H(x, \alpha, s) = (\tilde{\gamma}(x, \alpha, s), (t \mapsto \alpha(s + t - st))).$$

□

Given a map $f : X \to Y$, it is common to be sloppy and say "F is the fiber of f", or "$F \hookrightarrow X \to Y$ is a fibration" to mean that after replacing X by the homotopy equivalent space P_f and the map f by the fibration $P_f \to Y$, the fiber is a space of the homotopy type of F.

6.7. Cofibrations

Definition 6.19. A map $i : A \to X$ is called a *cofibration*, or *satisfies the homotopy extension property (HEP)*, if the following diagram has a solution for any space Y

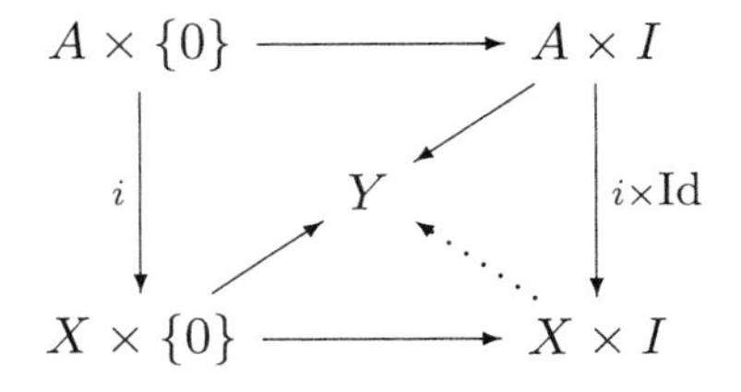

Cofibration is a "dual" notion to fibration, using the adjointness of the functors $- \times I$ and $-^I$ and reversing the arrows. To see this, note that since a map $A \times I \to B$ is the same as a map $A \to B^I$, the diagram defining a fibration $f : X \to Y$ can be written

$$\begin{array}{ccc} X & \xleftarrow{\text{eval. at } 0} & X^I \\ \uparrow & \nearrow & \downarrow f \\ Z & \longrightarrow & Y^I \end{array}$$

The diagram defining a cofibration $f : Y \to X$ can be written as

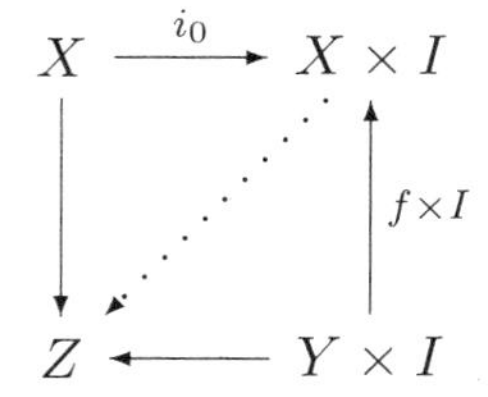

For reasonable spaces, any cofibration $i : A \to X$ can be shown to be an embedding whose image is closed in X. We will deal only with cofibrations given by a pair (X, A) with A a closed subspace. In that case one usually says that $A \hookrightarrow X$ is a cofibration if the problem

$$\begin{array}{ccc} X \times \{0\} \cup A \times I & \xrightarrow{f \cup h} & Y \\ \downarrow i & \nearrow & \\ X \times I & & \end{array}$$

has a solution for all spaces Y, maps $f : X \to Y$ and homotopies $h : A \times I \to Y$ extending $f|_A$. Hence the name homotopy extension property.

Definition 6.20. Let X be compactly generated, $A \subset X$ a subspace. Then (X,A) is called an *NDR-pair* (NDR stands for "neighborhood deformation retract") if there exist continuous maps $u : X \to I$ and $h : X \times I \to X$ so that:

1. $A = u^{-1}(0)$,
2. $h(-, 0) = \mathrm{Id}_X$,
3. $h(a, t) = a$ for all $t \in I$, $a \in A$, and
4. $h(x, 1) \in A$ for all $x \in X$ such that $u(x) < 1$.

In particular the neighborhood $U = \{x \in X | u(x) < 1\}$ of A deformation retracts to A.

Definition 6.21. A pair (X,A) is called a *DR–pair* (DR stands for "deformation retract") if (1), (2), and (3) hold, but also

4′. $h(x,1) \in A$ for all $x \in X$.

(This is slightly stronger than the usual definition of deformation retracts, because of the requirement that there exists a function $u : X \to I$ such that $u^{-1}(0) = A$.)

Theorem 6.22 (Steenrod). *Equivalent are:*

1. (X,A) *is an NDR pair.*
2. $(X \times I, X \times 0 \cup A \times I)$ *is a DR pair.*
3. $X \times 0 \cup A \times I$ *is a retract of* $X \times I$.
4. $i : A \hookrightarrow X$ *is a cofibration.*

For a complete proof see Steenrod's paper [**38**].

Proof of some implications.

$(4 \Rightarrow 3)$ Let $Y = X \times 0 \cup A \times I$. Then the solution of

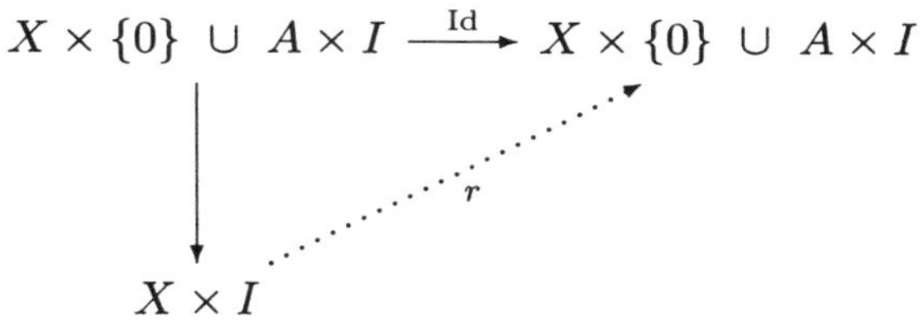

$$\begin{array}{ccc} X \times \{0\} \cup A \times I & \xrightarrow{\mathrm{Id}} & X \times \{0\} \cup A \times I \\ \downarrow & \overset{r}{\nearrow} & \\ X \times I & & \end{array}$$

is a retraction of $X \times I$ to $X \times 0 \cup A \times I$.

$(3 \Rightarrow 4)$ The problem

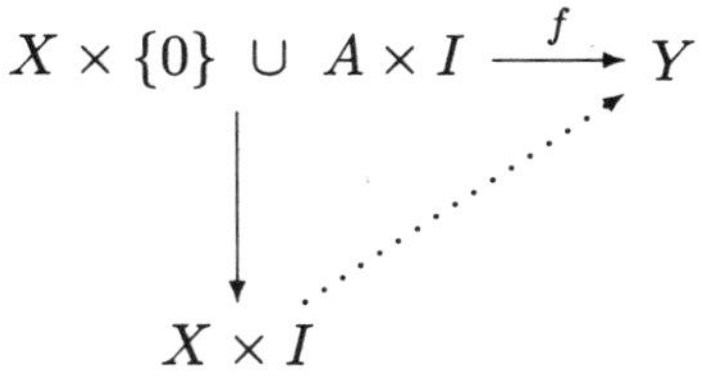

$$\begin{array}{ccc} X \times \{0\} \cup A \times I & \xrightarrow{f} & Y \\ \downarrow & \nearrow & \\ X \times I & & \end{array}$$

has a solution $f \circ r$, where $r : X \times I \to X \times \{0\} \cup A \times I$ is the retraction.

$(1 \Rightarrow 3)$ (This implication says that NDR pairs satisfy the homotopy extension property. This is the most important property of NDR pairs.)

The map $R : X \times I \to X \times \{0\} \cup A \times I$ given by

$$R(x,t) = \begin{cases} (x,t) & \text{if } x \in A \text{ or } t = 0, \\ (h(x,1), t - u(x)) & \text{if } t \geq u(x) \text{ and } t > 0, \text{ and} \\ (h(x, \frac{t}{u(x)}), 0) & \text{if } u(x) \geq t \text{ and } u(x) > 0 \end{cases}$$

is a well defined and continuous retraction. □

The next result should remind you of the result that fiber bundles over paracompact spaces are fibrations.

Theorem 6.23. *If X is a CW-complex and $A \subset X$ a subcomplex, then (X,A) is an NDR pair.*

Sketch of proof. The complex X is obtained from A by adding cells. Use a collar $S^{n-1} \times [0,1] \subset D^n$ given by $(\vec{v}, t) \mapsto (1 - \frac{t}{2})\vec{v}$ to define u and h cell-by-cell. □

Exercise 92. If (X, A) and (Y, B) are cofibrations, so is their product

$$(X, A) \times (Y, B) = (X \times Y, X \times B \cup A \times Y).$$

We next establish that a pushout of a cofibration is a cofibration; this is dual to the fact that pullback of a fibration is a fibration. The word dual here is used in the sense of reversing arrows.

Definition 6.24. A *pushout of maps $f : A \to B$ and $g : A \to C$* is a commutative diagram

$$\begin{array}{ccc} A & \xrightarrow{f} & B \\ {\scriptstyle g}\downarrow & & \downarrow \\ C & \longrightarrow & D \end{array}$$

which is initial among all such commutative diagrams; i.e. any problem of the form

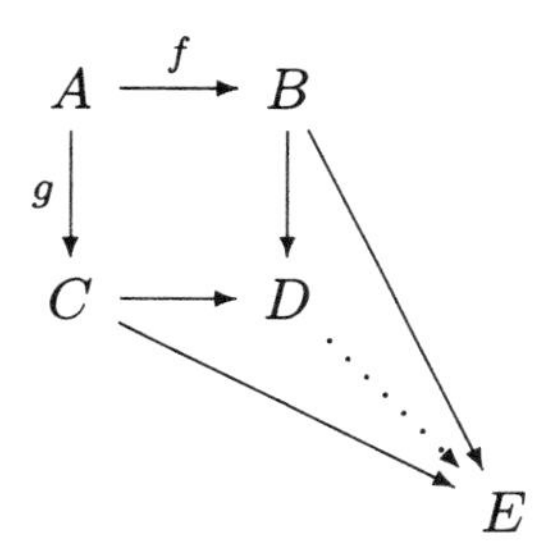

has a unique solution.

Pushouts are unique up to homeomorphism; this is proved using an "initial object" argument. Sometimes we just say D is the pushout, and sometimes we write $D = f_*C$, the pushout of g along f.

Pushouts always exist. They are constructed as follows.

When A is empty the pushout is the *disjoint union* $B \amalg C$. A concrete realization is given by choosing base points $b_0 \in B$ and $c_0 \in C$ and setting

$$B \amalg C = \{(b, c_0, 0) \in B \times C \times I \mid b \in B\} \cup \{(b_0, c, 1) \in B \times C \times I \mid c \in C\}.$$

In general, a concrete realization for the pushout of $f : A \to B$ and $g : A \to C$ is

$$\frac{B \amalg C}{f(a) \sim g(a)}.$$

Note that this is a quotient of a sum, just like the pushout in the category of abelian groups.

Theorem 6.25. *If $g : A \to C$ is a cofibration and*

$$\begin{array}{ccc} A & \xrightarrow{f} & B \\ {\scriptstyle g}\big\downarrow & & \big\downarrow \\ C & \longrightarrow & f_*C \end{array}$$

is a pushout diagram, then $B \to f_*C$ is a cofibration. □

The proof is obtained by reversing the arrows in the dual argument for fibrations. We leave it as an exercise.

Exercise 93. Prove Theorem 6.25.

6.8. Replacing a map by a cofibration

Let $f : A \to X$ be a continuous map. We will replace X by a homotopy equivalent space M_f and obtain a map $A \to M_f$ which is a cofibration. In short, every map is equivalent to a cofibration. If f is a cofibration to begin with, then the construction gives a homotopy equivalent cofibration relative to A.

Definition 6.26. The *mapping cylinder* of a map $f : A \to X$ is the space

$$M_f = \frac{(A \times I) \amalg X}{(a, 1) \sim f(a)}.$$

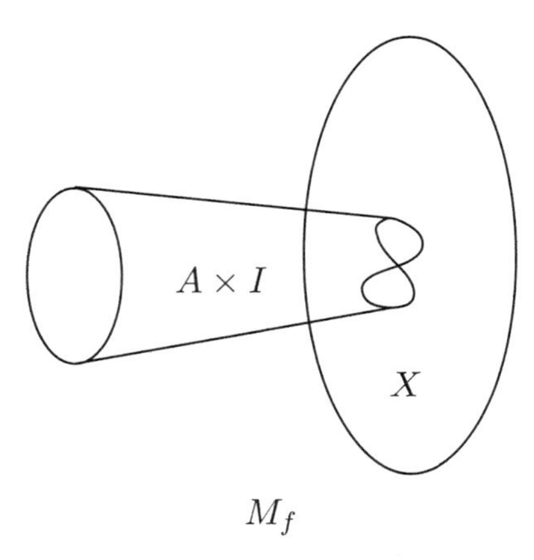

M_f

The *mapping cone* of $f : A \to X$ is

$$C_f = \frac{M_f}{A \times \{0\}}.$$

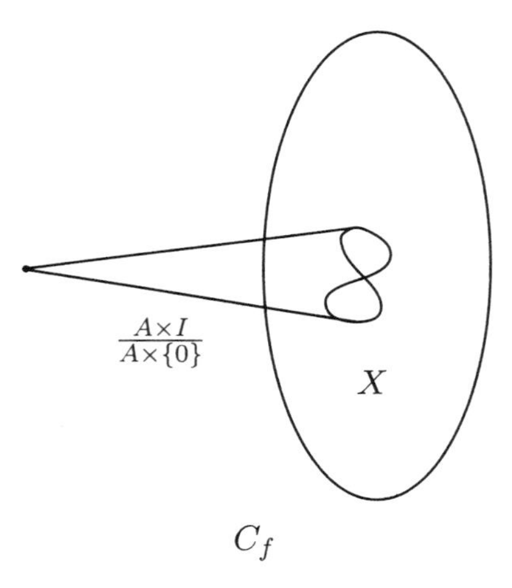

C_f

Note that the mapping cylinder M_f can also be defined as the pushout of

$$\begin{array}{ccc} A \times \{1\} & \longrightarrow & X \times \{1\} \\ \downarrow & & \\ A \times I & & \end{array}$$

This shows the analogue with the mapping path fibration P_f more clearly. Sometimes P_f is called the *mapping cocylinder* by those susceptible to categorical terminology.

The "dual" result to Theorem 6.18 is the following.

Theorem 6.27. *Let $f : A \to X$ be a map. Let $i : A \to M_f$ be the inclusion $i(a) = [a, 0]$.*

1. *There exists a homotopy equivalence $h : M_f \to X$ so that the diagram*

$$\begin{array}{ccc} & A & \\ {}^{f}\swarrow & & \searrow^{i} \\ X & \xleftarrow{h} & M_f \end{array}$$

commutes.

2. *The inclusion $i : A \to M_f$ is a cofibration.*
3. *If $f : A \to X$ is a cofibration, then h is a homotopy equivalence rel A. In particular h induces a homotopy equivalence of the cofibers $C_f \to X/f(A)$.*

Proof. 1. Let $h : M_f \to X$ be the map

$$h[a,s] = f(a), \qquad h[x] = x.$$

Then $f = h \circ i$ so the diagram commutes. The homotopy inverse of h is the inclusion $j : X \to M_f$. In fact, $h \circ j = \mathrm{Id}_X$, and the homotopy from Id_{M_f} to $j \circ h$ squashes the mapping cylinder onto X and is given by

$$F([a,s],t) = [a, s+t-st]$$
$$F([x],t) = [x].$$

2. By the implication ($3 \Rightarrow 4$) from Steenrod's theorem (Theorem 6.22), we need to construct a retraction $R : M_f \times I \to M_f \times 0 \cup A \times I$.

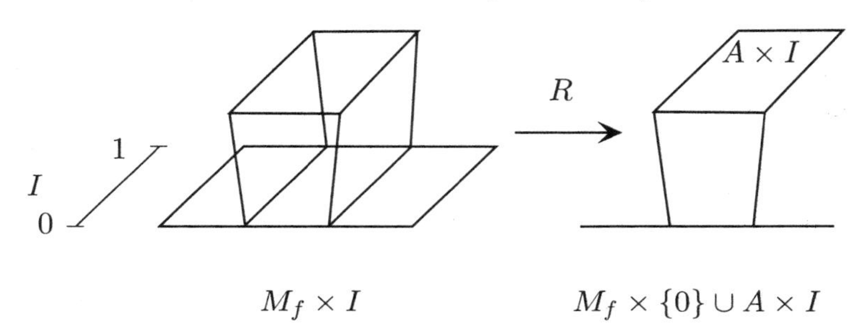

Let

$$r : I \times I \to I \times 0 \cup 0 \times I$$

be a retraction so that $r(1 \times I) = \{(1,0)\}$. (First retract the square onto 3 sides and then contract a side to a point.) Define $R([a,s],t) = [a, r(s,t)]$ and $R([x],t) = ([x],0)$. Thus $i : A \to M_f$ is a cofibration.

3. If $f : A \hookrightarrow X$ is a cofibration, by Steenrod's theorem there is a retraction

$$r : X \times I \to X \times 1 \cup f(A) \times I$$

and an obvious homeomorphism

$$q : X \times 1 \cup f(A) \times I \to M_f.$$

Define $g : X \to M_f$ by $g(x) = q(r(x,0))$. We will show that g and h are homotopy inverses rel A (recall $h[a,s] = f(a)$ and $h[x] = x$).

Define the homotopy

$$H : X \times I \to X$$

as $H = h \circ r$. Then $H(x,0) = h \circ g(x)$, $H(x,1) = x$, and $H(f(a),t) = f(a)$. Define the homotopy

$$F : M_f \times I \to M_f$$

by $F([x],t) = q(r(x,t))$ and $F([a,s],t) = q(r(f(a),st))$. Then $F(x,0) = g \circ h(x)$, $H(-,1) = \mathrm{Id}_{M_f}$, and $F(i(a),t) = i(a)$. The reader is encouraged to verify these formulae or to draw the motivating pictures. □

6.9. Sets of homotopy classes of maps

We introduce the following notation. If X, Y are spaces, then $[X,Y]$ denotes the set of homotopy classes of maps from X to Y, i.e.

$$[X,Y] = \text{Map}(X,Y)/\sim$$

where $f \sim g$ if f is homotopic to g.

Notice that if Y is path-connected, then the set $[X,Y]$ contains a distinguished class of maps, namely the unique class containing all the constant maps. We will use this as a base point for $[X,Y]$ if one is needed.

If X has a base point x_0 and Y has a base point y_0, let $[X,Y]_0$ denote the *based homotopy classes of based maps*, where a *based map* is a map $f : (X,x_0) \to (Y,y_0)$. Then $[X,Y]_0$ has a distinguished class, namely the class of the constant map at y_0. (In the based context, it is not necessary to assume Y is path-connected to have this distinguished class.) Given a map $f : X \to Y$ let $[f]$ denote its homotopy class in $[X,Y]$ or $[X,Y]_0$. Notice that if X and Y are based spaces, there is a forgetful map $[X,Y]_0 \to [X,Y]$. This map need not be injective or surjective.

The notion of an exact sequence of sets is a useful generalization of the corresponding concept for groups.

Definition 6.28. A sequence of functions

$$A \xrightarrow{f} B \xrightarrow{g} C$$

of *sets* (not spaces or groups) *with base points* is called *exact at* B if

$$f(A) = g^{-1}(c_0)$$

where c_0 is the base point of C.

All that was necessary here was that C be based. Notice that if A, B, C are groups, with base points the identity element, and f, g homomorphisms, then $A \to B \to C$ is exact *as a sequence of sets* if and only if it is exact as a sequence of groups.

The following two theorems form the cornerstone of constructions of exact sequences in algebraic topology.

Theorem 6.29 (basic property of fibrations). *Let $p : E \to B$ be a fibration, with fiber $F = p^{-1}(b_0)$ and B path-connected. Let Y be any space. Then the sequence of sets*

$$[Y,F] \xrightarrow{i_*} [Y,E] \xrightarrow{p_*} [Y,B]$$

is exact.

Proof. Clearly $p_*(i_*[g]) = 0$.

Suppose $f : Y \to E$ so that $p_*[f] = [\text{const}]$; i.e. $p \circ f : Y \to B$ is null homotopic. Let $G : Y \times I \to B$ be a null homotopy, and then let $H : Y \times I \to E$ be a solution to the lifting problem

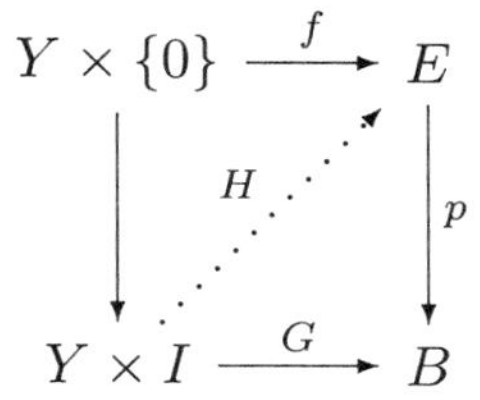

$$\begin{array}{ccc} Y \times \{0\} & \xrightarrow{f} & E \\ \downarrow & \overset{H}{\nearrow} & \downarrow p \\ Y \times I & \xrightarrow{G} & B \end{array}$$

Since $p \circ H(y,1) = G(y,1) = b_0$, $H(y,1) \in F = p^{-1}(b_0)$. Thus f is homotopic into the fiber, so $[f] = i_*[H(-,1)]$. □

Theorem 6.30 (basic property of cofibrations). *Let $i : A \hookrightarrow X$ be a cofibration, with cofiber X/A. Let $q : X \to X/A$ denote the quotient map. Let Y be any path-connected space. Then the sequence of sets*

$$[X/A,Y] \xrightarrow{q^*} [X,Y] \xrightarrow{i^*} [A,Y]$$

is exact.

Proof. Clearly $i^*(q^*([g])) = [g \circ q \circ i] = [\text{const}]$.

Suppose $f : X \to Y$ is a map and suppose that $f_{|A} : A \to Y$ is nullhomotopic. Let $h : A \times I \to Y$ be a null homotopy. The solution F to the problem

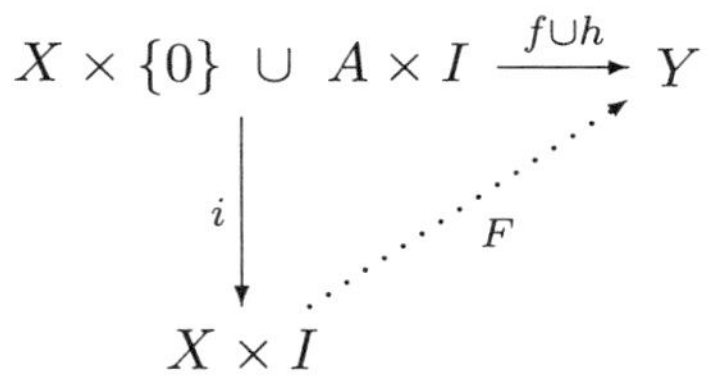

$$\begin{array}{ccc} X \times \{0\} \cup A \times I & \xrightarrow{f \cup h} & Y \\ i \downarrow & \overset{F}{\nearrow} & \\ X \times I & & \end{array}$$

defines a map $f' = F(-,1)$ homotopic to f whose restriction to A is constant, i.e. $f'(A) = y_0$. Therefore the diagram

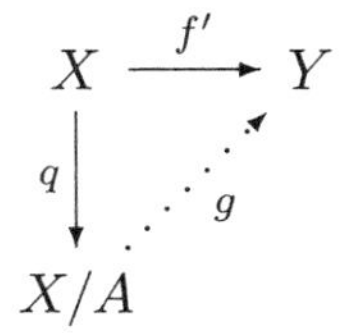

$$\begin{array}{ccc} X & \xrightarrow{f'} & Y \\ q \downarrow & \overset{g}{\nearrow} & \\ X/A & & \end{array}$$

can be completed, by the definition of quotient topology. Thus $[f] = [f'] = q^*[g]$. □

6.10. Adjoint of loops and suspension; smash products

Definition 6.31. Define $\mathcal{K}_*$ to be the category of compactly generated spaces with a *non-degenerate base point*, i.e. (X,x_0) is an object of $\mathcal{K}_*$ if the inclusion $\{x_0\} \subset X$ is a cofibration. The morphisms in $\mathcal{K}_*$ are the base point preserving continuous maps.

Exercise 94. Prove the base point versions of the previous two theorems:

1. If $F \hookrightarrow E \to B$ is a base point preserving fibration, then for any $Y \in \mathcal{K}_*$
$$[Y,F]_0 \to [Y,E]_0 \to [Y,B]_0$$
is exact.
2. If $A \hookrightarrow X \to X/A$ is a base point preserving cofibration, then for any $Y \in \mathcal{K}_*$
$$[X/A,Y]_0 \to [X,Y]_0 \to [A,Y]_0$$
is exact.

Most exact sequences in algebraic topology can be derived from Theorems 6.29, 6.30, and Exercise 94. We will soon use this exercise to establish exact sequences of homotopy groups. To do so, we need to be careful about base points and adjoints. Recall that if (X, x_0) and (Y, y_0) are based spaces, then $\mathrm{Map}(X, Y)_0$ is the set of maps of pairs $(X, x_0) \to (Y, y_0)$ with the compactly generated topology.

Definition 6.32. The *smash product* of based spaces is
$$X \wedge Y = \frac{X \times Y}{X \vee Y} = \frac{X \times Y}{X \times \{y_0\} \cup \{x_0\} \cup Y}.$$

Note that the smash product $X \wedge Y$ is a based space. Contrary to popular belief, the smash product is *not* the product in the category $\mathcal{K}_*$, although the *wedge product*
$$X \vee Y = (X \times \{y_0\}) \cup (\{x_0\} \times Y) \subset X \times Y$$
is the sum in $\mathcal{K}_*$. The smash product is the adjoint of the based mapping space. The following theorem follows from the unbased version of the adjoint theorem (Theorem 6.5), upon restricting to based maps.

Theorem 6.33 (adjoint theorem). *There is a (natural) homeomorphism*
$$\mathrm{Map}(X \wedge Y, Z)_0 \cong \mathrm{Map}(X, \mathrm{Map}(Y, Z)_0)_0.$$

□

Definition 6.34. The *(reduced) suspension* of a based space (X, x_0) is $SX = S^1 \wedge X$. The *(reduced) cone* is $CX = I \wedge X$. Here the circle is based by $1 \in S^1 \subset \mathbf{C}$ and the interval by $0 \in I$.

Using the usual identification $I/\{0,1\} = S^1$ via $t \mapsto e^{2\pi i t}$, one sees

$$SX = \frac{X \times I}{X \times \{0,1\} \cup \{x_0\} \times I}.$$

In other words, if ΣX is the unreduced suspension and cone(X) is the unreduced cone ($= \Sigma X / X \times \{0\}$), then there are quotient maps

$$\Sigma X \to SX \qquad \text{cone}(X) \to CX$$

given by collapsing $\{x_0\} \times I$, as indicated in the following figure.

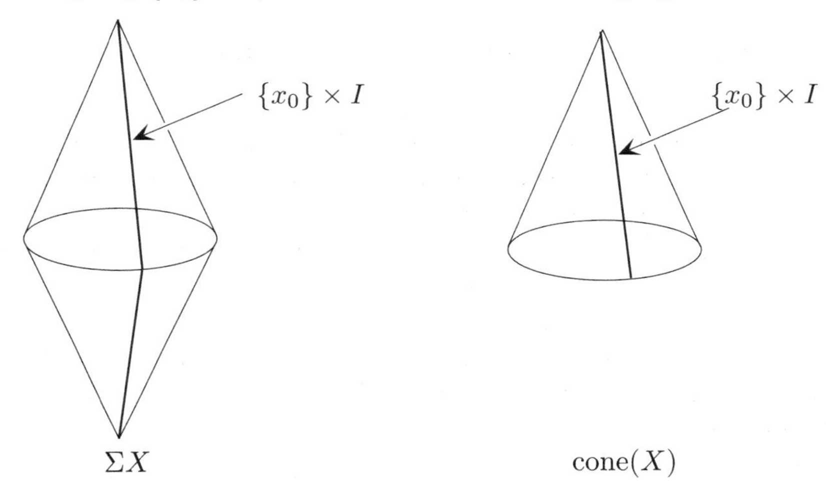

Notice that taking reduced suspensions and reduced cones is functorial. Reduced suspensions and cones are more useful than the unreduced variety since they have canonical base points and satisfy adjoint properties. Nonetheless, it is reassuring to connect them with the more familiar unreduced versions.

Exercise 95. If $X \in \mathcal{K}_*$, then the quotient maps $\Sigma X \to SX$ and cone$(X) \to CX$ are homotopy equivalences.

Proposition 6.35. *The reduced suspension SS^n is homeomorphic to S^{n+1}, and the reduced cone CS^n is homeomorphic to D^{n+1}.* □

Exercise 96. Prove Proposition 6.35. This shows in a special case that the smash product is associative. Prove associativity of the smash product in general.

Corollary 6.36. *$S^i \wedge S^j$ is homeomorphic to S^{i+j}.* □

We defined loop spaces by $\Omega_{x_0}X = \text{Map}(I, \{0,1\}; X, \{x_0\})$, but by using the identification of the circle as a quotient space of the interval, one sees

$$\Omega_{x_0}X = \text{Map}(S^1, X)_0.$$

Then a special case of Theorem 6.33 shows the following.

Theorem 6.37 (loops and suspension are adjoints). *The spaces*

$$\text{Map}(SX, Y)_0$$

and

$$\text{Map}(X, \Omega Y)_0$$

are naturally homeomorphic. □

6.11. Fibration and cofibration sequences

We will see eventually that the homotopy type of a fiber of a fibration measures how far the fibration is from being a homotopy equivalence. (For example, if the fiber is contractible, then the fibration is a homotopy equivalence.) More generally, given a map $f : X \to Y$, one can turn it into a fibration $P_f \to Y$ as above; the fiber of this fibration measures how far f is from a homotopy equivalence.

After turning $f : X \to Y$ into a fibration $P_f \to Y$, one then has an inclusion of the fiber $F \subset P_f$. Why not turn this into a fibration and see what happens? Now take the fiber of the resulting fibration and continue the process ....

Similar comments apply to cofibrations. Theorem 6.39 below identifies the resulting iterated fibers and cofibers. We first introduce some terminology.

Definition 6.38. If $f : X \to Y$ is a map, the *homotopy fiber of f* is the fiber of the fibration obtained by turning f into a fibration. The homotopy fiber is a space, well defined up to homotopy equivalence. Usually one is lazy and just calls this the *fiber of f*.

Similarly, the *homotopy cofiber of $f : X \to Y$* is the mapping cone C_f, the cofiber of $X \to M_f$.

Theorem 6.39.

1. *Let $F \hookrightarrow E \to B$ be a fibration. Let Z be the homotopy fiber of $F \hookrightarrow E$, so $Z \to F \to E$ is a fibration (up to homotopy). Then Z is homotopy equivalent to the loop space ΩB.*
2. *Let $A \hookrightarrow X \to X/A$ be a cofibration sequence. Let W be the homotopy cofiber of $X \to X/A$, so that $X \to X/A \to W$ is a cofibration (up*

to homotopy). Then W is homotopy equivalent to the (unreduced) suspension ΣA.

Proof. 1. Let $f : E \to B$ be a fibration with fiber $F = f^{-1}(b_0)$. Choose a base point $e_0 \in F$. In Section 6.6 we constructed a fibration $p : P_f \to B$ with

$$P_f = \{(e,\alpha) \in E \times B^I | f(e) = \alpha(0)\}$$

and $p(e,\alpha) = \alpha(1)$, and such that the map $h : E \to P_f$ given by $h(e) = (e, \text{const}_{f(e)})$ is a fiber homotopy equivalence.

Let $(P_f)_0 = p^{-1}(b_0)$, so $(P_f)_0 \hookrightarrow P_f \xrightarrow{p} B$ is a fibration equivalent to $F \hookrightarrow E \xrightarrow{f} B$.

Define $\pi : (P_f)_0 \to E$ by $\pi(e,\alpha) = e$. Notice that

$$(P_f)_0 = \{(e,\alpha) | f(e) = \alpha(0),\ \alpha(1) = b_0\}.$$

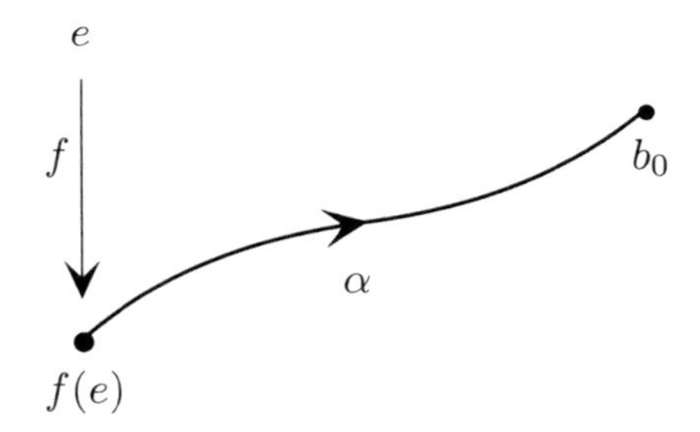

Claim. $\pi : (P_f)_0 \to E$ is a fibration with fiber $\Omega_{b_0} B$.

Proof of claim. Clearly $\pi^{-1}(e_0) = \{(e_0,\alpha) | \alpha(0) = \alpha(1) = b_0\}$ is homeomorphic to the loop space, so we just need to show π is a fibration. Given the problem

$$\begin{array}{ccc} A \times \{0\} & \xrightarrow{g} & (P_f)_0 \\ \downarrow & \overset{\widetilde{H}}{\nearrow} & \downarrow \pi \\ A \times I & \xrightarrow[H]{} & E \end{array}$$

the picture is

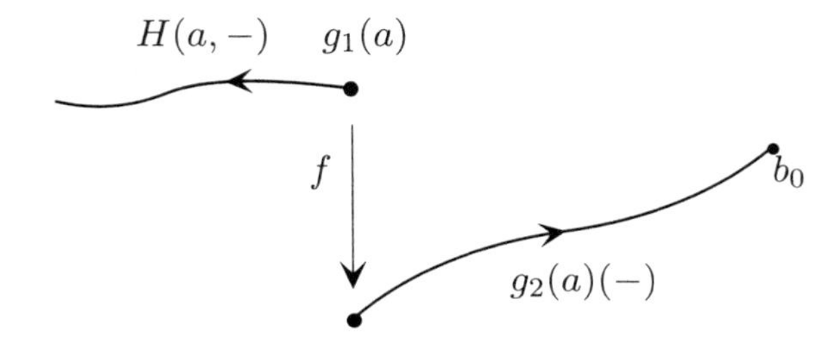

Hence we can set $\widetilde{H}(a,s) = (H(a,s), \widetilde{H}_2(a,s))$ where $\widetilde{H}_2(a,s))(-)$ has the picture

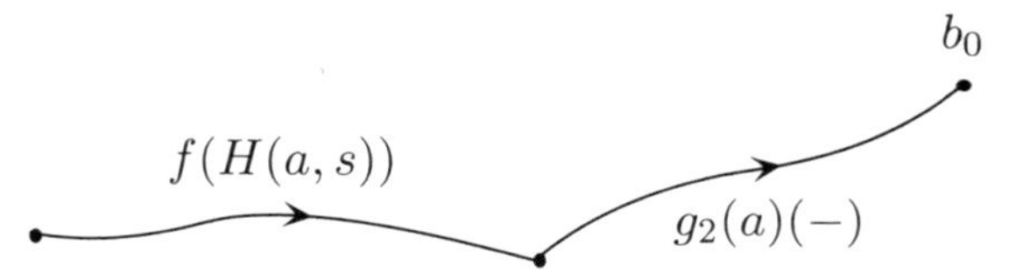

and is defined by

$$\widetilde{H}_2(a,s))(t) = \begin{cases} f(H(y, -(1+s)t+s)) & \text{if } 0 \le t \le s/(s+1), \\ g_2(a)((s+1)t-s) & \text{if } \ s/(s+1) \le t \le 1. \end{cases}$$

The map $F \hookrightarrow (P_f)_0$ is a homotopy equivalence, since $E \to P_f$ is a fiber homotopy equivalence. Thus the diagram

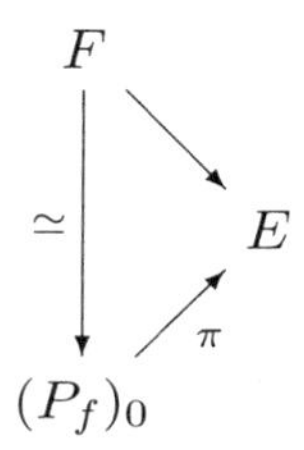

shows that the fibration $\pi : (P_f)_0 \to E$ is obtained by turning $F \hookrightarrow E$ into a fibration, and the homotopy fiber is $\Omega_{b_0} B$.

2. The map $X \to X/A$ is equivalent to $X \hookrightarrow C_i = X \cup \text{cone}(A)$ where $i : A \hookrightarrow X$. The following picture makes clear that $C_i/X = \Sigma A$. The fact that $X \to C_i$ is a cofibration is left as an exercise.

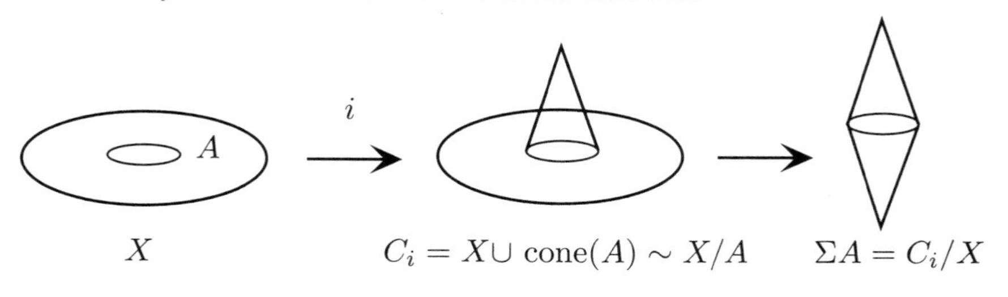

□

Exercise 97. Show that $X \hookrightarrow C_i = X \cup \text{cone}(A)$ is a cofibration.

We have introduced the notion of the loop space ΩX of a based space X as the space of paths in X which start and end at the base point. The loop space is itself a based space with base point the constant loop at the base point of X. Let $\Omega^n X$ denote the n-fold loop space of X. Similarly the reduced suspension SX of X is a based space. Let $S^n X$ denote the n-fold suspension of X.

The previous theorem can be restated in the following convenient form.

Theorem 6.40.

1. *Let $A \hookrightarrow X$ be a cofibration. Then any two consecutive maps in the sequence*

$$A \to X \to X/A \to \Sigma A \to \Sigma X \to \cdots \to \Sigma^n A \to \Sigma^n X \to \Sigma^n(X/A) \to \cdots$$

have the homotopy type of a cofibration followed by projection onto the cofiber.

1′. *Let $A \hookrightarrow X$ be a base point preserving cofibration. Then any two consecutive maps in the sequence*

$$A \to X \to X/A \to SA \to SX \to \cdots \to S^n A \to S^n X \to S^n(X/A) \to \cdots$$

have the homotopy type of a cofibration followed by projection onto the cofiber.

2. *Let $E \to B$ be a fibration with fiber F. Then any two consecutive maps in the sequence*

$$\cdots \to \Omega^n F \to \Omega^n E \to \Omega^n B \to \cdots \to \Omega F \to \Omega E \to \Omega B \to F \to E \to B$$

have the homotopy type of a fibration preceded by the inclusion of its fiber.

□

To prove (1′), one must use reduced mapping cylinders and reduced cones.

6.12. Puppe sequences

Lemma 6.41. *Let X and Y be spaces in $\mathcal{K}_*$.*

1. $[X,\Omega Y]_0 = [SX,Y]_0$ *is a group.*
2. $[X,\Omega(\Omega Y)]_0 = [SX,\Omega Y]_0 = [S^2X,Y]_0$ *is an abelian group.*

Sketch of proof. The equalities follow from Theorem 6.37, the adjointness of loops and suspension. The multiplication can be looked at in two ways: first on $[SX,Y]_0$ as coming from the map

$$\nu : SX \to SX \vee SX$$

given by collapsing out the "equator" $X \times 1/2$. Then define

$$fg \underset{\text{def}}{:=} \nu(f \vee g).$$

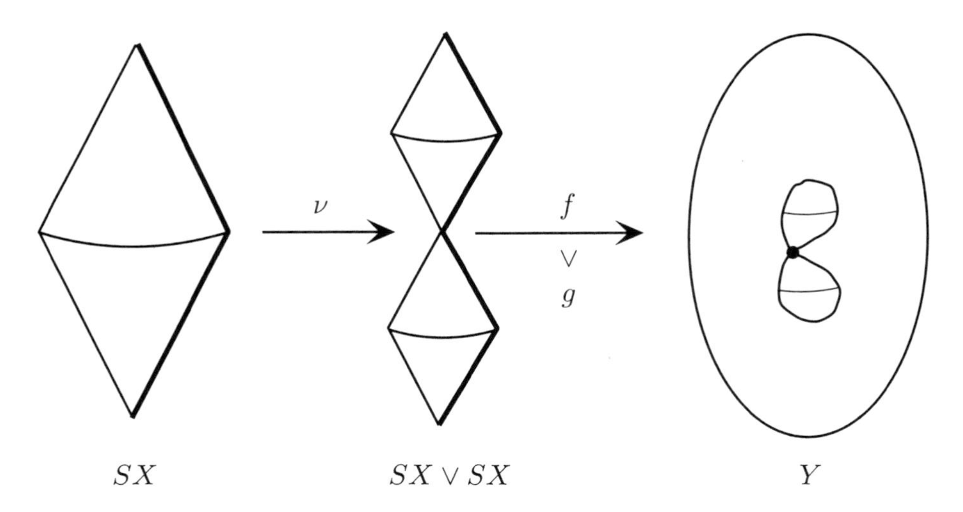

The second interpretation of multiplication is on $[X, \Omega Y]_0$ and comes from composition of loops

$$* : \Omega Y \times \Omega Y \to \Omega Y$$

with $(fg)x = f(x) * g(x)$.

The proof of 2 is obtained by meditating on the following sequence of pictures.

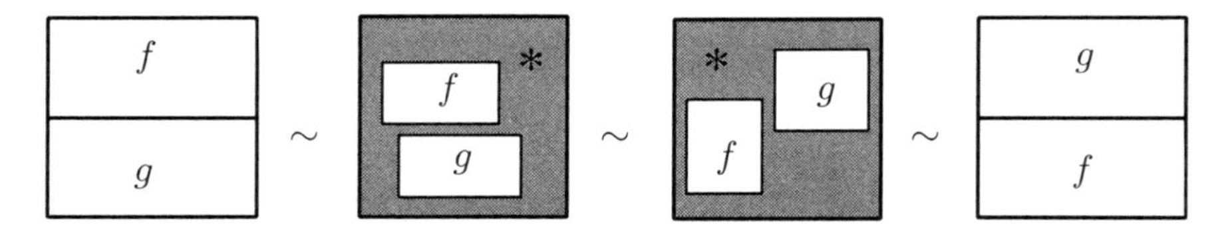

□

Exercise 98. Convince yourself that the two definitions of multiplication on $[X,\Omega Y]_0 = [SX,Y]_0$ are the same and that $\pi_1(Y, y_0) = [SS^0, Y]_0$.

The last lemma sits in a more general context. A loop space is an example of an H-group and a suspension is an example of a co-H-group. See [**36**] or [**43**] for precise definitions, but here is the basic idea. An H-group Z is a based space with a "multiplication" map $\mu : Z \times Z \to Z$ and an "inversion" map $\varphi : X \to X$ which satisfy the axioms of a group up to homotopy (e.g. is associative up to homotopy). For a topological group G and any space X, $\mathrm{Map}(X, G)$ is a group; similarly for an H-group Z, $[X, Z]_0$ is a group. To define a co-H-group, one reverses all the arrows in the definition of H-group, so there is a co-multiplication $\nu : W \to W \vee W$ and a co-inversion $\psi : W \to W$. Then $[W, X]_0$ is a group. Finally, there is a formal, but occasionally very useful result. If W is a co-H-group and Z is an H-group, then the two multiplications on $[W, Z]_0$ agree and are abelian. Nifty, huh? One consequence of this is that $\pi_1(X, x_0)$ of an H-group (e.g. a topological group) is abelian.

Combining Lemma 6.41 with Theorem 6.40 and Exercise 94 yields the proof of the following fundamental theorem.

Theorem 6.42 (Puppe sequences). *Let $Y \in \mathcal{K}_*$.*

1. *If $F \to E \to B$ is a fibration, the following sequence is a long exact sequence of sets ($i \geq 0$), groups ($i \geq 1$), and abelian groups ($i \geq 2$).*
$$\cdots \to [Y, \Omega^i F]_0 \to [Y, \Omega^i E]_0 \to [Y, \Omega^i B]_0 \to$$
$$\cdots \to [Y, \Omega B]_0 \to [Y, F]_0 \to [Y, E]_0 \to [Y, B]_0$$
where $\Omega^i Z$ denotes the iterated loop space
$$\Omega(\Omega(\cdots(\Omega Z)\cdots)).$$
2. *If (X, A) is a cofibration, the following sequence is a long exact sequence of sets ($i \geq 0$), groups ($i \geq 1$), and abelian groups ($i \geq 2$).*
$$\cdots \to [S^i(X/A), Y]_0 \to [S^i X, Y]_0 \to [S^i A, Y]_0 \to$$
$$\cdots \to [SA, Y]_0 \to [X/A, Y]_0 \to [X, Y]_0 \to [A, Y]_0.$$

□

This theorem is used as the basic tool for constructing exact sequences in algebraic topology.

6.13. Homotopy groups

We now define the homotopy groups of a based space.

Definition 6.43. Suppose that X is a space with base point x_0. Then the nth *homotopy group of X based at x_0* is the group (set if $n = 0$, abelian group if $n \geq 2$)
$$\pi_n(X, x_0) = [S^n, X]_0.$$

(We will usually only consider $X \in \mathcal{K}_*$.)

Notice that

$$\pi_n(X, x_0) = [S^n, X]_0 = [S^k \wedge S^{n-k}, X]_0 = \pi_{n-k}(\Omega^k(X)). \tag{6.2}$$

In particular,
$$\pi_n X = \pi_1(\Omega^{n-1} X).$$

There are other ways of looking at homotopy groups which are useful. For example, to get a handle on the group structure for writing down a proof, use $\pi_n(X, x_0) = [(I^n, \partial I^n), (X, x_0)]$. For the proof of the exact sequence of a pair (coming later) use $\pi_n(X, x_0) = [(D^n, S^{n-1}), (X, x_0)]$. For finding a geometric interpretation of the boundary map in the homotopy long exact sequence of a fibration given below, use

$$\pi_n(X, x_0) = [(S^{n-1} \times I, (S^{n-1} \times \partial I) \cup (* \times I)), (X, x_0)].$$

A useful observation is that the set $\pi_0(X, x_0)$ is in bijective correspondence with the path components of X. A based map $f : S^0 = \{\pm 1\} \to X$ corresponds to the path component of $f(-1)$. In general π_0 is just a based set, unless X is an H-space, e.g. a loop space or a topological group.

Also useful is the fact that $[X, Y]_0 = \pi_0(\mathrm{Map}(X, Y)_0)$, the set of path components of the function space $\mathrm{Map}(X, Y)_0$. In particular, Equation (6.2) shows that $\pi_n(X, x_0)$ is the set of path components of the n-fold loop space of X.

Homotopy groups are the most fundamental invariant of algebraic topology. For example, we will see below that a CW-complex is contractible if and only if all its homotopy groups vanish. More generally we will see that a map $f : X \to Y$ is a homotopy equivalence if and only if it induces an isomorphism on all homotopy groups. Finally, the homotopy type of a CW-complex X is determined by the homotopy groups of X together with a cohomological recipe (the k-invariants) for assembling these groups. (The homotopy groups by themselves do not usually determine the homotopy type of a space.)

Exercise 99. Show that $\pi_n(X \times Y) = \pi_n(X) \oplus \pi_n(Y)$.

As an application of the Puppe sequences (Theorem 6.42) we immediately get the extremely useful long exact sequence of homotopy groups associated to any fibration.

Corollary 6.44 (long exact sequence of a fibration). *Let $F \hookrightarrow E \to B$ be a fibration. Then the sequence*

$$\begin{aligned}\cdots \to \pi_n F \to \pi_n E \to \pi_n B \to \pi_{n-1} F \to \pi_{n-1} E \to \cdots\\ \to \pi_1 F \to \pi_1 E \to \pi_1 B \to \pi_0 F \to \pi_0 E \to \pi_0 B\end{aligned}$$

is exact. □

In Corollary 6.44, one must be careful with exactness at the right end of this sequence since $\pi_1 F$, $\pi_1 E$, and $\pi_1 B$ are non-abelian groups and $\pi_0 F$, $\pi_0 E$, and $\pi_0 B$ are merely sets.

Taking F discrete in Corollary 6.44 and using the fact that covering spaces are fibrations, one concludes the following important theorem.

Theorem 6.45. *Let $\tilde{X} \to X$ be a connected covering space of a connected space X. Then the induced map*

$$\pi_n(\tilde{X}) \to \pi_n(X)$$

is injective if $n = 1$, and an isomorphism if $n > 1$. □

Exercise 100. Give a covering space proof of Theorem 6.45.

6.14. Examples of fibrations

Many examples of fibrations and fiber bundles arise naturally in mathematics. Getting a feel for this material requires getting one's hands dirty. For that reason many facts are left as exercises. We will use the following theorem from equivariant topology to conclude that certain maps are fibrations. This is a special case of Theorem 4.5.

Theorem 6.46 (Gleason). *Let G be a compact Lie group acting freely on a compact manifold X. Then*

$$X \to X/G$$

is a principal fiber bundle with fiber G.

6.14.1. Hopf fibrations.

6.14.1. Hopf fibrations. The first class of examples we give are the famous *Hopf fibrations.* These were invented by Hopf to prove that there are non–nullhomotopic maps $S^n \to S^m$ when $n > m$.

There are four Hopf fibrations (these are fiber bundles):

$$S^0 \hookrightarrow S^1 \to S^1$$

$$S^1 \hookrightarrow S^3 \to S^2$$

$$S^3 \hookrightarrow S^7 \to S^4$$

and

$$S^7 \hookrightarrow S^{15} \to S^8.$$

These are constructed by looking at the various division algebras over $\mathbf{R}$.

Let $K = \mathbf{R}, \mathbf{C}, \mathbf{H}$, or $\mathbf{O}$ (the real numbers, complex numbers, quaternions, and octonions). Each of these has a *norm* $N : K \to \mathbf{R}_+$ so that

$$N(xy) = N(x)N(y)$$

and $N(x) > 0$ for $x \neq 0$.

More precisely,

1. If $K = \mathbf{R}$, then $N(x) = |x| = \sqrt{x\overline{x}}$ where $\overline{x} = x$.
2. If $K = \mathbf{C}$, then $N(x) = \sqrt{x\overline{x}}$ where $\overline{a + ib} = a - ib$.
3. If $K = \mathbf{H}$, then $N(x) = \sqrt{x\overline{x}}$, where $\overline{a + ib + jc + kd} = a - ib - jc - kd$.

4. The octonions are defined to be $\mathbf{O} = \mathbf{H} \oplus \mathbf{H}$. The conjugation is defined by the rule: if $p = (a, b)$, then $\overline{p} = (\overline{a}, -b)$. Multiplication is given by the rule
$$(a, b)(c, d) = (ac - \overline{d}b, b\overline{c} + da),$$
and the norm is defined by
$$N(p) = \sqrt{p\overline{p}}.$$

Let $E_K = \{(x, y) \in K \oplus K | N(x)^2 + N(y)^2 = 1\}$. Let $G_K = \{x \in K | N(x) = 1\}$.

Exercise 101. G_K is a compact Lie group homeomorphic to S^r for $r = 0, 1, 3$. For $K = \mathbf{O}$, G_K is homeomorphic to S^7, but it is not a group; associativity fails.

Let G_K act on E_K by $g \cdot (x, y) = (gx, gy)$. (Note $N(gx)^2 + N(gy)^2 = N(x)^2 + N(y)^2$ if $N(g) = 1$.)

This action is free. This is easy to show for $K = \mathbf{R}, \mathbf{C}$, or $\mathbf{H}$, since K is associative. Hence if $g(x, y) = (x, y)$, one of x or y is non-zero (since $N(x)$ and $N(y)$ are not both zero), and so if $x \neq 0$, $gx = x$ implies that $1 = xx^{-1} = (gx)x^{-1} = g(xx^{-1}) = g$. This argument does not work for $K = \mathbf{O}$ since G_K is not a group; in this case one defines an equivalence relation on E_K by $(x, y) \sim (gx, gy)$ for $g \in G_K$. The resulting quotient map $E_K \to E_k/\sim$ is a fiber bundle.

It is also easy to see that E_K consists of the unit vectors in the corresponding $\mathbf{R}^n$ and so $E_K = S^{2r+1}$ for $r = 0, 1, 3, 7$. Moreover, $G_K \cong S^r$, and so the fiber bundle $G_K \hookrightarrow E_K \to E_K/G_K$ can be rewritten
$$S^r \hookrightarrow S^{2r+1} \to Y = S^{2r+1}/S^r.$$

Exercise 102. Prove that Y is homeomorphic to the $(r+1)$-sphere S^{r+1} in the 4 cases. In fact, prove that the quotient map $S^{2r+1} \to Y$ can be written in the form $f : S^{2r+1} \to S^{r+1}$ where
$$f(z_1, z_2) = (2\overline{z}_1 z_2, N(z_1)^2 - N(z_2)^2).$$

Using these fibrations and the long exact sequence of a fibration (Corollary 6.44), one obtains exact sequences
$$\cdots \to \pi_n S^1 \to \pi_n S^3 \to \pi_n S^2 \to \pi_{n-1} S^1 \to \cdots$$
$$\cdots \to \pi_n S^3 \to \pi_n S^7 \to \pi_n S^4 \to \pi_{n-1} S^3 \to \cdots$$
$$\cdots \to \pi_n S^7 \to \pi_n S^{15} \to \pi_n S^8 \to \pi_{n-1} S^7 \to \cdots .$$

Since $\pi_n S^1 = 0$ for $n > 1$ (the universal cover of S^1 is contractible and so this follows from Theorem 6.45), it follows from the first sequence that

$\pi_n S^3 = \pi_n S^2$ for $n > 2$. The Hopf degree Theorem (Corollary 6.67 and a project for Chapter 3) implies that $\pi_n S^n = \mathbf{Z}$. In particular,

$$\pi_3 S^2 = \mathbf{Z}.$$

This is our second non-trivial calculation of $\pi_m S^n$ (the first being $\pi_n S^n = \mathbf{Z}$).

The quickest way to obtain information from the other sequences is to use the cellular approximation theorem. This is an analogue of the simplicial approximation theorem. Its proof is one of the projects for Chapter 1.

Theorem 6.47 (cellular approximation theorem). *Let (X, A) and (Y, B) be relative CW-complexes, and let $f : (X, A) \to (Y, B)$ be a continuous map. Then f is homotopic rel A to a cellular map.* □

Applying this theorem with $(X, A) = (S^n, x_0)$ and $(Y, B) = (S^m, y_0)$, one concludes that

$$\pi_n S^m = 0 \text{ if } n < m.$$

Returning to the other exact sequences, it follows from the cellular approximation theorem that $\pi_n S^4 = \pi_{n-1} S^3$ for $n \leq 6$ (since $\pi_n(S^7) = 0$ for $n \leq 6$) and that $\pi_n S^8 = \pi_{n-1} S^7$ for $n \leq 14$. We will eventually be able to say more.

6.14.2. Projective spaces.

The Hopf fibrations can be generalized by taking G_K acting on K^n for $n > 2$ at least for $K = \mathbf{R}, \mathbf{C}$, and $\mathbf{H}$.

For $K = \mathbf{R}$, $G_K = \mathbf{Z}/2$ acts on S^n with quotient real projective space $\mathbf{R}P^n$. The quotient map $S^n \to \mathbf{R}P^n$ is a covering space and in particular a fibration.

Let S^1 act on

$$S^{2n-1} = \{(z_1, \dots, z_n) \in \mathbf{C}^n \mid \Sigma |z_i|^2 = 1\}$$

by

$$t(z_1, \cdots, z_n) = (tz_1, \cdots, tz_n)$$

if $t \in S^1 = \{z \in \mathbf{C} \mid |z| = 1\}$.

Exercise 103. Prove that S^1 acts freely.

The orbit space is denoted by $\mathbf{C}P^{n-1}$ and is called *complex projective space*. The projection $S^{2n-1} \to \mathbf{C}P^{n-1}$ is a fibration with fiber S^1. (Can you prove directly that this is a fiber bundle?) In fact, if one uses the map $p : S^{2n-1} \to \mathbf{C}P^{n-1}$ to adjoin a $2n$-cell, one obtains $\mathbf{C}P^n$. Thus complex projective space is a CW-complex.

Notice that $\mathbf{C}P^n$ is a subcomplex of $\mathbf{C}P^{n+1}$, and in fact $\mathbf{C}P^{n+1}$ is obtained from $\mathbf{C}P^n$ by adding a single $2n+2$-cell. One defines infinite complex projective space $\mathbf{C}P^\infty$ to be the union of the $\mathbf{C}P^n$, with the CW-topology.

Exercise 104. Using the long exact sequence for a fibration, show that $\mathbf{C}P^\infty$ is an Eilenberg–MacLane space of type $K(\mathbf{Z}, 2)$, i.e. a CW-complex with π_2 the only non-zero homotopy group and $\pi_2 \cong \mathbf{Z}$.

Similarly, there is a fibration

$$S^3 \hookrightarrow S^{4n-1} \to \mathbf{H}P^{n-1}$$

using quaternions in the previous construction. The space $\mathbf{H}P^{n-1}$ is called *quaternionic projective space*.

Exercise 105.

1. Calculate the cellular chain complexes for $\mathbf{C}P^k$ and $\mathbf{H}P^k$.
2. Compute the *ring* structure of $H^*(\mathbf{C}P^k; \mathbf{Z})$ and $H^*(\mathbf{H}P^k; \mathbf{Z})$ using Poincaré duality.
3. Examine whether $\mathbf{O}P^k$ can be defined this way, for $k > 1$.
4. Show these reduce to Hopf fibrations for $k = 1$.

6.14.3. More general homogeneous spaces and fibrations.

Definition 6.48.

1. The *Stiefel manifold* $V_k(\mathbf{R}^n)$ is the space of orthonormal k-frames in $\mathbf{R}^n$:
$$V_k(\mathbf{R}^n) = \{(v_1, v_2, \dots, v_k) \in (\mathbf{R}^n)^k \mid v_i \cdot v_j = \delta_{ij}\}$$
given the topology as a subspace of $(\mathbf{R}^n)^k = \mathbf{R}^{nk}$.
2. The *Grassmann manifold* or *grassmannian* $G_k(\mathbf{R}^n)$ is the space of k-dimensional subspaces (a.k.a. k-planes) in $\mathbf{R}^n$. It is given the quotient topology using the surjection $V_k(\mathbf{R}^n) \to G_k(\mathbf{R}^n)$ taking a k-frame to the k-plane it spans.

Let G be a compact Lie group. Let $H \subset G$ be a closed subgroup (and hence a Lie group itself). The quotient G/H is called a *homogeneous space.* The (group) quotient map $G \to G/H$ is a principal H-bundle since H acts freely on G by right translation. If H has a closed subgroup K, then H acts on the homogeneous space H/K. Changing the fiber of the above bundle results in a fiber bundle $G/K \to G/H$ with fiber H/K.

For example, if $G = O(n)$ and $H = O(k) \times O(n-k)$ with $H \hookrightarrow G$ via

$$(A, B) \mapsto \begin{pmatrix} A & 0 \\ 0 & B \end{pmatrix},$$

let $K \subset O(n)$ be $O(n-k)$, with

$$A \mapsto \begin{pmatrix} I & 0 \\ 0 & A \end{pmatrix}.$$

Exercise 106. Identify G/H with the grassmannian and G/K with the Stiefel manifold. Conclude that the map taking a frame to the plane it spans defines a principal $O(k)$ bundle $V_k(\mathbf{R}^n) \to G_k(\mathbf{R}^n)$.

Let

$$\gamma_k(\mathbf{R}^n) = \{(p, V) \in \mathbf{R}^n \times G_k(\mathbf{R}^n) \mid p \text{ is a point in the } k\text{-plane } V\}.$$

There is a natural map $\gamma_k(\mathbf{R}^n) \to G_k(\mathbf{R}^n)$ given by projection on the second coordinate. The fiber bundle so defined is a vector bundle with fiber $\mathbf{R}^k$ (a *k-plane bundle*)

$$\mathbf{R}^k \hookrightarrow \gamma_k(\mathbf{R}^n) \to G_k(\mathbf{R}^n).$$

It is called the canonical (or tautological) vector bundle over the grassmannian.

Exercise 107. Identify the canonical bundle with the bundle obtained from the principal $O(k)$ bundle $V_k(\mathbf{R}^n) \to G_k(\mathbf{R}^n)$ by changing the fiber to $\mathbf{R}^k$.

Exercise 108. Show there are fibrations

$$O(n-k) \hookrightarrow O(n) \to V_k(\mathbf{R}^n)$$

$$O(n-1) \hookrightarrow O(n) \to S^{n-1}$$

taking a matrix to its last k columns. Deduce that

$$\pi_i(O(n-1)) \cong \pi_i(O(n)) \quad \text{for} \quad i < n-2, \tag{6.3}$$

and

$$\pi_i(V_k(\mathbf{R}^n)) = 0 \quad \text{for} \quad i < n-k-1.$$

The isomorphism of Equation (6.3) is an example of "stability" in algebraic topology. In this case it leads to the following construction. Consider the infinite orthogonal group

$$O = \operatorname*{colim}_{n \to \infty} O(n) = \bigcup_{n=1}^{\infty} O(n),$$

where $O(n) \subset O(n+1)$ is given by the continuous monomorphism

$$A \to \begin{pmatrix} A & 0 \\ 0 & 1 \end{pmatrix}.$$

Topologize O as the expanding union of the $O(n)$. Then any compact subset of O is contained in $O(n)$ for some n; hence $\pi_i O = \operatorname*{colim}_{n\to\infty} \pi_i(O(n)) = \pi_i(O(n))$ for any $n > i+2$.

A famous theorem of Bott says:

Theorem 6.49 (Bott periodicity).

$$\pi_k O \cong \pi_{k+8} O \quad \textit{for } k \in \mathbf{Z}_+.$$

Moreover the homotopy groups of O are computed to be

$k \pmod 8$	0	1	2	3	4	5	6	7
$\pi_k O$	$\mathbf{Z}/2$	$\mathbf{Z}/2$	0	$\mathbf{Z}$	0	0	0	$\mathbf{Z}$

□

An element of $\pi_k O$ is given by an element of $\pi_k(O(n))$, for some n, which by clutching (see Section 4.3.3) corresponds to a bundle over S^{k+1} with structure group $O(n)$. (Alternatively, one can show that $\pi_{k+1}(BO(n)) \cong \pi_k(O(n))$ by using the long exact sequence of homotopy groups of the fibration $O(n) \hookrightarrow EO(n) \to BO(n)$). The generators of the first eight homotopy groups of O are given by Hopf bundles.

Similarly one can consider stable Stiefel manifolds and stable grassmanians. Let $V_k(\mathbf{R}^\infty) = \operatorname{colim}_{n\to\infty} V_k(\mathbf{R}^n)$ and $G_k(\mathbf{R}^\infty) = \operatorname{colim}_{n\to\infty} G_k(\mathbf{R}^n)$. Then $\pi_i(V_k(\mathbf{R}^\infty)) = \operatorname{colim}_{n\to\infty} \pi_i(V_k(\mathbf{R}^n))$ and $\pi_i(G_k(\mathbf{R}^\infty)) = \operatorname{colim}_{n\to\infty} \pi_i(G_k(\mathbf{R}^n))$. In particular $\pi_i(V_k(\mathbf{R}^\infty)) = 0$.

A project for Chapter 4 was to show that for every topological group G, there is a principal G-bundle $EG \to BG$ where EG is contractible.

This bundle classifies principal G-bundles in the sense that given a principal G-bundle $p : G \hookrightarrow E \to B$ over a CW-complex B (or more generally a paracompact space), there is a map of principal G-bundles

$$\begin{array}{ccc} E & \xrightarrow{\tilde{f}} & EG \\ {\scriptstyle p}\downarrow & & \downarrow \\ B & \xrightarrow[f]{} & BG \end{array}$$

and the homotopy class $[f] \in [B, BG]$ is uniquely determined. It follows that the (weak) homotopy type of BG is uniquely determined.

Corollary 6.50. *The infinite grassmannian $G_k(\mathbf{R}^\infty)$ is a model for $BO(k)$. The principal $O(k)$ bundle*

$$O(k) \hookrightarrow V_k(\mathbf{R}^\infty) \to G_k(\mathbf{R}^\infty)$$

is universal and classifies principal $O(k)$-bundles. The canonical bundle

$$\mathbf{R}^k \hookrightarrow \gamma_k(\mathbf{R}^\infty) \to G_k(\mathbf{R}^\infty)$$

classifies $\mathbf{R}^k$-vector bundles with structure group $O(k)$ (i.e. $\mathbf{R}^k$-vector bundles equipped with a metric on each fiber which varies continuously from fiber to fiber). □

The fact that the grassmannian classifies orthogonal vector bundles makes sense from a geometric point of view. If $M \subset \mathbf{R}^n$ is a k-dimensional smooth submanifold, then for any point $p \in M$, the tangent space T_pM

defines a k-plane in $\mathbf{R}^n$, and hence a point in $G_k(\mathbf{R}^n)$. Likewise a tangent vector determines a point in the canonical bundle $\gamma_k(\mathbf{R}^n)$. Thus there is a bundle map

$$\begin{array}{ccc} TM & \xrightarrow{\tilde{f}} & \gamma_k(\mathbf{R}^n) \\ {\scriptstyle p}\downarrow & & \downarrow \\ M & \xrightarrow[f]{} & G_k(\mathbf{R}^n) \end{array}$$

Moreover, $G_k(\mathbf{R}^\infty)$ is also a model for $BGL_k(\mathbf{R})$ and hence is a classifying space for k-plane bundles over CW-complexes. This follows either by redoing the above discussion, replacing k-frames by sets of k-linearly independent vectors, or by using the fact that $O(k) \hookrightarrow GL_k(\mathbf{R})$ is a homotopy equivalence, with the homotopy inverse map being given by the Gram-Schmidt process.

Similar statements apply in the complex setting to unitary groups $U(n)$. Let

$$G_k(\mathbf{C}^n) = \text{complex } k\text{-planes in } \mathbf{C}^n$$
$$G_k(\mathbf{C}^n) = U(n)/(U(k) \times U(n-k)), \quad \text{the complex grassmannian}$$

$$V_k(\mathbf{C}^n) = U(n)/U(n-k), \quad \text{the unitary Stiefel manifold.}$$

There are principal fiber bundles

$$U(n-k) \hookrightarrow U(n) \to V_k(\mathbf{C}^n)$$

and

$$U(k) \hookrightarrow V_k(\mathbf{C}^n) \to G_k(\mathbf{C}^n).$$

Moreover, $V_1(\mathbf{C}^n) \cong S^{2n-1}$. Therefore

$$\pi_k(U(n)) \cong \pi_k(U(n-1)) \text{ if } k < 2n-2.$$

So letting

$$U = \operatorname*{colim}_{n\to\infty} U(n),$$

we conclude that

$$\pi_k U = \pi_k(U(n)) \text{ for } n > 1 + \frac{k}{2}.$$

Bott periodicity holds for the unitary group; the precise statement is the following.

Theorem 6.51 (Bott periodicity).

$$\pi_k U \cong \pi_{k+2} U \text{ for } k \in \mathbf{Z}_+.$$

Moreover,

$$\pi_k U = \begin{cases} \mathbf{Z} & \text{if } k \text{ is odd, and} \\ 0 & \text{if } k \text{ is even.} \end{cases}$$

Exercise 109. Prove that $\pi_1 U = \mathbf{Z}$ and $\pi_2 U = 0$.

Taking determinants gives fibrations $SO(n) \hookrightarrow O(n) \xrightarrow{\det} \{\pm 1\}$ and $SU(n) \hookrightarrow U(n) \xrightarrow{\det} S^1$. In particular, $SO(n)$ is the identity path–component of $O(n)$, so $\pi_k(SO(n)) = \pi_k(O(n))$ for $k \geq 1$. Similarly, since $\pi_k(S^1) = 0$ for $k > 1$, $\pi_1(SU(n)) = 0$ and $\pi_k SU(n) = \pi_k(U(n))$ for $k > 1$.

Exercise 110. Prove that $SO(2) = U(1) = S^1$, $SO(3) \cong \mathbf{R}P^3$, $SU(2) \cong S^3$, and that the map $p : S^3 \times S^3 \to SO(4)$ given by $(a, b) \mapsto (v \mapsto av\bar{b})$ where $a, b \in S^3 \subset \mathbf{H}$ and $v \in \mathbf{H} \cong \mathbf{R}^4$ is a 2-fold covering map.

Exercise 111. Using Exercise 110 and the facts:

1. $\pi_n S^n = \mathbf{Z}$ (Hopf degree Theorem).
2. $\pi_k S^n = 0$ for $k < n$ (Hurewicz theorem).
3. $\pi_k S^n \cong \pi_{k+1} S^{n+1}$ for $k < 2n - 1$ (Freudenthal suspension theorem).
4. There is a covering $\mathbf{Z} \hookrightarrow \mathbf{R} \to S^1$.
5. $\pi_n S^{n-1} = \mathbf{Z}/2$ for $n > 3$ (this theorem is due to V. Rohlin and G. Whitehead; see Corollary 9.27).

Compute as many homotopy groups of S^n's, $O(n)$, Grassmann manifolds, Stiefel manifolds, etc., as you can.

6.15. Relative homotopy groups

Let (X, A) be a pair, with base point $x_0 \in A \subset X$. Let $p = (1, 0, \cdots, 0) \in S^{n-1} \subset D^n$.

Definition 6.52. The *relative homotopy group (set if $n = 1$) of the pair* (X, A) is

$$\pi_n(X, A, x_0) = [D^n, S^{n-1}, p; X, A, x_0],$$

the set of based homotopy classes of base point preserving maps from the pair (D^n, S^{n-1}) to (X, A). This is a functor from pairs of spaces to sets $(n = 1)$, groups $(n = 2)$, and abelian groups $(n > 2)$.

Thus, representatives for $\pi_n(X, A, x_0)$ are maps $f : D^n \to X$ such that $f(S^{n-1}) \subset A$, $f(p) = x_0$ and f is equivalent to g if there exists a homotopy $F : D^n \times I \to X$ so that for each $t \in I$, $F(-,t)$ is base point preserving and takes S^{n-1} into A, and $F(-,0) = f$, $F(-,1) = g$.

(Technical note: associativity is easier to see if instead one takes

$$\pi_n(X, A, x_0) = [D^n, S^{n-1}, P; X, A, x_0]$$

where P is one-half of a great circle, running from p to $-p$, e.g.

$$P = \{(\cos\theta, \sin\theta, 0, \cdots, 0) \mid \theta \in [0, \pi]\}.$$

This corresponds to the previous definition since the reduced cone on the sphere is the disk.)

Theorem 6.53 (homotopy long exact sequence of a pair). *The homotopy set $\pi_n(X, A)$ is a group for $n \geq 2$ and is abelian for $n \geq 3$. Moreover, there is a long exact sequence*

$$\cdots \to \pi_n A \to \pi_n X \to \pi_n(X, A) \to \pi_{n-1} A \to \cdots \to \pi_1(X, A) \to \pi_0 A \to \pi_0 X.$$

Proof. The proof that $\pi_n(X, A)$ is a group is a standard exercise, with multiplication based on the idea of the following picture.

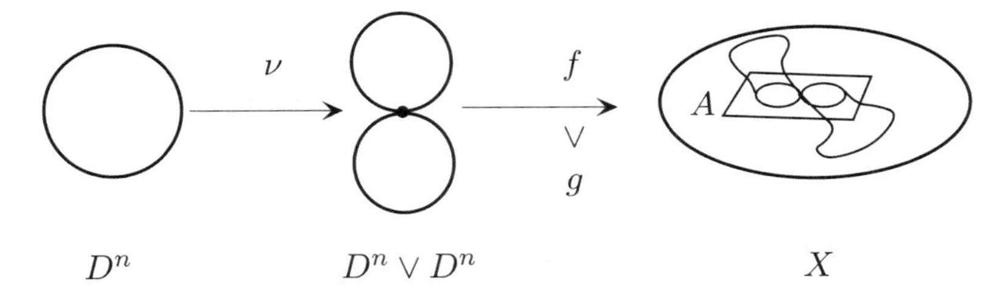

Exercise 112. Concoct an argument from this picture and use it to figure out why $\pi_1(X, A)$ is not a group. Also use it to prove that the long exact sequence is exact.

□

Lemma 6.54. *Let $f : E \to B$ be a fibration with fiber F. Let $A \subset B$ be a subspace, and let $G = f^{-1}(A)$, so that $F \hookrightarrow G \xrightarrow{f} A$ is a fibration. Then f induces isomorphims $f_* : \pi_k(E, G) \to \pi_k(B, A)$ for all k. In particular, taking $A = \{b_0\}$ one obtains the commuting ladder*

$$\begin{array}{ccccccccccc}
\cdots & \longrightarrow & \pi_k F & \longrightarrow & \pi_k E & \longrightarrow & \pi_k(E, F) & \longrightarrow & \pi_{k-1}(F) & \longrightarrow & \cdots \\
 & & \downarrow{\scriptstyle \mathrm{Id}} & & \downarrow{\scriptstyle \mathrm{Id}} & & \downarrow{\scriptstyle f_*} & & \downarrow{\scriptstyle \mathrm{Id}} & & \\
\cdots & \longrightarrow & \pi_k F & \longrightarrow & \pi_k E & \longrightarrow & \pi_k(B) & \longrightarrow & \pi_{k-1}(F) & \longrightarrow & \cdots
\end{array}$$

with all vertical maps isomorphisms, taking the long exact sequence of the pair (E, F) *to the long exact sequence in homotopy for the fibration* $F \hookrightarrow E \to B$.

Proof. This is a straightforward application of the homotopy lifting property. Suppose that $h_0 : (D^k, S^{k-1}) \to (B, A)$ is a map. Viewed as a map $D^k \to B$ it is nullhomotopic, i.e. homotopic to the constant map $c_{b_0} = h_1 : D^k \to B$. Let H be a homotopy, and let $\tilde{h}_1 : D^k \to G \subset E$ be the constant map at the base point of G. Since $f \circ \tilde{h}_1 = h_1 = H(-, 1)$, the homotopy lifting property implies that there is a lift $\widetilde{H} : D^k \times I \to E$ with $f \circ \widetilde{H}(-, 0) = h_0$. This proves that $f_* : \pi_k(E, G) \to \pi_k(B, A)$ is surjective. A similar argument shows that $f_* : \pi_k(E, G) \to \pi_k(B, A)$ is injective.

The only square in the diagram for which commutativity is not obvious is

$$\begin{array}{ccc} \pi_k(E, F) & \longrightarrow & \pi_{k-1}(F) \\ \downarrow{\scriptstyle f_*} & & \downarrow{\scriptstyle \mathrm{Id}} \\ \pi_k(B) & \longrightarrow & \pi_{k-1}(F) \end{array} \tag{6.4}$$

We leave this as an exercise. □

Exercise 113. Prove that the diagram (6.4) commutes. You will find the constructions in the proof of Theorem 6.39 useful. Notice that the commutativity of this diagram and the fact that f_* is an isomorphism give an alternative definition of the connecting homomorphism $\pi_k(B) \to \pi_{k-1}(F)$ in the long exact sequence of the fibration $F \hookrightarrow E \to B$.

An alternative and useful perspective on Theorem 6.53 is obtained by replacing a pair by a fibration as follows.

Turn $A \hookrightarrow X$ into a fibration, with A' replacing A and $L(X, A)$ the fiber. Using the construction of Section 6.6 we see that

$$\begin{aligned} L(X, A) &= \{(a, \alpha) \mid \alpha : I \to X, \alpha(0) = a \in A, \alpha(1) = x_0\} \\ &= \mathrm{Map}((I, 0, 1), (X, A, x_0)). \end{aligned}$$

This shows that if $\Omega X \hookrightarrow PX \xrightarrow{e} X$ denotes the path space fibration, then $L(X, A) = PX|_A = e^{-1}(A)$. Thus Lemma 6.54 shows that e induces an isomorphism $e_* : \pi_k(PX, L(X, A)) \to \pi_k(X, A)$ for all k. Since PX is contractible, using the long exact sequence for the pair $(PX, L(X, A))$ gives an isomorphism $\partial : \pi_k(PX, L(X, A)) \xrightarrow{\cong} \pi_{k-1}(L(X, A))$. Therefore the composite

$$\pi_{k-1}(L(X, A)) \xrightarrow{e_* \circ \partial^{-1}} \pi_k(X, A)$$

is an isomorphism which makes the diagram

$$\begin{array}{ccccccccc}
\cdots \longrightarrow & \pi_{k+1}X & \longrightarrow & \pi_k(L(X,A)) & \longrightarrow & \pi_k A & \longrightarrow & \pi_k X & \longrightarrow \cdots \\
& \downarrow{\scriptstyle \mathrm{Id}} & & \downarrow & & \downarrow{\scriptstyle \mathrm{Id}} & & \downarrow{\scriptstyle \mathrm{Id}} & \\
\cdots \longrightarrow & \pi_{k+1}X & \longrightarrow & \pi_{k+1}(X,A) & \longrightarrow & \pi_k A & \longrightarrow & \pi_k X & \longrightarrow \cdots
\end{array}$$

commute, where the top sequence is the long exact sequence for the fibration $L(X,A) \hookrightarrow A \to X$ and the bottom sequence is the long exact sequence of the pair (X,A).

Homotopy groups are harder to compute and deal with than homology groups, essentially because excision fails for relative homotopy groups. In Chapter 8 we will discuss stable homotopy and generalized homology theories, in which (properly interpreted) excision does hold. Stabilization is a procedure which looks at a space X only in terms of what homotopy information remains in $S^n X$ as n gets large. The fiber $L(X,A)$ and cofiber X/A are stably homotopy equivalent.

6.16. The action of the fundamental group on homotopy sets

The question which arises naturally when studying based spaces is, what is the difference between the based homotopy classes $[X,Y]_0$ and the unbased classes $[X,Y]$? Worrying about base points can be a nuisance. It turns out that for simply connected spaces one need not worry; the based and unbased homotopy sets are the same. In general, the fundamental group acts on the based set as we will now explain.

Let X be in $\mathcal{K}_*$; i.e. it is a based space with a non-degenerate base point x_0. Suppose Y is a based space.

Definition 6.55. Let $f_0, f_1 : X \to Y$. Let $u : I \to Y$ be a path and suppose there is a homotopy $F : X \times I \to Y$ from f_0 to f_1 so that $F(x_0,t) = u(t)$. Then we say f_0 *is freely homotopic to* f_1 *along* u, and write

$$f_0 \underset{u}{\simeq} f_1.$$

Notice that if $f_0, f_1 : (X,x_0) \to (Y,y_0)$, then u is a loop. Thus a *free* homotopy of *based maps* gives rise to an element of $\pi_1(Y,y_0)$.

Lemma 6.56.

1. *(Existence) Given a map* $f_0 : X \to Y$ *and a path* u *in* Y *starting at* $f_0(x_0)$, $f_0 \underset{u}{\simeq} f_1$ *for some* f_1.
2. *(Uniqueness) Suppose* $f_0 \underset{u}{\simeq} f_1$, $f_0 \underset{v}{\simeq} f_2$ *and* $u \simeq v$ *(rel* ∂I*). Then* $f_1 \underset{\text{const}}{\simeq} f_2$.
3. *(Multiplicativity)* $f_0 \underset{u}{\simeq} f_1$, $f_1 \underset{v}{\simeq} f_2 \Longrightarrow f_0 \underset{uv}{\simeq} f_2$.

Proof. 1. There exists a free homotopy $F : X \times I \to Y$ with $F(x_0, t) = u(t)$, $F(-, 0) = f_0$, since (X, x_0) is a cofibration:

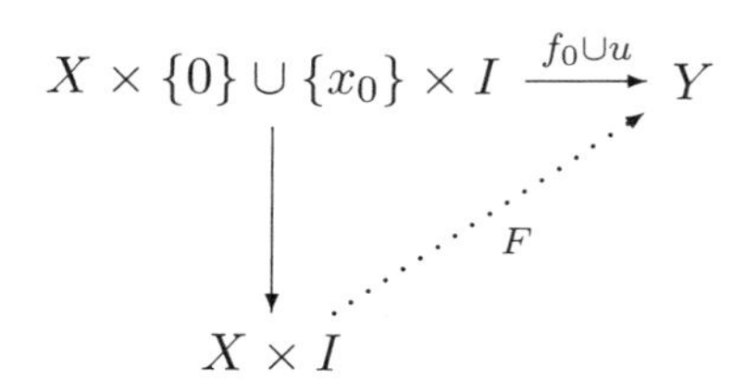

2. Since $(I, \partial I), (X, x_0)$ are cofibrations, so is their product $(X \times I, X \times \partial I \cup x_0 \times I)$ (See Exercise 92), and so the following problem has a solution:

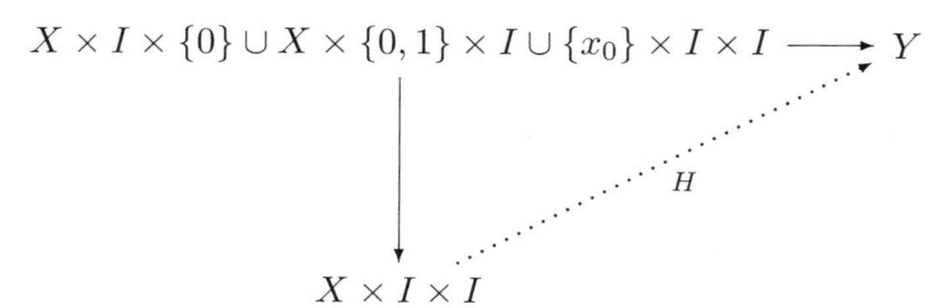

In this diagram,

1. $X \times I \times \{0\} \to Y$ is the map $(x, s, 0) \mapsto f_0(x)$.
2. $X \times \{0\} \times I \to Y$ is the homotopy of f_0 to f_1 along u.
3. $X \times \{1\} \times I \to Y$ is the homotopy of f_0 to f_2 along v.
4. $\{x_0\} \times I \times I \to Y$ is the path homotopy of u to v.

The situation is represented in the following picture of a cube $X \times I \times I$.

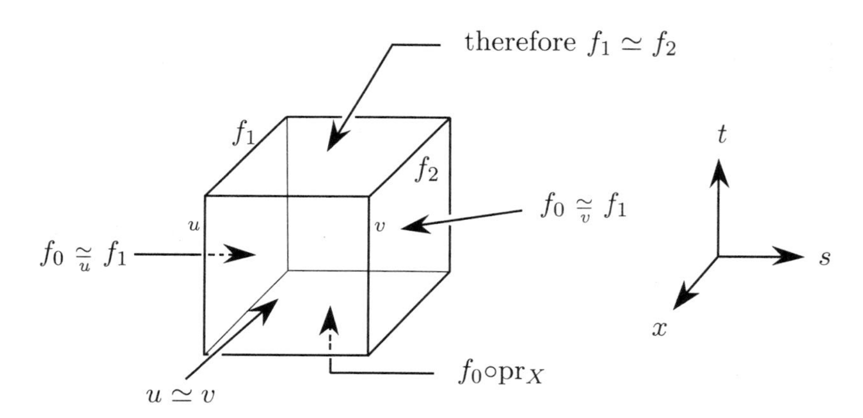

Then $H(-, -, 1)$ is a homotopy of f_1 to f_2 along a constant path.

3. This is clear. □

In light of Lemma 6.56, we can define an action of $\pi_1(Y, y_0)$ on $[X, Y]_0$ by the following recipe.

For $[u] \in \pi_1(Y, y_0)$ and $[f] \in [X, Y]_0$, define $[u][f]$ to be $[f_1]$, where f_1 is any map so that $f \underset{u}{\simeq} f_1$.

Theorem 6.57. *This defines an action of $\pi_1(Y, y_0)$ on the based set $[X, Y]_0$, and $[X, Y]$ is the quotient set of $[X, Y]_0$ by this action if Y is path connected.*

Proof. We need to verify that this action is well defined. It is independent of the choice of representative of $[u]$ by Lemma 6.56, Part 2. Suppose now $[f] = [g] \in [X, Y]_0$ and $g \underset{u}{\simeq} g_1$. Then

$$f_1 \underset{u^{-1}}{\simeq} f \underset{\text{const}}{\simeq} g \underset{u}{\simeq} g_1$$

so that f_1 and g_1 are based homotopic by Lemma 6.56, Parts 2 and 3.

This is an action of the group $\pi_1(Y, y_0)$ on the set $[X, Y]_0$ by Lemma 6.56, Part 3. Let

$$\Phi : [X, Y]_0 \to [X, Y]$$

be the forgetful functor. Clearly $\Phi([u][f]) = [f]$, and if $\Phi[f_0] = \Phi[f_1]$, then there is a u so that $[u][f_0] = [f_1]$. Finally Φ is onto by Lemma 6.56, Part 3, and the fact that Y is path-connected. □

Corollary 6.58. *A based map of path connected spaces is null-homotopic if and only if it is based null-homotopic.*

Proof. If c denotes the constant map, then clearly $c \underset{u}{\simeq} c$ for any $u \in \pi_1 Y$. Thus $\pi_1 Y$ fixes the class in $[X, Y]_0$ containing the constant map. □

Corollary 6.59. *Let $X, Y \in \mathcal{K}_*$. If Y is a path connected and simply-connected space then the forgetful functor $[X, Y]_0 \to [X, Y]$ is bijective.* □

6.16.1. Alternative description in terms of covering spaces. Suppose Y is path connected, and X is *simply connected.* Then covering space theory says that any map $f : (X, x_0) \to (Y, y_0)$ lifts to a *unique* map $\tilde{f} : (X, x_0) \to (\tilde{Y}, \tilde{y}_0)$, where $\tilde{Y}$ denotes the universal cover of Y. Moreover based homotopic maps lift to based homotopic maps. Thus the function

$$p_* : [X, \tilde{Y}]_0 \to [X, Y]_0$$

induced by the cover $p : (\tilde{Y}, \tilde{y}_0) \to (Y, y_0)$ is a *bijection.* On the other hand, since $\tilde{Y}$ is path connected and simply connected, Corollary 6.59 shows that the function $[X, \tilde{Y}]_0 \to [X, \tilde{Y}]$ induced by the inclusion is a bijection.

Now $\pi_1(Y, y_0)$ can be identified with the group of covering transformations of $\tilde{Y}$. Thus, $\pi_1(Y, y_0)$ acts on $[X, \tilde{Y}]$ by post composition; i.e. $\alpha : \tilde{Y} \to \tilde{Y}$ acts on $f : X \to \tilde{Y}$ by $\alpha \circ f$. (Note: one must be careful with left and right actions: by convention $\pi_1(Y, y_0)$ acts on $\tilde{Y}$ on the right, so $\alpha \circ f$ means the function $x \mapsto f(x) \cdot \alpha$.)

A standard exercise in covering space theory shows that if $\alpha \in \pi_1(Y, y_0)$, the diagram

$$\begin{array}{ccccc} [X,Y]_0 & \xleftarrow{\cong} & [X,\tilde{Y}]_0 & \xrightarrow{\cong} & [X,\tilde{Y}] \\ \downarrow{\scriptstyle \alpha} & & & & \downarrow{\scriptstyle \alpha} \\ [X,Y]_0 & \xleftarrow{\cong} & [X,\tilde{Y}]_0 & \xrightarrow{\cong} & [X,\tilde{Y}] \end{array}$$

commutes, where the action on the left is via an α-homotopy, and the action on the right is the action induced by the covering translation corresponding to α, and the two left horizontal bijections are induced by the covering projection. Thus the two notions of action agree.

Since $\pi_n Y = [S^n, Y]_0$, we have the following corollary.

Corollary 6.60. *For any space Y, $\pi_1(Y, y_0)$ acts on $\pi_n(Y, y_0)$ for all n with quotient $[S^n, Y]$, the set of free homotopy classes.* □

One could restrict to simply connected spaces Y and never worry about the distinction between based and unbased homotopy classes of maps into Y. This is not practical in general, and so instead one can make a dimension-by-dimension definition.

Definition 6.61. We say Y is *n-simple* if $\pi_1 Y$ acts trivially on $\pi_n Y$. We say Y is *simple* if Y is n-simple for all n.

Thus, simply connected spaces are simple.

Proposition 6.62. *If F is n-simple, then the fibration $F \hookrightarrow E \to B$ defines a local coefficient system over B with fiber $\pi_n F$.*

(A good example to think about is the Klein bottle mapping onto the circle.)

Proof. Theorem 6.12 shows that given any fibration, $F \hookrightarrow E \to B$, there is a well defined homomorphism

$$\pi_1 B \to \left\{ \begin{array}{l} \text{Homotopy classes of self-homotopy} \\ \text{equivalences } F \to F \end{array} \right\}.$$

A homotopy equivalence induces a bijection

$$[S^n, F] \xrightarrow{\cong} [S^n, F].$$

But, since we are assuming that F is n-simple, this is the same as an automorphism

$$\pi_n F \longrightarrow \pi_n F.$$

Thus, we obtain a homomorphism

$$\rho : \pi_1 B \longrightarrow \mathrm{Aut}(\pi_n(F)),$$

i.e. a local coefficient system over B. $\square$

Exercise 114. Prove that the action of $\pi_1(Y, y_0)$ on itself is just given by conjugation, so that Y is 1-simple if and only if $\pi_1 Y$ is abelian.

Exercise 115. Show that a topological group is simple. (In fact H-spaces are simple.)

Theorem 6.63. *The group $\pi_1 A$ acts on $\pi_n(X, A)$, $\pi_n X$, and $\pi_n A$ for all n. Moreover, the long exact sequence of the pair*

$$\cdots \longrightarrow \pi_n A \longrightarrow \pi_n X \longrightarrow \pi_n(X, A) \longrightarrow \pi_{n-1} A \longrightarrow \cdots$$

is $\pi_1 A$-equivariant.

Proof. Let $h : (I, 0, 1) \longrightarrow (A, x_0, x_0)$ represent $u \in \pi_1(A, x_0)$. Let $f : (D^n, S^{n-1}, p) \longrightarrow (X, A, x_0)$. Then since (S^{n-1}, p) is an NDR–pair, the problem

$$\begin{array}{ccc} S^{n-1} \times \{0\} \cup \{p\} \times I & \xrightarrow{f_{|S^{n-1}} \cup h} & A \\ \downarrow & \nearrow_{h} & \\ S^{n-1} \times I & & \end{array}$$

has a solution h. Since (D^n, S^{n-1}) is a cofibration, the problem

$$\begin{array}{ccc} D^n \times \{0\} \cup S^{n-1} \times I & \xrightarrow{f \cup h} & X \\ \downarrow & \nearrow_{F} & \\ D^n \times I & & \end{array}$$

has a solution F. By construction, $F(x, 0) = f(x)$, and also $F(-, 1)$ takes the triple (D^n, S^{n-1}, p) to (X, A, x_0). Taking $u \cdot [f] = [F(-, 1)]$ defines the action of $\pi_1(A, x_0)$ on $\pi_n(X, A; x_0)$. It follows immediately from the definitions that the maps in the long exact sequence are $\pi_1 A$-equivariant. $\square$

Definition 6.64. A *pair* (X, A) is *n-simple* if $\pi_1 A$ acts trivially on $\pi_n(X, A)$ for all n.

6.17. The Hurewicz and Whitehead theorems

Perhaps the most important result of homotopy theory is the Hurewicz Theorem. We will state the general relative version of the Hurewicz theorem and its consequence, the Whitehead theorem, in this section.

Recall that D^n is oriented as a submanifold of $\mathbf{R}^n$; i.e., the chart $D^n \hookrightarrow \mathbf{R}^n$ determines the local orientation at any $x \in D^n$ via the excision isomorphism $H_n(D^n, D^n - \{x\}) \cong H_n(\mathbf{R}^n, \mathbf{R}^n - \{x\})$. This determines the fundamental class $[D^n, S^{n-1}] \in H_n(D^n, S^{n-1})$. The sphere S^{n-1} is oriented as the boundary of D^n; i.e. the fundamental class $[S^{n-1}] \in H_{n-1}(S^{n-1})$ is defined by $[S^{n-1}] = \delta([D^n, S^{n-1}])$ where $\delta : H_n(D^n, S^{n-1}) \xrightarrow{\cong} H_{n-1}(S^{n-1})$ is the connecting homomorphism in the long exact sequence for the pair (D^n, S^{n-1}).

Definition 6.65. The *Hurewicz map* $\rho : \pi_n X \to H_n X$ is defined by

$$\rho([f]) = f_*([S^n]),$$

where $f : S^n \to X$ represents an element of $\pi_n X, [S^n] \in H_n S^n \cong \mathbf{Z}$ is the generator (given by the natural orientation of S^n) and $f_* : H_n S^n \to H_n X$ is the induced map.

There is also a *relative Hurewicz map* $\rho : \pi_n(X, A) \to H_n(X, A)$ defined by

$$\rho([f]) = f_*([D^n, S^{n-1}]).$$

Here $[D^n, S^{n-1}] \in H_n(D^n, S^{n-1}) \cong \mathbf{Z}$ is the generator given by the natural orientation, and $f_* : H_n(D^n, S^{n-1}) \to H_n(X, A)$ is the homomorphism induced by $f : (D^n, S^{n-1}, *) \to (X, A, x_0) \in \pi_n(X, A; x_0)$.

Since the connecting homomorphism $\partial : H_n(D^n, S^{n-1}) \to H_{n-1}(S^{n-1})$ takes $[D^n, S^{n-1}]$ to $[S^{n-1}]$, the map of exact sequences

$$\begin{array}{ccccccccc} \cdots \longrightarrow & \pi_n(A) & \longrightarrow & \pi_n(X) & \longrightarrow & \pi_n(X,A) & \longrightarrow & \pi_{n-1}(A) & \longrightarrow \cdots \\ & \downarrow{\scriptstyle\rho} & & \downarrow{\scriptstyle\rho} & & \downarrow{\scriptstyle\rho} & & \downarrow{\scriptstyle\rho} & \\ \cdots \longrightarrow & H_n(A) & \longrightarrow & H_n(X) & \longrightarrow & H_n(X,A) & \longrightarrow & H_{n-1}(A) & \longrightarrow \cdots \end{array}$$

commutes.

Let $\pi_n^+(X, A)$ be the quotient of $\pi_n(X, A)$ by the normal subgroup generated by

$$\{x(\alpha(x))^{-1} | x \in \pi_n(X, A),\ \alpha \in \pi_1 A\}.$$

(Thus $\pi_n^+(X, A) = \pi_n(X, A)$ if $\pi_1 A = \{1\}$, or if (X, A) is n-simple.)

Clearly ρ factors through $\pi_n^+(X, A)$, since $f_*([D^n, S^n])$ depends only on the free homotopy class of f. The following theorem is the subject of one of the projects for this chapter. It says that for simply connected spaces, the first non-vanishing homotopy and homology groups coincide. We will give a proof of the Hurewicz theorem for simply connected spaces in Chapter 10.

Theorem 6.66 (Hurewicz theorem).

1. *Let $n > 0$. Suppose that X is path-connected. If $\pi_k(X, x_0) = 0$ for all $k < n$, then $H_k(X) = 0$ for all $0 < k < n$, and the Hurewicz map*
$$\rho : \pi_n X \to H_n X$$
is an isomorphism if $n > 1$, and a surjection with kernel the commutator subgroup of $\pi_1 X$ if $n = 1$.
2. *Let $n > 1$. Suppose X and A are path-connected. If $\pi_k(X, A) = 0$ for all $k < n$ then $H_k(X, A) = 0$ for all $k < n$, and*
$$\rho : \pi_n^+(X, A) \to H_n(X, A)$$
is an isomorphism. In particular $\rho : \pi_n(X, A) \to H_n(X, A)$ is an epimorphism. □

Corollary 6.67 (Hopf degree theorem). *The Hurewicz map $\rho : \pi_n S^n \to H_n S^n$ is an isomorphism. Hence a degree zero map $f : S^n \to S^n$ is null-homotopic.* □

Although we have stated this as a corollary of the Hurewicz theorem, it can be proven directly using only the (easy) simplicial approximation theorem. (The Hopf degree theorem was covered as a project in Chapter 5.)

Definition 6.68.

1. A space X is called *n-connected* if $\pi_k X = 0$ for $k \leq n$. (Thus "simply connected" is synonymous with 1-connected.)
2. A pair (X, A) is called *n-connected* if $\pi_k(X, A) = 0$ for $k \leq n$.
3. A map $f : X \to Y$ is called *n-connected* if the pair (M_f, X) is n-connected, where M_f = mapping cylinder of f.

Using the long exact sequence for (M_f, X) and the homotopy equivalence $M_f \sim Y$, we see that f is *n-connected* if and only if
$$f_* : \pi_k X \to \pi_k Y$$
is an isomorphism for $k < n$ and an epimorphism for $k = n$. Replacing the map $f : X \to Y$ by a fibration and using the long exact sequence for the homotopy groups of a fibration one concludes that f is n-connected if and only if the homotopy fiber of f is $(n-1)$-connected.

Corollary 6.69 (Whitehead theorem).

1. *If $f : X \to Y$ is n-connected, then $f_* : H_q X \to H_q Y$ is an isomorphism for all $q < n$ and an epimorphism for $q = n$.*
2. *If X, Y are 1-connected, and $f : X \to Y$ is a map such that*
$$f_* : H_q X \to H_q Y$$
is an isomorphism for all $q < n$ and an epimorphism for $q = n$, then f is n-connected.
3. *If X, Y are 1-connected spaces and $f : X \to Y$ is a map inducing an isomorphism on $\mathbf{Z}$-homology, then f induces isomorphisms $f_* : \pi_k X \xrightarrow{\cong} \pi_k Y$ for all k.*

Exercise 116. Prove Corollary 6.69.

A map $f : X \to Y$ inducing an isomorphism of $\pi_k X \to \pi_k Y$ for all k is called a *weak homotopy equivalence.* Thus a map inducing a homology isomorphism between simply connected spaces is a weak homotopy equivalence. Conversely a weak homotopy equivalence between two spaces gives a homology isomorphism.

We will see later (Theorem 7.34) that if X, Y are CW-complexes, then $f : X \to Y$ is a weak homotopy equivalence if and only if f is a homotopy equivalence. As a consequence,

Corollary 6.70. *A continuous map $f : X \to Y$ between simply connected CW-complexes inducing an isomorphism on all $\mathbf{Z}$-homology groups is a homotopy equivalence.* □

This corollary does not imply that if X, Y are two simply connected spaces with the same homology, then they are homotopy equivalent; one needs a *map* inducing the homology equivalence.

For example, $X = S^4 \vee (S^2 \times S^2)$ and $Y = \mathbf{C}P^2 \vee \mathbf{C}P^2$ are simply connected spaces with the same homology. They are not homotopy equivalent because their cohomology rings are different. In particular, there does not exist a continuous map from X to Y inducing isomorphisms on homology.

The Whitehead theorem for non-simply connected spaces involves homology with local coefficients: If $f : X \to Y$ is a map, let $\tilde{f} : \tilde{X} \to \tilde{Y}$ be the corresponding lift to universal covers. Recall from Shapiro's lemma (Exercise 76) that
$$H_k(\tilde{X}; \mathbf{Z}) \cong H_k(X, \mathbf{Z}[\pi_1 X]) \quad \text{for all } k$$
and
$$\pi_k \tilde{X} \cong \pi_k X \quad \text{for } k > 1$$

(and similarly for Y).

We obtain (with $\pi = \pi_1 X \cong \pi_1 Y$):

Theorem 6.71. *If $f : X \to Y$ induces an isomorphism $f_* : \pi_1 X \to \pi_1 Y$, then f is n-connected if and only if it induces isomorphisms*

$$H_k(X; \mathbf{Z}[\pi]) \to H_k(Y; \mathbf{Z}[\pi])$$

for $k < n$ and an epimorphism

$$H_n(X; \mathbf{Z}[\pi]) \to H_n(Y; \mathbf{Z}[\pi]).$$

In particular, f is a weak homotopy equivalence (homotopy equivalence if X, Y are CW-complexes) if and only if $f_* : H_k(X; A_\rho) \to H_k(Y; A_\rho)$ is an isomorphism for all local coefficient systems $\rho : \pi \to \mathrm{Aut}(A)$. □

Thus, in the presence of a map $f : X \to Y$, *homotopy equivalences can be detected by homology.*

6.18. Projects for Chapter 6

6.18.1. The Hurewicz theorem. Give or outline a proof of Theorem 6.66. A reference is [**43**, §IV.4-IV.7]. Another possibility is to give a spectral sequence proof. Chapter 10 contains a spectral sequence proof of the Hurewicz theorem.

6.18.2. The Freudenthal suspension theorem. Give or outline a proof of Theorem 8.7. A good reference for the proof is [**43**, §VII.6-VII.7]. You can find a spectral sequence proof in Section 10.3.

Chapter 7

Obstruction Theory and Eilenberg-MacLane Spaces

7.1. Basic problems of obstruction theory

Obstruction theory addresses the following types of problems. Let (X, A) be a CW-pair, Y an arbitrary space, and $p : E \to B$ a fibration.

1. *Extension problem.* Suppose $f : A \to Y$ is a continuous map. When does f extend to all of X? The problem is stated in the following diagram.

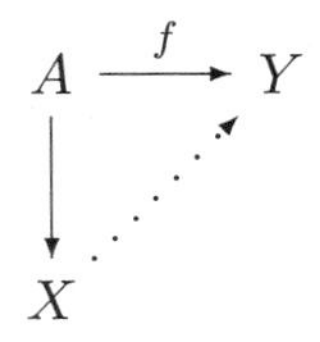

(Given the two solid arrows, can one find a dotted arrow so that the diagram commutes?)

2. *Homotopy problem.*

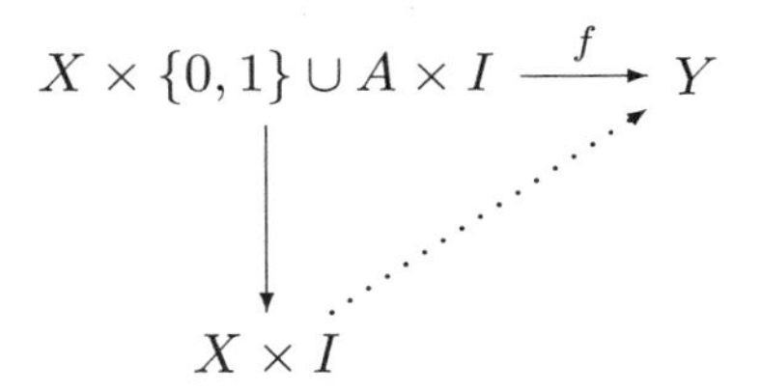

In words, given two maps $f_0, f_1 : X \to Y$ and a homotopy of the restrictions $f_{0|_A} : A \to Y$ to $f_{1|_A} : A \to Y$, can one find a homotopy from f_0 to f_1?

If A is empty, this is just the question of whether two maps f_0 and f_1 are homotopic. This problem is different from the homotopy extension problem (which is always solvable in our context) since in this case f_1 is *specified.*

Notice that the homotopy problem is a special case of the extension problem.

3. *Lifting problem.*

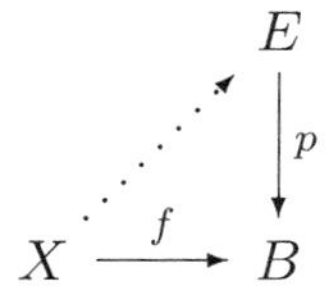

If $f : X \to B$ is given, can we find a lift of f to E? This is a special case of the *relative lifting problem*

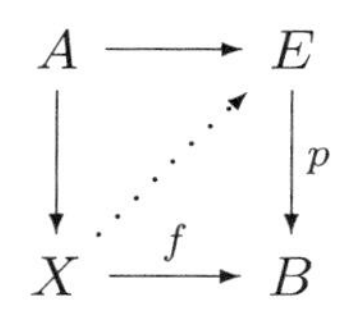

4. *Cross section problem.* This is just a special case of the relative lifting problem in the case when $X = B$ and $f : X \to B$ is the identity map.

The cross section and relative lifting problems are equivalent since the relative lifting problem reduces to finding a cross section of the pullback bundle $f^*E \to X$.

Fibrations and cofibrations are easier to work with than arbitrary maps since they have fibers and cofibers. Although we have required that (X, A) be a CW-pair and $p : E \to B$ be a fibration, the methods of Chapter 6 show how to work in complete generality. Suppose that X and A are arbitrary CW-complexes and $g : A \to X$ is an arbitrary continuous map. The cellular approximation theorem (Theorem 6.47) implies that g is homotopic to a cellular map; call it $h : A \to X$. The mapping cylinder M_h is then a CW-complex containing A as a subspace, and (M_h, A) is a CW-pair. Similarly if $p : E \to B$ is not a fibration, replace E by the mapping path space P_p of p to obtain a fibration $P_p \to B$.

Then the following exercise is an easy consequence of the homotopy lifting property, the homotopy extension property, and the method of turning maps into fibrations or cofibrations.

Exercise 117. Each of the four problems stated above is solvable for arbitrary continuous maps $g : A \to X$ and $p : E \to B$ between CW-complexes if and only if it is solvable for the CW-pair (M_h, A) and the fibration $P_p \to E$.

To solve the following exercise, work cell-by-cell one dimension at a time. Obstruction theory is a formalization of this geometric argument.

Exercise 118. (Motivating exercise of obstruction theory) Any map $f : X \to Y$ from an n-dimensional CW-complex to an n-connected space is null-homotopic.

It turns out that if Y is only assumed to be $(n-1)$-connected, there is a single *obstruction* $\theta(f) \in H^n(X; \pi_n Y)$ which vanishes if and only if the map f is null-homotopic.

The strategy of obstruction theory is to solve the four problems cell-by-cell and skeleton-by-skeleton. Thus, if the problem is solved over the n-skeleton X_n of X and e^{n+1} is an $(n+1)$-cell on X, some map is defined on ∂e^{n+1} and so the problem is to extend it over e^{n+1}. The obstruction to extending this map is that it be nullhomotopic or, more formally, that the element of $\pi_n Y$ represented by the composite $S^n \to \partial e^{n+1} \to X_n \to Y$ equals zero.

In this way we obtain a *cellular cochain* which assigns to $e^{n+1} \in X$ the element in $\pi_n Y$. If this cochain is the zero cochain, then the map can be extended over the $(n+1)$-skeleton of X. It turns out this cochain is in fact a cocycle and so represents a cohomology class in $H^{n+1}(X; \pi_n Y)$.

The remarkable result is that if this cocycle represents the zero cohomology class, then by *redefining* the map on the n-skeleton one can then extend it over the $(n+1)$-skeleton of X (if you take one step backwards, then you will be able to take two steps forward).

We will deal with the extension and homotopy problems first. The homotopy problem can be viewed as a relative form of the extension problem; just take

$$(X', A') = (X \times I, X \times \partial I \cup A \times I).$$

Hence the problem of finding a homotopy between $f : X \to Y$ and $g : X \to Y$ is obstructed by classes in $H^{n+1}(X \times I, X \times \{0,1\}; \pi_n Y)$, which is isomorphic to $H^n(X; \pi_n Y)$.

We end this introduction to obstruction theory with some comments to indicate how the results of obstruction theory lead to a major conceptual shift in perspective on what homology is.

Let A be an abelian group and $K(A, n)$ be a space such that

$$\pi_k(K(A,n)) = \begin{cases} A & \text{if } k = n, \\ 0 & \text{otherwise.} \end{cases}$$

Such a space is called an *Eilenberg-MacLane space of type* (A, n). Then the solution to the homotopy problem for maps into $Y = K(A, n)$ shows that there is a single obstruction $\theta \in H^n(X; \pi_n(K(A, n)))$ to homotoping a map $f : X \to K(A, n)$ to another map $g : X \to K(A, n)$. With a little more work one shows that this sets up an isomorphism

$$[X, K(A, n)] \xrightarrow{\cong} H^n(X; A).$$

This suggests one could *define* $H^n(X; A)$ to be $[X, K(A, n)]$. This observation forms the basic link between the homological algebra approach to cohomology and homotopy theory. From this perspective the Puppe sequence (Theorem 6.42) immediately gives the long exact sequence in cohomology, and the other Eilenberg–Steenrod axioms are trivial to verify. But more importantly, it suggests that one could find generalizations of cohomology by replacing the sequence of spaces $K(A, n)$ by some other sequence E_n and defining functors from spaces to sets (or groups, or rings, depending on how much structure one has on the sequence E_n) by

$$X \mapsto [X, E_n].$$

This indeed works and leads to the notion of a *spectrum* $\{E_n\}$ and its corresponding generalized homology theory, one of the subjects of Chapter 8.

7.2. The obstruction cocycle

Suppose that (X, A) is a relative CW-complex. We refer the reader to Definition 1.3 for the precise definition.

Notice that X/A is a CW-complex. The dimension of (X, A) is defined to be the highest dimension of the cells attached (we allow the dimension to be infinite).

Suppose that $g : A \to Y$ is a continuous map with Y path connected. We wish to study the question of whether g can be extended to map $X \to Y$.

Since Y is path connected, the map $g : A \to Y$ extends over the 1-skeleton X_1. Thus the zeroth and first step in extending $g : A \to Y$ to X is always possible (when Y is path connected).

Suppose that $g : A \to Y$ has been extended to $g : X_n \to Y$ for some $n \geq 1$.

We now make a simplifying assumption.

Assumption. Y is n-simple, so that $[S^n, Y] = \pi_n Y$.

(We will indicate later how to avoid this assumption by using local coefficients.)

Theorem 7.1 (main theorem of obstruction theory). *Let (X, A) be a relative CW-complex, $n \geq 1$, and Y a path–connected n-simple space. Let $g : X_n \to Y$ be a continuous map.*

1. *There is a cellular cocycle $\theta(g) \in C^{n+1}(X, A; \pi_n Y)$ which vanishes if and only if g extends to a map $X_{n+1} \to Y$.*
2. *The cohomology class $[\theta(g)] \in H^{n+1}(X, A; \pi_n Y)$ vanishes if and only if the restriction $g_{|X_{n-1}} : X_{n-1} \to Y$ extends to a map $X_{n+1} \to Y$.*

The proof of this theorem will occupy several sections.

7.3. Construction of the obstruction cocycle

Recall that if J_n indexes the n-cells of (X, A),

$$\begin{aligned} C^{n+1}(X, A; \pi_n Y) &= \mathrm{Hom}_{\mathbf{Z}}(C_{n+1}(X, A), \pi_n Y) \\ &= \mathrm{Funct}(\{e_i^{n+1} | i \in J_{n+1}\}, \pi_n Y). \end{aligned}$$

Each $(n+1)$-cell e_i^{n+1} admits a characteristic map

$$\phi_i : (D^{n+1}, S^n) \to (e_i^{n+1}, \partial e_i^{n+1}) \subset (X_{n+1}, X_n)$$

whose restriction to S^n we call the *attaching map*

$$f_i = \phi_{i|S^n} : S^n \to X_n.$$

Composing f_i with $g : X_n \to Y$ defines a map

$$S^n \xrightarrow{f_i} X_n \xrightarrow{g} Y.$$

This defines an element $[g \circ f_i] \in [S^n, Y]$, which equals $[S^n, Y]_0 = \pi_n Y$, since Y is assumed to be n-simple.

Definition 7.2. Define the *obstruction cochain* $\theta^{n+1}(g) \in C^{n+1}(X, A; \pi_n Y)$ on the basis of $(n+1)$-cells by the formula

$$\theta^{n+1}(g)(e_i^{n+1}) = [g \circ f_i]$$

and extend by linearity.

A map $h : S^n \to Y$ is homotopically trivial if and only if h extends to a map $D^{n+1} \to Y$. The following lemma follows from this fact and the definition of adjoining cells.

Lemma 7.3. $\theta(g) = 0$ *if and only if g extends to a map $X_{n+1} \to Y$.* □

We gave a geometric definition of the obstruction cochain and came to a geometric conclusion. Next we give an algebraic definition.

Recall from Section 1.1.3 that the cellular chain complex is defined by taking the chain groups to be $C_n(X, A) = H_n(X_n, X_{n-1})$. The differential is defined to be the composite

$$H_n(X_n, X_{n-1}) \xrightarrow{\partial} H_{n-1}(X_{n-1}) \xrightarrow{i} H_{n-1}(X_{n-1}, X_{n-2}).$$

By the cellular approximation theorem, $\pi_1 X_n \to \pi_1 X_{n+1}$ is onto if $n = 1$, and an isomorphism if $n > 1$, so $\pi_1(X_{n+1}, X_n) = 0$. Similarly, the cellular approximation theorem implies that $\pi_k(X_{n+1}, X_n) = 0$ for $k \leq n$; hence the (relative) Hurewicz theorem implies that $H_k(X_{n+1}, X_n) = 0$ for $k \leq n$ and the Hurewicz map

$$\rho : \pi_{n+1}(X_{n+1}, X_n) \to H_{n+1}(X_{n+1}, X_n)$$

is onto, with kernel the subgroup of $\pi_{n+1}(X_{n+1}, X_n)$ generated by the set

$$K = \{x(\alpha(x))^{-1} | x \in \pi_{n+1}(X_{n+1}, X_n),\ \alpha \in \pi_1 X_n\}.$$

Recall that $\pi^+_{n+1}(X_{n+1}, X_n)$ denotes $\pi_{n+1}(X_{n+1}, X_n)/K$.

Lemma 7.4. *There is a factorization:*

$$\begin{array}{ccccc} \pi_{n+1}(X_{n+1}, X_n) & \xrightarrow{\partial} & \pi_n(X_n) & \xrightarrow{g_*} & \pi_n(Y) \\ \downarrow & & & \nearrow_{\overline{g_* \circ \partial}} & \\ \pi^+_{n+1}(X_{n+1}, X_n) & & & & \end{array}$$

Proof. If $\alpha \in \pi_1 X_n$, then $\partial(\alpha \cdot x) = \alpha \cdot \partial x$. Moreover, $g_*(\alpha \cdot z) = g_*(\alpha) \cdot (g_*(z)) = g_*(z)$ for $z \in \pi_n X_n$ since Y is n-simple. $\square$

Using Lemma 7.4 one can form the composite

$$C_{n+1}(X, A) = H_{n+1}(X_{n+1}, X_n) \xrightarrow[\rho^{-1}]{\cong} \pi^+_{n+1}(X_{n+1}, X_n) \xrightarrow{\overline{g_* \circ \partial}} \pi_n Y \tag{7.1}$$

where $\rho : \pi^+_{n+1}(X_{n+1}, X_n) \to H_{n+1}(X_{n+1}, X_n)$ denotes the Hurewicz isomorphism. The composite map of Equation (7.1) defines another cochain in $\mathrm{Hom}_{\mathbf{Z}}(C_{n+1}(X, A)); \pi_n Y)$ which we again denote by $\theta^{n+1}(g)$.

Proposition 7.5. *The two definitions of $\theta^{n+1}(g)$ agree.*

Proof. We first work on the second definition. Given an $(n+1)$-cell e^{n+1}_i, let $\phi_i : (D^{n+1}, S^n) \to (X_{n+1}, X_n)$ be the characteristic map for the cell. We construct a map $(\phi_i \vee u) \circ q : (D^{n+1}, S^n, p) \to (X_{n+1}, X_n, x_0)$ as the composite of a map $q : (D^{n+1}, S^n, p) \to (D^{n+1} \vee I, S^n \vee I, p)$, illustrated in the next figure,

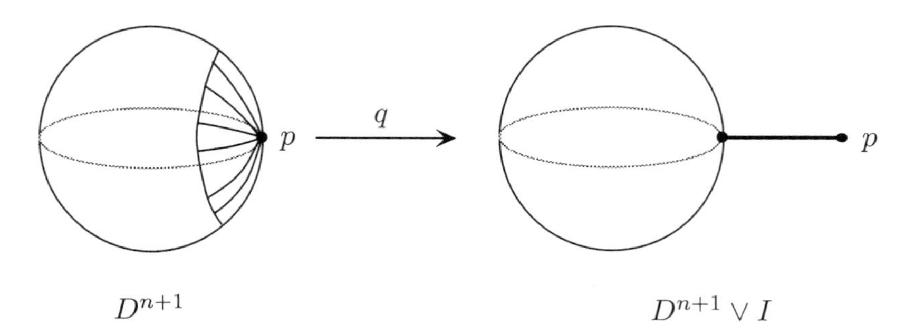

and a map $D^{n+1} \vee I \xrightarrow{\phi_i \vee u} X_{n+1}$ where u is a path in X_n to the base point x_0. Clearly $(\phi_i \vee u) \circ q$ is homotopic to the characteristic map ϕ_i. Thus $\rho((\phi_i \vee u) \circ q)$ is the generator of $H_{n+1}(X_{n+1}, X_n)$ represented by the cell e_i^{n+1}. Hence $(\phi_i \vee u) \circ q$ represents the element $\rho^{-1}(e_i^{n+1})$ in $\pi_{n+1}^+(X_{n+1}, X_n)$.

By definition, $\partial((\phi_i \vee u) \circ q) \in \pi_n X_n$ is represented by the composite of the map $\overline{q} : S^n \to S^n \vee I$ (obtained by restricting the map q of the previous figure to the boundary) and the attaching map $f_i = \phi_{i|S^n}$ for the cell e_i^{n+1} together with a path u to x_0:

$$\partial((\phi_i \vee u) \circ q) = (f_i \vee u) \circ \overline{q} : S^n \to X_n.$$

Hence by the second definition, $\theta(g)(e_i^{n+1}) = g \circ (f_i \vee u) \circ \overline{q}$. But this is $(g \circ f_i \vee g \circ u) \circ \overline{q}$ which equals $[f_i] \in [S^n, Y] = \pi_n Y$, which in turn is the first definition of $\theta(g)(e_i^{n+1})$. □

Theorem 7.6. *The obstruction cochain $\theta^{n+1}(g)$ is a cocycle.*

Proof. Consider the following commutative diagram:

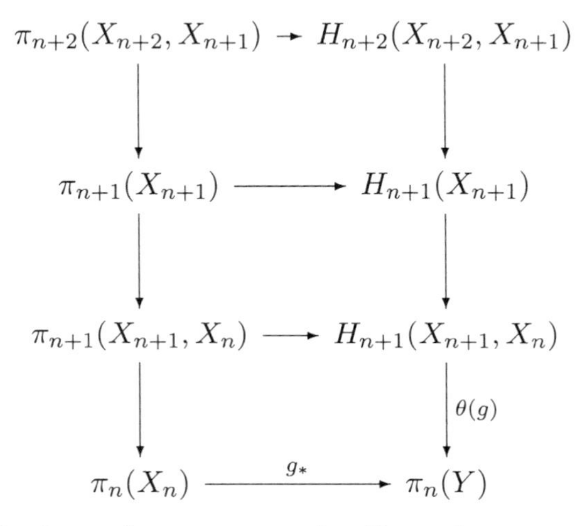

The unlabeled horizontal arrows are the Hurewicz maps. The unlabeled vertical arrows come from homotopy or homology exact sequences of the pairs (X_{n+2}, X_{n+1}) and (X_{n+1}, X_n).

The theorem follows by noting that the $\delta\theta(g)$ is the composite of all the right vertical maps, that the top horizontal arrow is onto by the Hurewicz Theorem, and that the composite of the bottom two vertical maps on the left are zero, because they occur in the homotopy exact sequence of the pair (X_{n+1}, X_n). □

7.4. Proof of the extension theorem

Lemma 7.3 says that $\theta^{n+1}(g)$ is the zero cocycle if and only if g extends over X_{n+1}. Theorem 7.1 says that if $\theta^{n+1}(g)$ is *cohomologous to* 0, then the restriction of g to the $(n-1)$-skeleton X_{n-1} extends over X_{n+1}. Thus to prove Theorem 7.1 it must be shown that if $\theta^{n+1}(g) = \delta d$ for some cochain $d \in C^n(X, A; \pi_n Y)$, then g can be redefined on the n-skeleton relative to the $(n-1)$-skeleton, then extended over the $(n+1)$-skeleton.

Theorem 7.7. *If $\theta^{n+1}(g)$ is cohomologous to 0, then the restriction of $g : X_n \to Y$ to the $(n-1)$-skeleton, $g_{|X_{n-1}} : X_{n-1} \to Y$ extends over the $(n+1)$-skeleton X_{n+1}.*

Before we prove this, we prove some preliminary lemmas which will be useful for the homotopy classification as well.

Lemma 7.8. *Let $f_0, f_1 : X_n \to Y$ be two maps so that $f_{0|X_{n-1}}$ is homotopic to $f_{1|X_{n-1}}$. Then a choice of homotopy defines a* difference *cochain $d \in C^n(X, A; \pi_n Y)$ with*

$$\delta d = \theta^{n+1}(f_0) - \theta^{n+1}(f_1).$$

Proof. Let $\hat{X} = X \times I$, $\hat{A} = A \times I$. Then $(\hat{X}, \hat{A})$ is a relative CW-complex, with $\hat{X}_k = X_k \times \partial I \cup X_{k-1} \times I$. Hence a map $\hat{X}_n \to Y$ is a pair of maps $f_0, f_1 : X_n \to Y$ and a homotopy $G : X_{n-1} \times I \to Y$ of the restrictions of f_0, f_1 to X_{n-1}.

Thus one obtains the cocycle

$$\theta(f_0, G, f_1) \in C^{n+1}(\hat{X}, \hat{A}; \pi_n Y)$$

which obstructs finding an extension of $f_0 \cup G \cup f_1$ to $\hat{X}_{n+1}$. From this cocycle one obtains the *difference cochain*

$$d(f_0, G, f_1) \in C^n(X, A; \pi_n Y)$$

by restricting to cells of the form $e^n \times I$; that is,

$$d(f_0, G, f_1)(e_i^n) = (-1)^{n+1}\theta(f_0, G, f_1)(e_i^n \times I) \tag{7.2}$$

for each e_i^n an n-cell of X. (The reason for the sign will be apparent shortly.)

Using the fact that $\theta(f_0, G, f_1)$ is a cocycle, for all $(n+1)$-cells e_i^{n+1},

$$\begin{aligned}
0 &= (\delta\theta(f_0, G, f_1))(e_i^{n+1} \times I) \\
&= \theta(f_0, G, f_1)(\partial(e_i^{n+1} \times I)) \\
&= \theta(f_0, G, f_1)(\partial(e_i^{n+1}) \times I) \\
&\quad +(-1)^{n+1}(\theta(f_0, G, f_1)(e_i^{n+1} \times \{1\}) - \theta(f_0, G, f_1)(e^{n+1} \times \{0\})) \\
&= (-1)^{n+1}(\delta(d(f_0, G, f_1))(e_i^{n+1}) + \theta(f_1)(e_i^{n+1}) - \theta(f_0)(e_i^{n+1})).
\end{aligned}$$

Therefore

$$\delta d(f_0, G, f_1) = \theta^{n+1}(f_0) - \theta^{n+1}(f_1).$$

□

There is a geometric interpretation for the difference cochain. Identify S^n with $\partial(D^n \times I)$. For an n-cell e_i^n, let $\varphi_i : (D^n, S^{n-1}) \to (X_n, X_{n-1})$ be its characteristic map. Then $\pm d(f_0, G, f_1)(e_i^n) \in \pi_n Y$ is $f_0 \cup G \cup f_1$ composed with the attaching map of $e_i^n \times I$ as indicated in the next figure.

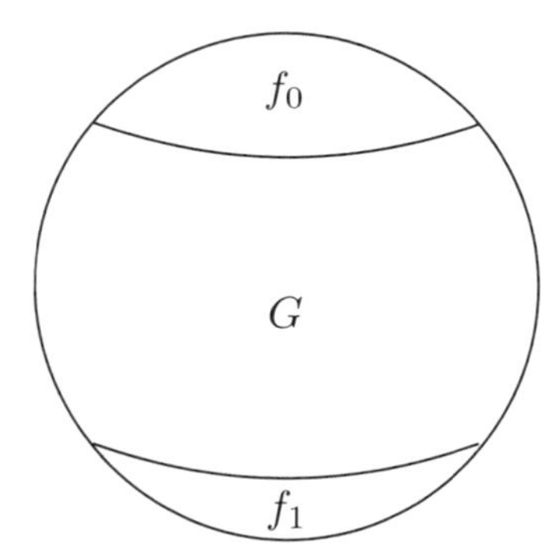

Lemma 7.8 immediately implies the following.

Corollary 7.9. *Given $g : X_n \to Y$, the obstruction cocycle $\theta^{n+1}(g)$ is null homologous if the restriction $g_{|X_{n-1}} : X_{n-1} \to Y$ is homotopic to a map which extends over X_{n+1}.* □

What we want is a converse. What we need is a *realization theorem*, i.e. a proof that for any coboundary $\delta d \in C^{n+1}(X, A; \pi_n Y)$, any map g, and any homotopy G, there is a map f_1 so that the difference cochain $d(g, G, f_1) = d$. In particular, G could be a homotopy constant in time in which case if $\theta^{n+1}(g) = \delta d = \delta d(g, g_{|X_{n-1}} \times \mathrm{Id}_I, f_1)$, the previous lemma shows that $\theta^{n+1}(f_1) = 0$, so that f_1 extends over X_{n+1}. Since $g'_{|X_{n-1}} = g_{|X_{n-1}}$ this would finish this step in the extension process. The realization result is the following.

Proposition 7.10. *Given a map $f_0 : X_n \to Y$ and a homotopy $G : X_{n-1} \times I \to Y$, so that $G_0 = f_{0|X_{n-1}}$ and an element $d \in C^n(X, A; \pi_n Y)$, there is a map $f_1 : X_n \to Y$ so that $G(-,1) = f_{1|X_{n-1}}$ and $d = d(f_0, G, f_1)$.*

Given the previous geometric description of the difference cochain, all we really need to prove is:

Lemma 7.11. *For any map $f : D^n \times \{0\} \cup S^{n-1} \times I \to Y$ and for any element $\alpha \in [\partial(D^n \times I), Y]$, there is a map $F : \partial(D^n \times I) \to Y$ so that F represents α and restricts to f.*

Proof. The proof is easy. Let $K : \partial(D^n \times I) \to Y$ be any map representing α and let $D = D^n \times \{0\} \cup S^{n-1} \times I$. Since D is contractible, both $K_{|D}$ and f are null-homotopic, hence homotopic to each other; the map K gives an extension of one end of this homotopy. The homotopy extension property of the pair $(\partial(D^n \times I), D)$ gives a homotopy $H : \partial(S^{n-1} \times I) \times I \to Y$ and, $F = H_1$ is the required map. □

Proof of Proposition 7.10. Given an n-cell e_i^n of X_n, let

$$\varphi_i : (D^n, S^{n-1}) \to (X_n, X_{n-1})$$

be the characteristic map. Apply Lemma 7.11 with $f = f_0 \circ \varphi_i \cup G \circ (\varphi_{i|S^{n-1}} \times \mathrm{Id}_I)$ and $\alpha = d(e_i^n)$ and let F_i be the map provided by the conclusion of Lemma 7.11. Define $f_1 : X_n \to Y$ on the n-cells by $f_1(\varphi_i(x)) = F_i(x, 1)$. The geometric interpretation of the difference cochain shows $d(f_0, G, f_1)(e_i^n) = d(e_i^n)$ as desired. □

Proof of Theorem 7.1. The only thing left to show is that if $g : X_n \to Y$ and if $\theta(g)$ is a coboundary δd, then $g_{|X_{n-1}}$ extends to X_{n+1}. Apply the realization proposition to g, d, and the stationary homotopy $((x,t) \mapsto g(x))$ from $g_{|X_{n-1}}$ to itself . One obtains a map $g' : X_n \to Y$ which agrees with g upon restriction to X_{n-1} and satisfies

$$\delta d = \theta(g) - \theta(g').$$

Then $\theta(g') = 0$ so g' extends to X_{n+1}. □

Exercise 119. Find examples of (X, A), Y, and g where:

1. $\theta^{n+1}(g) = 0$.
2. $\theta^{n+1}(g) \neq 0$, but $[\theta^{n+1}(g)] = 0$.
3. $[\theta^{n+1}(g)] \neq 0$.

It is conceivable (and happens frequently) that finding an extension of $g : X_n \to Y$ to X_{n+1} may require g to be redefined not just on the n-cells, but maybe even the $(n-1)$-cells, or perhaps even on the $(n-k)$-cells for $k = 1, \cdots, r$ for some r.

This suggests that there may be theorems which state "given g, the restriction $g_{|X_{n-k}}$ extends to X_{n+1} if and only if some obstruction vanishes." Such theorems exist, and working them out leads to the definition of secondary and higher obstructions.

To get a feel for where such obstructions may lie, notice that the obstruction cochain θ is the obstruction to extending $g : X_n \to Y$ to X_{n+1}, and that its cohomology class $[\theta]$ is the obstruction to extending $g_{|X_{n-1}}$ to X_{n+1}. The cohomology group is a subquotient of the cochain group $C^n(X, A; \pi_n Y)$. It turns out that the obstructions live in further subquotients, that is, in subquotients of cohomology.

7.5. Obstructions to finding a homotopy

We now turn to the construction of obstructions to finding a homotopy between $f_0 : X \to Y$ and $f_1 : X \to Y$ extending a fixed homotopy on A.

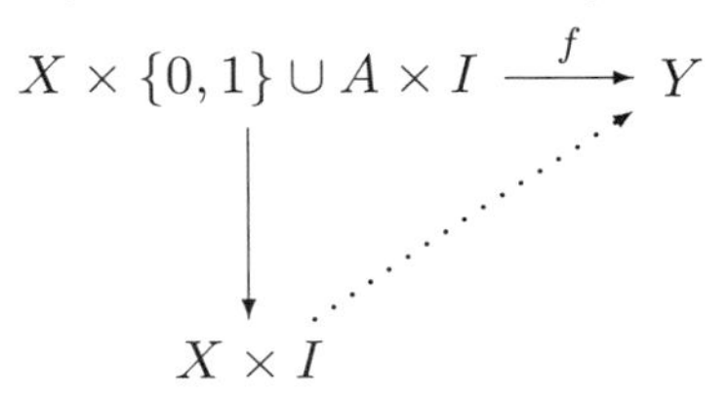

This is accomplished by viewing the homotopy problem as an extension problem and then applying the Künneth theorem.

Consider the product relative CW-complex:

$$(X^*, A^*) = (X, A) \times (I, \partial I) = (X \times I, X \times \partial I \cup A \times I).$$

Then a map $F : X_n^* \to Y$ is a pair of maps $f_0, f_1 : X \to Y$ and a homotopy of $f_{0|X_{n-1}}$ to $f_{1|X_{n-1}}$. Therefore the obstruction class $[\theta^{n+1}(F)] \in H^{n+1}(X^*, A^*; \pi_n Y)$ is defined. This group is isomorphic to $H^n(X, A; \pi_n Y)$ by the Künneth theorem (this is really the suspension isomorphism). Call the corresponding element $d^n(f_0, f_1) \in H^n(X, A; \pi_n Y)$. Then one gets the following theorem.

Theorem 7.12. *Let (X, A) be a relative CW-complex, Y an n-simple space, $f_0, f_1 : X \to Y$ two maps which agree on A, and $F : X_{n-1} \times I \to Y$ a homotopy from $f_{0|X_{n-1}}$ to $f_{1|X_{n-1}}$ (rel A). Then the cohomology class of $d^n(f_0, f_1)$ equals 0 if and only if the restriction of F to $X_{n-2} \times I$ extends to a homotopy of $f_{0|X_n}$ to $f_{1|X_n}$.* □

An interesting special case occurs when f_1 is constant (see Exercise 118).

Corollary 7.13. *Any continuous map from an n-dimensional CW-complex to an n-connected space is null-homotopic.* □

7.6. Primary obstructions

A case where obstruction theory is easy to use occurs if $H^{n+1}(X, A; \pi_n Y) = 0$ for all n. This occurs quite frequently. For example, if (X, A) has dimension a and Y is $(a-1)$-connected, then any map from A to Y extends to X.

The next interesting case occurs when $H^{k+1}(X, A; \pi_k Y)$ is non-zero in only one dimension. Then there is a single obstruction to extending g, and this obstruction sets up a correspondence between extensions and the corresponding cohomology group. As a first step in understanding this correspondence, we have the following theorem.

Theorem 7.14. *Let (X, A) be a relative CW-complex, $n \geq 1$, and Y an $(n-1)$-connected space (if $n = 1$, assume $\pi_1 Y$ is abelian). Let $f : A \to Y$ be a map. Then f extends to a map $g : X_n \to Y$. If g_0, g_1 are extensions of f, then $g_{0|X_{n-1}} \simeq g_{1|X_{n-1}}$ (rel A) and the obstructions $\theta^{n+1}(g_0)$ and $\theta^{n+1}(g_1)$ are cohomologous.*

Proof. Since Y is path connected, f can be extended over X_1. Since the obstructions to extending f lie in $H^{r+1}(X, A; \pi_r Y)$, f can be extended to X_{n-1}. Since the obstructions to finding a homotopy between maps lie in $H^r(X, A; \pi_r Y)$, any two extensions of f are homotopic over X_{n-1}, and as we saw, the difference cochain has coboundary equal to the difference of the obstruction cocycles. □

Definition 7.15. Let (X, A) be a relative CW-complex, $n \geq 1$, and Y an $(n-1)$-connected space (if $n = 1$, assume $\pi_1 Y$ is abelian). Let $f : A \to Y$ be a map. The obstruction to extending f to X_{n+1} is denoted by

$$\gamma^{n+1}(f) \in H^{n+1}(X, A; \pi_n Y).$$

It is called the *primary obstruction to extending f*.

Theorem 7.14 says that the primary obstruction is well-defined and vanishes if and only if f extends over X_{n+1}. We next show that it is homotopy invariant.

Theorem 7.16. *Let (X, A) be a relative CW-complex, $n \geq 1$, and Y an $(n-1)$-connected space (if $n = 1$, assume $\pi_1 Y$ is abelian). Let $f : A \to Y$ be a map. Suppose f' is homotopic to f. Then $\gamma^{n+1}(f') = \gamma^{n+1}(f)$.*

Proof. By Theorem 7.14, f extends to a map $g : X_n \to Y$ and f' extends to a map $g' : X_n \to Y$. Likewise $g_{|X_{n-1}} \simeq g'_{|X_{n-1}}$, since Y is highly connected. Call the homotopy F. Then the difference cochain satisfies

$$\delta d(g, F, g') = \theta(g) - \theta(g')$$

by Lemma 7.8. This shows $\gamma^{n+1}(f)$ and $\gamma^{n+1}(f')$ are cohomologous. □

In the situation of the above theorems, if the primary obstruction vanishes, then the map f extends to $g : X_{n+1} \to Y$. However the next obstruction class $[\theta^{n+2}(g)]$ may depend on the choice of g. So it is usually only the primary obstruction which is computable. Obstruction theory ain't all it's cracked up to be.

To define the primary obstruction for two maps to be homotopic, we apply the above theorems to $(X \times I, X \times \partial I \cup A \times I)$ and obtain the following theorem.

Theorem 7.17. *Let (X, A) be a relative CW-complex, $n \geq 1$, and Y an $(n-1)$-connected space (if $n = 1$ assume $\pi_1 Y$ is abelian). Let $f_0, f_1 : X \to Y$ be two functions which agree on A. Then $f_{0|X_{n-1}} \simeq f_{1|X_{n-1}}$ rel A, and the cohomology class in $H^n(X, A; \pi_n Y)$ of the obstruction to extending this homotopy to X_n is independent of the choice of homotopy on X_{n-1} and depends only on the homotopy classes of f_0 and f_1 relative to A.* □

In light of this theorem, one can make the following definition.

Definition 7.18. Let (X, A) be a relative CW-complex, $n \geq 1$, and Y an $(n-1)$-connected space (if $n = 1$, assume $\pi_1 Y$ is abelian). Let $f_0, f_1 : X \to Y$ two functions which agree on A. The obstruction to constructing a homotopy $f_{0|X_n} \simeq f_{1|X_n}$ rel A is denoted

$$d^n(f_0, f_1) \in H^n(X, A; \pi_n Y)$$

and is called the *primary obstruction to homotoping f_0 to f_1*. It depends only on the homotopy classes of f_0 and f_1 relative to A.

7.7. Eilenberg-MacLane spaces

An important class of spaces is the class of those spaces Y satisfying $\pi_k Y = 0$ for all $k \neq n$.

Definition 7.19. Let n be a positive integer and let π be a group, with π abelian if $n > 1$. A CW-complex Y is called a *$K(\pi, n)$-space* if

$$\pi_k Y = \begin{cases} 0 & \text{if } k \neq n \\ \pi & \text{if } k = n. \end{cases}$$

We will see later that (π, n) determines the homotopy type of Y; that is, for a fixed pair (π, n), any two $K(\pi, n)$ spaces are homotopy equivalent. A $K(\pi, n)$-space is called an *Eilenberg–MacLane space*.

Theorem 7.20. *Given any $n > 0$ and any group π with π abelian if $n > 1$, there exists a $K(\pi, n)$-CW-complex.*

Sketch of proof. Let $\langle x_i, i \in I | r_j, \ j \in J \rangle$ be a presentation (abelian if $n > 1$) of π. Let K_n be the wedge $\vee_{i \in I} S^n$ of n-spheres, one for each generator of π. Then the Van Kampen and Hurewicz theorems imply that $\pi_k K_n = 0$ for $k < n$, $\pi_1 K_1$ is the free group on the generators of π when $n = 1$, and $\pi_n K_n$ is the free abelian group on the generators of π when $n > 1$.

For each relation, attach an $(n + 1)$-cell using the relation to define the homotopy class of the attaching map. This defines a complex K_{n+1} with

$$\pi_k(K_{n+1}) = \begin{cases} 0 & \text{if } k < n \\ \pi & \text{if } k = n. \end{cases}$$

For $n = 1$, this follows from the Van Kampen theorem. For $n > 1$ and $k < n$, this follows from the cellular approximation theorem, and for $k = n$ from the Hurewicz theorem.

Attach $(n + 2)$-cells to kill $\pi_{n+1}(K_{n+1})$. More precisely, choose a set of generators for $\pi_{n+1}(K_{n+1})$ and attach one $(n + 2)$-cell for each generator, using the generator as the homotopy class of the attaching map. This gives a $(n+2)$-dimensional complex K_{n+2}. By the cellular approximation theorem, the homotopy groups in dimensions less than $n+1$ are unaffected, and there is a surjection $\pi_{n+1}(K_{n+1}) \to \pi_{n+1}(K_{n+2})$. Thus

$$\pi_k(K_{n+2}) = \begin{cases} 0 & \text{if } k < n \text{ or } k = n+1, \\ \pi & \text{if } k = n. \end{cases}$$

Now attach $(n + 3)$-cells to kill π_{n+2}, etc. The union of the K_r is a CW-complex and a $K(\pi, n)$ space. □

An important property of Eilenberg-MacLane spaces is that they possess fundamental cohomology classes. These classes are extremely useful. They allow us to set up a functorial correspondence between $H^n(X, \pi)$ and $[X, K(\pi, n)]$. They are used to define cohomology operations. They can be used to give the "fibering data" needed to decompose an arbitrary space into Eilenberg-MacLane spaces (Postnikov towers) and also to construct characteristic classes for fiber bundles.

Assume π is abelian, so that $K(\pi, n)$ is simple. Then

$$H^n(K(\pi, n); \pi) \cong \mathrm{Hom}(H_n(K(\pi, n); \mathbf{Z}), \pi) \cong \mathrm{Hom}(\pi, \pi), \tag{7.3}$$

where the first isomorphism is the adjoint of the Kronecker pairing (Exercise 11) and is an isomorphism by the universal coefficient theorem (Theorem 2.29). The second map is the Hurewicz isomorphism (Theorem 6.66).

Definition 7.21. The *fundamental class of the* $K(\pi, n)$,

$$\iota \in H^n(K(\pi, n); \pi),$$

is the class corresponding to the identity map $\mathrm{Id} : \pi \to \pi$ under the isomorphisms of Equation (7.3).

The fundamental class can be used to define a function

$$\Phi : [X, K(\pi, n)] \to H^n(X; \pi) \tag{7.4}$$

by the formula

$$\Phi([f]) = f^*[\iota]$$

for $f : X \to K(\pi, n)$.

The primary obstruction class can be used to define another function

$$\Psi : [X, K(\pi, n)] \to H^n(X; \pi) \tag{7.5}$$

by setting $\Psi[f]$ to be the primary obstruction to homotoping f to the constant map,

$$\Psi[f] = d^n(f, \mathrm{const}).$$

Theorem 7.14 shows that $\Psi[f]$ depends only on the homotopy class of f, and hence is well-defined.

Theorem 7.22. *The functions* Φ *and* Ψ

$$[X, K(\pi, n)] \to H^n(X; \pi)$$

of Equations (7.4) and (7.5) coincide, are bijections, and are natural with respect to maps $X \to X'$.

Proof. *Step 1.* Ψ *is injective.* Let $f : X \to K(\pi, n)$ be a continuous map. Obstruction theory says that if $d^n(f, \mathrm{const}) = 0$, then f and the constant map are homotopic over the n-skeleton. But all higher obstructions vanish since they live in zero groups. Hence if $\Psi[f] = 0$, f is nullhomotopic. In other words $\Psi^{-1}[0] = [\mathrm{const}]$.

If we knew that $[X, K(\pi, n)]$ were a group and Ψ a homomorphism, then we could conclude that Ψ is injective. (Of course, this follows from the present theorem.)

Instead, we will outline the argument proving the "addition formula"

$$d^n(f, g) = d^n(f, \mathrm{const}) - d^n(g, \mathrm{const});$$

i.e. $d^n(f, g) = \Psi[f] - \Psi[g]$ for any two functions $f, g : X \to K(\pi, n)$. This implies that Ψ is injective.

To see this, if F is a homotopy from $f_{|X_{n-1}}$ to the constant map and G is a homotopy from $g_{|X_{n-1}}$ to the constant map, then compose F and $\overline{G}$ to get a homotopy from $f_{|X_{n-1}}$ to $g_{|X_{n-1}}$. (Here $\overline{G}$ means the reverse homotopy, i.e. $\overline{G} = G \circ r$ where $r(x,t) = (x, 1-t)$.)

Write $F * \bar{G}$ for this homotopy from $f_{|X_{n-1}}$ to $g_{|X_{n-1}}$. Then on an n-cell $e \subset X$ the obstruction $d^n(f,g)(e) \in \pi_n(K(\pi,n))$ is defined to be the homotopy class of the map $S^n \to K(\pi,n)$ defined as follows. Decompose S^n as a neighborhood of the poles together with a neighborhood of the equator: $S^n = D_0^n \cup (S^{n-1} \times I) \cup D_1^n$. Then define $d = d^n(f,g)(e) : S^n \to K(\pi,n)$ to be the homotopy class of the map which equals f on D_0^n, $F * \bar{G}$ on $S^{n-1} \times I$, and g on D_1^n. Since this map is constant on the equator $S^{n-1} \times \frac{1}{2}$, the homotopy class of d is clearly the sum of two classes, the first representing $d^n(f, \text{const})(e)$ and the second representing $d^n(\text{const}, g)(e)$. Therefore $d^n(f,g) = d^n(f,\text{const}) - d^n(g,\text{const})$ and so $\Psi[f] = \Psi[g]$ if and only if f is homotopic to g.

Step 2. Ψ is surjective. We do this by proving a variant of the realization proposition (Proposition 7.10) for the difference cochain.

Given $[\alpha] \in H^n(X;\pi)$, choose a cocycle α representing $[\alpha]$. Since the quotient X_n/X_{n-1} is the wedge of n-spheres, one for each n-cell of X, α defines a function (up to homotopy)

$$g : X_n/X_{n-1} \to K(\pi,n)$$

by having the restriction of g to the i-th n-sphere be a function representing $\alpha(e_i^n) \in \pi = \pi_n(K(\pi,n))$.

The function g extends to $X_{n+1}/X_{n-1} \to K(\pi,n)$ because α is a cocycle. In fact, for each $(n+1)$-cell e_i^{n+1}

$$0 = (\delta\alpha)(e_i^{n+1}) = \alpha(\partial e_i^{n+1}),$$

which implies that the composite

$$S^n \to \partial e_i^{n+1} \to X_n \to X_n/X_{n-1} \xrightarrow{g} K(\pi,n)$$

of the attaching map of e_i^{n+1} and g is nullhomotopic. Thus g extends over the $(n+1)$-skeleton.

Since $H^{n+i+1}(X;\pi_{n+i}(K(\pi,n))) = 0$ for $i \geq 1$, obstruction theory and induction show that there exists an extension of $g : X_n/X_{n-1} \to K(\pi,n)$ to $\tilde{g} : X/X_{n-1} \to K(\pi,n)$. Composing with the quotient map, one obtains a map

$$f : X \to K(\pi,n),$$

constant on the $(n-1)$-skeleton, so that the characteristic map for an n-cell e_i^n, induces

$$\alpha(e^n) : D^n/\partial D^n \to X/X_{n-1} \to K(\pi, n).$$

But $d^n(f, \text{const})[e^n]$ was defined to be the map on S^n which equals f on the upper hemisphere and the constant map on the lower hemisphere. Therefore $d^n(f, \text{const})$ equals $\alpha(e^n)$; hence $\Psi[f] = [\alpha]$ and so Ψ is onto.

Step 3. Ψ is natural. This follows from the algebraic definition of the obstruction cocycle, but we leave the details as an exercise.

Exercise 120. Prove that Ψ is natural.

Step 4. $\Psi = \Phi$. We first prove this for the identity map $\text{Id} : K(\pi, n) \to K(\pi, n)$, and then use naturality. In other words, we need to show the primary obstruction to finding a null homotopy of Id is the fundamental class $\iota \in H^n(K(\pi, n); \pi))$. Since $K(\pi, n)$ is $(n-1)$-connected, Id is homotopic to a map, say Id', which is constant on the $(n-1)$-skeleton $K(\pi, n)_{n-1}$. By the universal coefficient and Hurewicz theorems,

$$H^n(K(\pi, n); \pi) \cong \text{Hom}(\rho(\pi_n(K(\pi, n))), \pi).$$

The definition of the fundamental class is equivalent to the formula

$$\langle \iota, \rho[g] \rangle = [g] \in \pi_n(K(\pi, n))$$

where $\langle\ ,\ \rangle$ denotes the Kronecker pairing. Thus what we need to show is:

$$\langle d^n(\text{Id}', \text{const}), \rho[g] \rangle = [g].$$

This is an equation which can be lifted to the cochain level; i.e. we need to show that if $[g] \in \pi_n(K(\pi, n)_n, K(\pi, n)_{n-1})$, then

$$d^n(\text{Id}', \text{const})(\rho[g]) = \text{Id}'_*[g].$$

In particular, we only need to verify this equation for the characteristic maps $\varphi_i : (D^n, S^{n-1}) \to (K(\pi, n)_n, K(\pi, n)_{n-1})$ of the n-cells. But the element $d^n(\text{Id}', \text{const})[\rho(\varphi_i)] \in \pi_n(K(\pi, n))$ is represented by the map $S^n \to K(\pi, n)$ given by the characteristic map composed with Id' on the upper hemisphere and the constant map on the lower hemisphere, and this map is homotopic to the characteristic map composed with the identity. Thus

$$d^n(\text{Id}', \text{const})[\rho(\varphi_i)] = \text{Id}'_*[\varphi_i]$$

as desired. Hence $\Phi[\text{Id}] = \Psi[\text{Id}]$.

Now suppose $[f] \in [X, K(\pi, n)]$. Then naturality of Ψ and Φ means that the diagram

$$\begin{array}{ccc} [X, K(\pi,n)] & \xleftarrow{f^*} & [K(\pi,n), K(\pi,n)] \\ \downarrow{\scriptstyle \Psi,\Phi} & & \downarrow{\scriptstyle \Psi,\Phi} \\ H^n(X;\pi) & \xleftarrow[f^*]{} & H^n(K(\pi,n);\pi) \end{array}$$

commutes when either both vertical arrows are labeled by Ψ or when both are labeled by Φ. Then we have

$$\begin{aligned} \Psi[f] &= \Psi f^*[\mathrm{Id}] \\ &= f^*\Psi[\mathrm{Id}] \\ &= f^*\Phi[\mathrm{Id}] \\ &= \Phi f^*[\mathrm{Id}] \\ &= \Phi[f]. \end{aligned}$$

□

Corollary 7.23. *For abelian groups π and π' and $n > 1$, there is a 1-1 correspondence*

$$[K(\pi,n), K(\pi',n)]_0 = [K(\pi,n), K(\pi',n)] \longleftrightarrow \mathrm{Hom}(\pi,\pi')$$

taking a map $K(\pi,n) \to K(\pi',n)$ to the induced map on homotopy groups.

Proof.

$$\begin{aligned} [K(\pi,n), K(\pi',n)] &\cong H^n(K(\pi,n);\pi') \\ &\cong \mathrm{Hom}(H_n(K(\pi,n)),\pi') \\ &\cong \mathrm{Hom}(\pi,\pi'). \end{aligned}$$

This follows from Theorem 7.22, the universal coefficient theorem, and the Hurewicz theorem. The composite is the map induced on the n-homotopy group. That the based and unbased homotopy sets are the same follows from Corollary 6.59 since $K(\pi,n)$ is simply connected for $n > 1$. □

Corollary 7.24. *Let $K(\pi,n)$ and $K'(\pi,n)$ be two Eilenberg-MacLane spaces of type (π,n) for $n > 1$. The identity map determines a canonical homotopy equivalence between them.*

Proof. The homotopy equivalence is simply the map $K(\pi,n) \to K'(\pi,n)$ corresponding to the identity $\pi \to \pi$. □

We shall see in the next section that Corollaries 7.23 and 7.24 continue to hold when $n = 1$ and π is non-abelian provided one uses based homotopy classes.

Computing the cohomology of Eilenberg-MacLane spaces is very important, because of connections to cohomology operations.

Definition 7.25. For positive integers n and m and abelian groups π and π', a *cohomology operation of type* (n, π, m, π') is a natural transformation of functors $\theta : H^n(-;\pi) \to H^m(-;\pi')$.

For example $u \mapsto u \cup u$ gives a cohomology operation of type $(n, \mathbf{Z}, 2n, \mathbf{Z})$.

Exercise 121. (Serre) Let $O(n, \pi, m, \pi')$ be the set of all cohomology operations of type (n, π, m, π'). Show that $\theta \leftrightarrow \theta(\iota)$ gives a 1-1 correspondence

$$O(n, \pi, m, \pi') \longleftrightarrow H^m(K(\pi, n); \pi') = [K(\pi, n), K(\pi', m)].$$

We will return to this subject in Section 10.4.

7.8. Aspherical spaces

It follows from our work above that for π abelian, $[X, K(\pi, 1)] = H^1(X; \pi) = \mathrm{Hom}(H_1 X, \pi) = \mathrm{Hom}(\pi_1 X, \pi)$.

For π non-abelian we have the following theorem.

Theorem 7.26. *For a based CW-complex X, taking induced maps on fundamental groups gives a bijection*

$$[X, K(\pi, 1)]_0 \to \mathrm{Hom}(\pi_1 X, \pi).$$

Sketch of proof. We will assume that the zero-skeleton of X is a single point. Then by the Van Kampen theorem, $\pi_1 X$ is presented with generators given by the characteristic maps of the 1-cells

$$\overline{\varphi_i^1} : D^1/S^0 \to X,$$

and relations given by the attaching maps of the 2-cells

$$\psi_j^2 : S^1 \to X.$$

We will discuss why the above correspondence is onto. Let $\gamma : \pi_1 X \to \pi$ be a group homomorphism. Construct a map $g : X_1 \to K(\pi, 1)$ by defining g on a 1-cell (a circle) e_i^1 to be a representative of $\gamma[\overline{\varphi_i^1}]$. The attaching maps ψ_j^2 are trivial in $\pi_1 X$, and hence $g_*[\psi_j^2] = \gamma[\psi_j^2] = \gamma(e) = e$. Thus g extends over the 2-skeleton. The attaching maps of the 3-cells of X are null-homotopic in $K(\pi, 1)$, so the map extends over the 3-cells. Continuing inductively, one obtains a map $X \to K(\pi, 1)$ realizing γ on the fundamental group.

The proof that if $g, h : X \to K(\pi, 1)$ are two maps inducing the same homomorphism (i.e. $g_* = h_* : \pi_1 X \to \pi$) then g is based point preserving homotopic to h (rel x_0) is similar in nature and will be omitted. □

Corollary 7.27. *Let $K(\pi, 1)$ and $K'(\pi, 1)$ be two Eilenberg-MacLane spaces of type $(\pi, 1)$. The identity map determines a canonical based homotopy equivalence between them.* □

Proposition 7.28. *Suppose that*

$$1 \to L \xrightarrow{\phi} \pi \xrightarrow{\gamma} H \to 1$$

is an exact sequence of (not necessarily abelian) groups. Then the homotopy fiber of the map $g : K(\pi, 1) \to K(H, 1)$ inducing γ as in Theorem 7.26 is $K(L, 1)$ and the inclusion of the fiber $K(L, 1) \to K(\pi, 1)$ induces the homomorphism ϕ.

If L, π, and H are abelian, the same assertions hold with $K(\pi, 1)$ replaced by $K(\pi, n)$ for any positive integer n. □

Thus short exact sequences of groups correspond exactly to fibrations of Eilenberg–MacLane spaces; the sequence of groups

$$1 \to L \to \pi \to H \to 1$$

is a short exact sequence of groups if and only if the corresponding sequence of spaces and maps

$$K(L, 1) \hookrightarrow K(\pi, 1) \to K(H, 1)$$

is a fibration sequence up to homotopy.

Similarly the sequence of abelian groups

$$0 \to L \to \pi \to H \to 0$$

is exact if and only if for any n the corresponding sequence of spaces and maps

$$K(L, n) \hookrightarrow K(\pi, n) \to K(H, n)$$

is a fibration sequence up to homotopy.

Exercise 122. Prove Proposition 7.28.

Definition 7.29. A space is *aspherical* if its universal cover is contractible.

Corollary 7.34 below implies that a CW-complex is aspherical if and only if it is a $K(\pi, 1)$.

Using $K(\pi, 1)$ spaces, one can define functors from groups to abelian groups by taking homology and cohomology. The group $H_n(K(\pi, 1))$ is called the *nth homology of the group* π and is often denoted by $H_n(\pi)$. Similarly the *nth cohomology of the group* π is defined by $H^n(\pi) = H^n(K(\pi, 1))$. We will study these functors in greater detail in Chapter 9. Other purely algebraic definitions of the (co)homology of groups can also be given.

Aspherical spaces are ubiquitous. Compact 2-manifolds other than the sphere and projective space are $K(\pi, 1)$'s. Also, $K(\mathbf{Z}/2, 1) = \mathbf{R}P^\infty$. More generally $K(\mathbf{Z}/n, 1) = L_n^\infty$, where L_n^∞ is the infinite lens space given as $S^\infty/(\mathbf{Z}/n)$ where $S^\infty \subset \mathbf{C}^\infty$ is the infinite dimensional sphere and the action is given by multiplication by a primitive n-th root of unity in every coordinate. Since $\pi_n(X \times Y) = \pi_n(X) \oplus \pi_n(Y)$, $K(\mathbf{Z}^n, 1) = (S^1)^n$, the *n-torus*. The Cartan–Hadamard Theorem states that if M is a complete Riemannian manifold with sectional curvature everywhere ≤ 0, then for every point $p \in M$, the exponential map

$$\exp : T_pM \to M$$

is a covering map. In particular M is aspherical. Here is an application.

Exercise 123. If M is a complete Riemannian manifold with sectional curvature everywhere ≤ 0, then $\pi_1 M$ is torsion-free.

We also mention the still open

Borel conjecture. *Compact aspherical manifolds with isomorphic fundamental groups are homeomorphic.*

The $K(\pi, 1)$-spaces are important for at least three reasons.

1. If M is a $\mathbf{Z}\pi$-module, then $H_*(K(\pi, 1); M)$ is an important algebraic invariant of the group and the module.
2. $K(\pi, 1) = B\pi$, and hence $[X, B\pi] = \mathrm{Hom}(\pi_1 X, \pi)/(\phi \sim g\phi g^{-1})$ classifies regular covers with deck transformations π.
3. In the study of flat bundles, that is, bundles whose structure group G reduces to a discrete group π, the classifying map $X \to BG$ factors through some $K(\pi, 1)$.

7.9. CW-approximations and Whitehead's theorem

Definition 7.30.

1. A *weak homotopy equivalence* is a map $f : X \to Y$ which induces isomorphisms $\pi_i(X, x) \to \pi_i(Y, f(x))$ for all i and for all base points x in X.

2. A *CW-approximation of a topological space* Y is a weak homotopy equivalence
$$X \to Y$$
where X is a CW-complex.

Theorem 7.31. *Any space* Y *has a CW-approximation.*

Proof. We may reduce to the case where Y is path-connected by approximating each path-component separately. We will inductively construct maps
$$g_n : X_n \to Y$$
which are n-connected, that is, give a surjection on π_n and a bijection on π_i for $i < n$. Also the restriction of g_n to the $(n-1)$-skeleton will be g_{n-1}.

Take X_0 to be a point. Assume inductively the existence of an n-connected map $g_n : X_n \to Y$, where X_n is an n-dimensional CW-complex. Attach an $(n+1)$-cell to X_n for every generator ker $g_{n*} : \pi_n X_n \to \pi_n Y$ to obtain a complex X'_{n+1}. Since the attaching maps are in the kernel, g_n extends to a map $g'_{n+1} : X'_{n+1} \to Y$. By cellular approximation and by construction $g'_{n+1*} : \pi_i X'_{n+1} \to \pi_i Y$ is an isomorphism for $i < n+1$. (One could alternatively use the relative Hurewicz theorem.) Finally, define $X_{n+1} = X'_{n+1} \vee (\bigvee S_i^{n+1})$ with an $(n+1)$-sphere for each generator of the cokernel of $g'_{n+1*} : \pi_{n+1} X'_{n+1} \to \pi_{n+1} Y$. Define the map $g_{n+1} : X_{n+1} \to Y$ by defining the map on S_i^{n+1} to be a representative of the corresponding element of the cokernel.

This shows how to construct the skeleta of $X = \cup X_n$. Topologize X as a CW-complex; see Definition 1.3. □

By the relative Hurewicz theorem, a CW-approximation induces an isomorphism on homology.

Milnor defined a *functorial* CW-approximation using simplicial methods [**26**]. This is done by defining a CW-complex X with an n-cell for each non-degenerate singular n-simplex in Y, where a non-degenerate simplex means that it does not factor through a face map. Milnor's construction gives a functor from topological spaces to CW-complexes; this functor takes a CW-complex to another complex of the same homotopy type.

A very useful theorem is given by the following.

Theorem 7.32 (cofibrant theorem). *A map* $f : Y \to Z$ *is a weak homotopy equivalence if and only if for all CW-complexes* X,
$$f_* : [X, Y] \to [X, Z] \qquad [g] \mapsto [f \circ g]$$
is a bijection.

Proof. ($\Longleftarrow$) When Y and Z are simple (e.g. simply-connected), then by choosing X to be the n-sphere $n = 0, 1, 2, \ldots$, one sees that f is a weak homotopy equivalence. We omit the proof without the simplicity hypothesis and refer the reader to Whitehead's book [**43**].

($\Longrightarrow$) Philosophically, f_* is a bijection, since it is for spheres and disks, and CW-complexes are built from spheres and disks. An easy proof along these lines can be given using the Puppe sequence for finite–dimensional CW-complexes, but for the general case we need a lemma similar to the motivating exercise (Exercise 118).

Lemma 7.33. *Let $g : (X, A) \to (Z, Y)$ be a map of pairs where (X, A) is a relative CW-complex and $Y \hookrightarrow Z$ is a weak homotopy equivalence. Then $g \simeq h$ (rel A) where $h(X) \subset Y$.*

Proof. We will construct a sequence of maps $h_n : X \to Z$ so that $h_n(X_n) \subset Y$ with $h_{-1} = g$ and $h_{n-1} \simeq h_n$ (rel X_{n-1}). (Slowly drag g into Y.) Then for a point x in an open n-cell, we define $h(x) = h_n(x)$, and the homotopy from g to h is defined by squeezing the homotopy $h_{n-1} \simeq h_n$ into the time interval $[1 - (1/2^n), 1 - (1/2^{n+1})] \subset [0, 1]$.

Assume inductively that $h_{n-1} : X \to Z$ has been constructed. Let

$$\varphi_i : (D^n, S^{n-1}) \to (X_n, X_{n-1})$$

be the characteristic map of an n-cell. Since $\pi_n(Z, Y) = 0$, the map

$$h_{n-1} \circ \varphi_i : (D^n, S^{n-1}) \to (Z, Y)$$

is homotopic rel S^{n-1} to a map $h_{n,i}$ whose image lies in Y. (See Exercise 124 below for this interpretation of the vanishing of the relative homotopy group.) Then define $h_n : X_n \to Y$ by

$$h_n(\varphi_i(e)) = h_{n,i}(e)$$

where $e \in D^n$. Using the above homotopy, one sees $h_{n-1|X_n} \simeq h_n : X_n \to Z$. Apply the homotopy extension theorem to extend this to a homotopy $H : X \times I \to Z$ and define $h_n(x)$ as $H(x, 1)$. □

We now return to the proof of the cofibrant theorem. We have a weak homotopy equivalence $f : Y \to Z$, which we may as well assume is the inclusion of a subspace by replacing Z by a mapping cylinder. We see

$$f_* : [X, Y] \to [X, Z]$$

is onto by applying the lemma to the pair (X, ϕ). We see f_* is injective by applying the lemma to the pair $(X \times I, X \times \{0, 1\})$. □

Exercise 124. Let $Y \subset Z$ be path-connected spaces. If $\pi_n(Z, Y, y_0) = 0$, show that any map $f : (D^n, S^{n-1}) \to (Z, Y)$ is homotopic rel S^{n-1} to a map whose image lies in Y.

Corollary 7.34 (Whitehead theorem). *A weak homotopy equivalence between CW-complexes is a homotopy equivalence.*

Proof. Let $f : Y \to Z$ be a weak homotopy equivalence between CW-complexes. By the surjectivity of $f_* : [Z, Y] \to [Z, Z]$, there is a $g : Z \to Y$ so that $[\mathrm{Id}_Z] = f_*[g] = [f \circ g]$. Then

$$f_*[g \circ f] = [f \circ g \circ f] = [\mathrm{Id}_Z \circ f] = [f \circ \mathrm{Id}_Y] = f_*[\mathrm{Id}_Y].$$

By the injectivity of f_*, $[g \circ f] = [\mathrm{Id}_Y]$, so f and g are homotopy inverses. □

Corollary 7.35. *Any n-connected CW-complex Y has the homotopy type of a CW-complex X whose n-skeleton is a point.*

Proof. Apply the proof of the CW-approximation to Y to find a weak homotopy equivalence $X \to Y$ where X_n is a point. By Corollary 7.34 it is a homotopy equivalence. □

Theorem 7.36. *Let $f : X \to Y$ be a continuous map. Suppose that CW-approximations $u : X' \to X$ and $v : Y' \to Y$ are given. Then there exists a map f' so that the diagram*

$$\begin{array}{ccc} X' & \xrightarrow{u} & X \\ {\scriptstyle f'}\downarrow & & \downarrow{\scriptstyle f} \\ Y' & \xrightarrow[v]{} & Y \end{array}$$

commutes up to homotopy. Furthermore, the map f' is unique up to homotopy. □

The theorem follows from the cofibrant theorem (v_* is a bijection). Applying it to the case where $f = \mathrm{Id}_X$, we see that CW-approximations are unique up to homotopy type. The theorem and the cofibrant theorem imply that for the purposes of homotopy theory, one may as well assume all spaces involved are CW-complexes. A relative version of the cofibrant theorem gives the same result for based homotopy theory.

7.10. Obstruction theory in fibrations

We next turn to the lifting and cross section problems.

Consider the lifting problem:

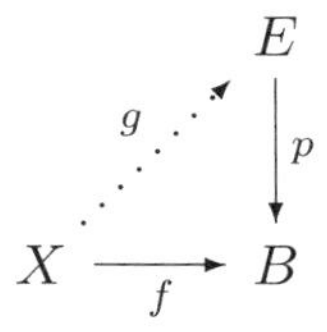

where $p : E \to B$ is a fibration. Note that if $f = \mathrm{Id}_B$, then the lifting problem is the same as constructing a cross section of p.

Suppose g has been defined over the n-skeleton of X. Given an $(n+1)$-cell e^{n+1} of X, the composite of the attaching map and g gives a map $S^n \to X \xrightarrow{g} E$. The composite $S^n \to X \xrightarrow{g} E \xrightarrow{p} B$ is null homotopic since it equals $S^n \to X \xrightarrow{f} B$, which extends over the cell $e^{n+1} \hookrightarrow X$.

The homotopy lifting property of fibrations implies that the composite

$$S^n \to X \xrightarrow{g} E$$

is homotopic to a map $S^n \to F$ by lifting the null-homotopy in the base (cf. Corollary 6.44).

Thus, to each $(n+1)$-cell of X we have defined, in a highly non-canonical way, a map $S^n \to F$. We would like to say that this defines a cochain on X with values in $\pi_n F$.

If we assume F is n-simple so that $\pi_n F = [S^n, F]$ (unbased maps), then any map $S^n \to F$ defines an element in $\pi_n F$.

However, if $\pi_1 B \neq 0$, then some ambiguity remains; namely it was not necessary that f preserved base points, and hence, *even if F is n-simple*, we do not obtain a cochain in $C^n(X; \pi_n F)$. However, one does get a cochain with local coefficients. Thus obstruction theory for fibrations requires the use of cohomology with local coefficients, as we will now see.

Recall from Proposition 6.62 that if F is n-simple, then the fibration $F \hookrightarrow E \to B$ defines a local coefficient system over B with fiber $\pi_n F$. In fact Proposition 6.62 shows how to associate to each $\alpha \in \pi_1 B$ a homotopy class h_α of self-homotopy equivalences of F. Then h_α induces an automorphism of $[S^n, F]$ by $f \mapsto h_\alpha \circ f$. Since we are assuming F is n-simple, $[S^n, F] = \pi_n F$, and so this shows how the fibration determines a representation $\rho : \pi_1 B \to \mathrm{Aut}(\pi_n F)$.

Now pull back this local coefficient system over X via $f : X \to B$ to obtain a local coefficient system over X. We continue to call it ρ, so

$$\rho : \pi_1 X \xrightarrow{f_*} \pi_1 B \xrightarrow{\rho} \mathrm{Aut}(\pi_n(F)).$$

With these hypotheses, one obtains an obstruction cocycle

$$\theta^{n+1}(g) \in C^{n+1}(X; \pi_n(F)_\rho) = \mathrm{Hom}_{\mathbf{Z}[\pi_1 X]}(C_{n+1}(\widetilde{X}), \pi_n(F)_\rho).$$

One then can prove the following theorem.

Theorem 7.37. *Let X be a CW-complex, $p : E \to B$ be a fibration with fiber F. Let $f : X \to B$ be a map and $g : X^n \to E$ a lift of f on the n-skeleton.*

If F is n-simple, then an obstruction class

$$[\theta^{n+1}(g)] \in H^{n+1}(X; \pi_n(F)_\rho)$$

is defined.

If $[\theta^{n+1}(g)]$ vanishes, then g can be redefined over the n-skeleton (rel the $(n-1)$-skeleton), then extended over the $(n+1)$-skeleton X^{n+1}. □

If the local coefficient system is trivial, for example if $\pi_1 X = 0$ or $\pi_1 B = 0$, then $[\theta^{n+1}(g)] \in H^{n+1}(X; \pi_n F)$ (untwisted coefficients). If $\pi_1 F = 0$, then F is k-simple for all k, so that the hypotheses of Theorem 7.37 hold.

If $\pi_k F = 0$ for $k \leq n-1$, then $[\theta^{n+1}(g)] \in H^{n+1}(X; \pi_n F_\rho)$ is called the *primary obstruction to lifting* f and is well-defined; i.e.

1. a lift over the n-skeleton always exists, and
2. $[\theta^{n+1}(g)]$ is independent of the choice of lift to the n-skeleton.

Henceforth we write $\gamma^{n+1}(f)$ for the primary obstruction to lifting f.

The proof of Theorem 7.37 is in many ways similar to the proofs given earlier. In certain important cases one can reduce this theorem to a special case of the extension problem by the following useful device.

Suppose there is a fibration $E \to B$, so that the fiber can be "delooped" in the following sense. Namely there exists a fibration $q : B \to Z$ with fiber E' so that the inclusion $E' \hookrightarrow B$ is equivalent to the fibration $E \to B$. Then we have seen that F is homotopy equivalent to the loop space ΩZ, and the sequence

$$[X, F] \to [X, E] \to [X, B] \xrightarrow{q_*} [X, Z]$$

is exact by Theorem 6.42.

This sequence shows that $f : X \to B$ can be lifted to $g : X \to E$ if and only if $q_*[f]$ is nullhomotopic. Thus the problem of lifting f is equivalent to the problem of nullhomotoping $q \circ f$.

As was explained above, there are obstructions

$$d^k(q \circ f, *) \in H^k(X; \pi_k(Z))$$

to nullhomotoping $q \circ f$ (provided Z is simple, etc.). But since

$$\pi_k(Z) = \pi_{k-1}(\Omega Z) = \pi_{k-1} F,$$

we can view $d^k(q \circ f, *)$ as an element of $H^k(X; \pi_{k-1}F)$. Thus the obstructions to finding cross sections are in this special case obtainable from the homotopy obstruction theorem.

This point of view works if $E \to B$ is, say, a principal G bundle, since one can take Z to be the classifying space BG.

7.11. Characteristic classes

One application of obstruction theory is to define characteristic classes. For example, suppose $p : E \to B$ is an oriented n-plane vector bundle, i.e. a bundle with fiber $\mathbf{R}^n$ and structure group $GL_n^+(\mathbf{R})$, the group of automorphisms of $\mathbf{R}^n$ with positive determinant. Then the *Euler class* $e(p) \in H^n(B; \mathbf{Z})$ is the primary obstruction to finding a section of the bundle

$$\mathbf{R}^n - \{0\} \hookrightarrow E_0 \to B$$

where $E_0 = E - i(B)$ is E minus the zero section.

Exercise 125. The primary obstruction to finding a cross section of $E_0 \to B$ lies in $H^n(B; \pi_{n-1}(\mathbf{R}^n - \{0\}))$. Show that $\mathbf{R}^n - \{0\}$ is n-simple, that $\pi_{n-1}(\mathbf{R}^n - \{0\}) = \mathbf{Z}$, and that $\pi_1 B$ acts trivially on $\pi_{n-1}(\mathbf{R}^n - \{0\})$; i.e. the local coefficient system is trivial (this uses the fact that the bundle is orientable).

In other words the Euler class is the primary obstruction to finding a nowhere zero section of p. The Euler class is a *characteristic class* in the sense that given a map of oriented n-plane vector bundles which is an isomorphism on fibers

$$\begin{array}{ccc} E' & \longrightarrow & E \\ {\scriptstyle p'}\downarrow & & \downarrow{\scriptstyle p} \\ B' & \xrightarrow[f]{} & B \end{array}$$

then

$$f^* e(p) = e(p').$$

If B is a CW-complex of dimension n, then the primary obstruction is the only obstruction, so there is a nowhere zero section if and only if the Euler class is zero. The Euler class is related to the Euler characteristic of a manifold by the following theorem (see [**30**]).

Theorem 7.38. *If $p : TB \to B$ is the tangent bundle of a closed, oriented n-manifold, then*

$$\langle e(p), [B] \rangle = \chi(B).$$

□

Corollary 7.39 (Poincaré–Hopf theorem). *A closed, oriented n-manifold has a nowhere-zero vector field if and only if its Euler characteristic is zero.* □

For example, you can't comb the hairy ball!

The mod 2 reduction of the Euler class of an $\mathbf{R}^n$-vector bundle $E \to B$ is called the *nth Stiefel–Whitney class* $w_n(E) \in H^n(B; \mathbf{Z}/2)$.

There are many other aspects of obstruction theory: for example the analogue of the homotopy problem in this setting is the problem of finding vertical homotopies between two cross sections. There are obstructions

$$d^n(g, g^1) \in H^n(X; \pi_n(F)_\rho).$$

Here are a few more examples to ponder.

Exercise 126. When we studied the extension problem, we did not come across local coefficient systems. This is because we assumed that Y was n-simple. Use Theorem 7.37 stated above together with the inverse of the delooping method outlined above to find a statement of a theorem about obstructions to extending maps into non-simple spaces.

Exercise 127. Write down a careful statement of a theorem about the obstruction to finding vertical homotopies between cross sections of a fibration.

7.12. Projects for Chapter 7

7.12.1. Postnikov systems. The decomposition of a CW-complex into its skeleta has a "dual" construction leading to *Postnikov decompositions* of a space. The word "dual" here is used in the same sense that cofibrations and fibrations are dual. The building blocks for CW-complexes are cells (D^n, S^{n-1}). These have homology $\mathbf{Z}$ in dimension n and zero in other dimensions. The building blocks for Postnikov decompositions are the Eilenberg-MacLane spaces $K(\pi, n)$. For CW-complexes, the attaching maps describe how the cells are put together. For Postnikov decompositions, spaces are described as iterated fibrations with fibers Eilenberg-MacLane spaces and the primary obstruction to finding cross sections determine how the space is to be assembled from its $K(\pi, n)$s.

For this project, show how to construct a Postnikov tower for a space X.

Theorem 7.40. *If X is a simple path connected space, there exists a "tower"*

$$\cdots \to X_n \xrightarrow{p_n} X_{n-1} \to \cdots \to X_1 \xrightarrow{p_1} X_0,$$

as well as maps $f_n : X \to X_n$ *so that the diagrams*

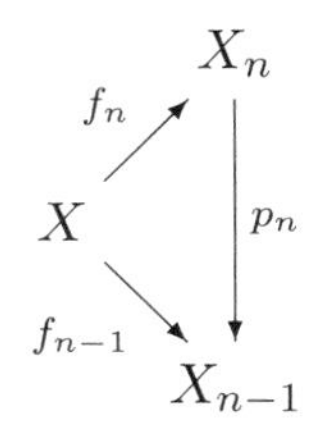

commute for each n.

For each n, *the map* $p_n : X_n \to X_{n-1}$ *is a fibration with fiber the Eilenberg-MacLane space* $K(\pi_n(X), n)$. *Moreover,* $\pi_k(X_n) = 0$ *for* $k > n$ *and* $(f_n)_* : \pi_k(X) \to \pi_k(X_n)$ *is an isomorphism for all* $k \leq n$.

To avoid complications you may assume that X is simply connected.

Each fibration in the tower $p_n : X_n \to X_{n-1}$ has fiber $K(\pi_n(X), n)$, and so there is a single obstruction (namely the primary obstruction) to finding a cross section. This obstruction lies in

$$H^{n+1}(X_{n-1}; \pi_n(K(\pi_n(X), n))) = H^{n+1}(X_{n-1}; \pi_n X)$$

and is called the $(n+1)$*st k-invariant of* X and is denoted by k^{n+1}. Using the identification $H^{n+1}(X_{n-1}; \pi_n X) = [X_{n-1}, K(\pi_n, n+1)]$, k^{n+1} can be thought of as a (homotopy class of a) map $X_{n-1} \to K(\pi_n, n+1)$ so that the fibration $K(\pi_n, n) \hookrightarrow X_n \to X_{n-1}$ is the pullback of the path space fibration $K(\pi_n, n) \hookrightarrow P \to K(\pi_n, n+1)$ via k^{n+1}.

Thus to a (simple) path connected space X this construction associates a collection $\{\pi_n, p_n, k^n\}$ where

1. π_n is an abelian group,
2. $p_n : X_n \to X_{n-1}$ is a fibration with fiber $K(\pi_n, n)$,
3. X_0 is contractible,
4. $k^n \in H^n(X_{n-2}; \pi_{n-1})$ classifies p_{n-1},
5. the inclusion of the fiber induces an isomorphism $\pi_n(K(\pi_n, n)) \to \pi_n(X_n)$.

This collection has the property that $\pi_n = \pi_n(X)$.

This data is called the *Postnikov system* or *Postnikov decomposition* for X.

Prove the main result about Postnikov systems.

Theorem 7.41. *The weak homotopy type of* X *is determined by its Postnikov system. More precisely, given the data* $\{\pi_n, p_n, k^n\}$ *satisfying conditions 1–5 above, there exists a space* X *with this data as its Postnikov* *decomposition. If Y is any space with this Postnikov decomposition, then X and Y are weakly homotopy equivalent.* □

Thus a space is completely determined up to homotopy inductively by the π_n and the k-invariants. More precisely, let $X_1 = K(\pi_1, 1)$ and let $p_1 : X_1 \to X_0 = \text{pt}$ be the constant map. Inductively, suppose $k^3, \cdots, k^n$ determine fibrations $p_\ell : X_\ell \to X_{\ell-1}$ for $\ell \leq n-1$, and suppose $k^{n+1} \in H^{n+1}(X_{n-1}; \pi_n)$ is given. Then define $p_n : X_n \to X_{n-1}$ to be the pullback of the path space fibration $K(\pi_n, n) \to P \to K(\pi_n, n+1)$ via the map $k^{n+1} : X_{n-1} \to K(\pi_n, n+1)$. If $\{\pi_n, k^n\}$ is the Postnikov system for a space X, then X is homotopy equivalent to $\lim X_n$.

A good reference for this material is [**43**, pp. 421–437] and [**15**, pp. 78-82].

If time permits, lecture on the dual exposition of obstruction theory from the point of view of Postnikov decompositions. Spanier's book [**36**] is one place to find this material.

Chapter 8

Bordism, Spectra, and Generalized Homology

This chapter contains a mixture of algebraic and differential topology and serves as an introduction to generalized homology theories. We will give a precise definition of a generalized homology theory later, but in the meantime you should think of a generalized homology theory as a functor from pairs of spaces to graded abelian groups (or graded R-modules) satisfying all Eilenberg–Steenrod axioms but the dimension axiom.

The material in this chapter will draw on the basic notions and theorems of differential topology, and you should re-familiarize yourself with the notion of smooth maps between smooth manifolds, submanifolds, tangent bundles, orientation of a vector bundle, the normal bundle of a submanifold, the Sard theorem, transversality and the tubular neighborhood theorem. One of the projects for this chapter is to prepare a lecture on these topics. A good reference for this material is Hirsch's book [**17**]; more elementary references include [**27**] and [**16**].

In this chapter (in contrast to the rest of this book), the word "manifold" will mean a *compact, smooth* manifold with or without boundary, and a submanifold $V \subset M$ will mean a compact submanifold whose boundary is contained in the boundary of M in such a way that V meets the boundary of M transversely. The normal bundle of a submanifold $i : V \hookrightarrow M$ is the quotient bundle $i^*(TM)/TV$, and we will use the notation $\nu(V \hookrightarrow M)$ or $\nu(i)$. If M is a submanifold of $\mathbf{R}^n$, or more generally if M has a Riemannian metric, then the normal bundle $\nu(V \hookrightarrow M)$ can be identified with the subbundle of $TM|_V$ consisting of all tangent vectors in T_pM which are perpendicular to T_pV, where $p \in V$. A *tubular neighborhood* of a submanifold

$i : V \hookrightarrow M$ is an embedding $f : \nu(i) \to M$ which restricts to the identity on (the zero section) V. Informally, we say that the open set $U = f(\nu(i)) \subset M$ is a tubular neighborhood of V.

8.1. Framed bordism and homotopy groups of spheres

Pontrjagin and Thom in the 1950's noted that in many situations there is a one-to-one correspondence between problems in geometric topology (= manifold theory) and problems in algebraic topology. Usually the algebraic problem is more tractable, and its solution leads to geometric consequences. In this section we discuss the quintessential example of this correspondence; a reference is the last section of Milnor's beautiful little book [**27**].

We start with an informal discussion of the passage from geometric topology to algebraic topology.

Definition 8.1. A *framing* of a submanifold V^{k-n} of a closed manifold M^k is a embedding ϕ of $V \times \mathbf{R}^n$ in M so that $\phi(p, 0) = p$ for all $p \in V$. If (W^{k+1-n}, ψ) is a framed submanifold of $M \times I$, then the two framed submanifolds of M given by intersecting W with $M \times \{0\}$ and $M \times \{1\}$ are *framed bordant*. Let $\Omega^{\mathrm{fr}}_{k-n,M}$ be the set of framed bordism classes of $(k-n)$-dimensional framed submanifolds of M.

A framed submanifold defines a *collapse map* $M \to S^n = \mathbf{R}^n \cup \{\infty\}$ by sending $\phi(p, v)$ to v and all points outside the image of ϕ to ∞. Note that $0 \in S^n$ is a regular value and the inverse image of 0 is V. A framed bordism gives a homotopy of the two collapse maps. A framed bordism from a framed submanifold to the empty set is a *null-bordism*. In the special case of a framed submanifold V^{k-n} of S^k, a null-bordism is given by an extension to a framed submanifold W^{k+1-n} of D^{k+1}.

Theorem 8.2. *The collapse map induces a bijection* $\Omega^{\mathrm{fr}}_{k-n,M} \to [M, S^n]$.

This method of translating between bordism and homotopy sets is called the *Pontrjagin–Thom construction.*

Here are some examples (without proof) to help your geometric insight. A (framed) point in S^k gives a map $S^k \to S^k$ which generates $\pi_k S^k \cong \mathbf{Z}$. Any framed circle in S^2 is null-bordant: for example the equator with the obvious framing is the boundary of the 2-disk in the 3-ball. However, a framed S^1 in S^3 so that the circle $\phi(S^1 \times \{(1,0)\})$ links the S^1 with linking number 1 represents the generator of $\pi_3(S^2) \cong \mathbf{Z}$. (Can you reinterpret this in terms of the Hopf map? Why can't one see the complexities of knot theory in framed bordism?) Now S^3 is naturally framed in S^4, S^4 in S^5, etc.

so we can suspend the linking number 1 framing of S^1 in S^3 to get a framing of S^1 in S^{k+1} for $k > 2$. This represents the generator of $\pi_{k+1}S^k \cong \mathbf{Z}_2$.

More generally, one can produce examples of framed manifolds by twisting and suspending. If (V^{k-n}, ϕ) is a framed submanifold of M^k and $\alpha : V \to O(n)$, then the *twist* is the framed submanifold $(V, \phi.\alpha)$ where $\phi.\alpha(p, v) = \phi(p, \alpha(p)v)$. The framed bordism class depends only on (V, ϕ) and the homotopy class of α. (See Exercise 133 below for more on this construction.) Next if (V^{k-n}, ϕ) is a framed submanifold of S^k, then the *suspension* of (V^{k-n}, ϕ) is the framed submanifold $(V^{k-n}, S\phi)$ of S^{k+1}, defined using the obvious framing of S^k in S^{k+1}, with $S^k \times \mathbf{R}_{>0}$ mapping to the upper hemisphere of S^{k+1}. Then the generator of $\pi_3(S^2)$ mentioned earlier can be described by first suspending the inclusion of a framed circle in the 2-sphere and then twisting by the inclusion of the circle in $O(2)$.

To prove Theorem 8.2 we first want to reinterpret $\Omega^{\text{fr}}_{k-n,M}$ in terms of normal framings. The key observation is that a framed submanifold determines n linearly independent normal vector fields on M.

Definition 8.3.

1. A *trivialization* of a vector bundle $p : E \to B$ with fiber $\mathbf{R}^n$ is a collection $\{\sigma_i : B \to E\}_{i=1}^n$ of sections which form a basis pointwise. Thus $\{\sigma_1(b), \dots, \sigma_n(b)\}$ is linearly independent and spans the fiber E_b for each $b \in B$.

   Equivalently, a trivialization is a specific bundle isomorphism $E \cong B \times \mathbf{R}^n$. A trivialization is also the same as a choice of section of the associated principal frame bundle.

2. A *framing* of a vector bundle is a homotopy class of trivializations, where two trivializations are called *homotopic* if there is a continuous 1-parameter family of trivializations joining them. In terms of the associated frame bundle this says the two sections are homotopic in the space of sections of the frame bundle.

A section of a normal bundle is called a normal vector field.

Definition 8.4. A *normal framing* of a submanifold V of M is a homotopy class of trivializations of the normal bundle $\nu(V \hookrightarrow M)$. If W is a normally framed submanifold of $M \times I$, then the two normally framed submanifolds of M given by intersecting W with $M \times \{0\}$ and $M \times \{1\}$ are *normally framed bordant*. (You should convince yourself that restriction of $\nu(W \hookrightarrow M \times I)$ to $V_0 = (M \times \{0\}) \cap W$ is canonically identified with $\nu(V_0 \hookrightarrow M)$.)

Exercise 128. Show that a framed submanifold (V, ϕ) of M determines a normal framing of V in M. Use notation from differential geometry and denote the standard coordinate vector fields on $\mathbf{R}^n$ by $\{\partial/\partial x_1, \dots, \partial/\partial x_n\}$.

Exercise 129. Define a map from the set of bordism classes of $(k-n)$-dimensional framed submanifolds of M to the set of bordism classes of $(k-n)$-dimensional normally framed submanifolds of M and show it is a bijection. (The existence part of the tubular neighborhood theorem will show the map is surjective, while the uniqueness part will show the map is injective.)

Henceforth we let $\Omega^{\text{fr}}_{k-n,M}$ denote both the bordism classes of framed submanifolds and the bordism classes of normally framed submanifolds of M.

Proof of Theorem 8.2. To define an inverse

$$d : [M, S^n] \to \Omega^{\text{fr}}_{k-n,M}$$

to the collapse map

$$c : \Omega^{\text{fr}}_{k-n,M} \to [M, S^n]$$

one must use differential topology; in fact, this was the original motivation for the development of transversality.

Any element of $[M, S^n]$ can be represented by a map $f : M \to S^n = \mathbf{R}^n \cup \{\infty\}$, which is smooth in a neighborhood of $f^{-1}(0)$ and transverse to 0 (i.e. 0 is a regular value). Thus:

1. The inverse image $f^{-1}(0) = V$ is a smooth submanifold of M^k of codimension n (i.e. of dimension $k-n$), and
2. The differential of f identifies the normal bundle of V in M^k with the pullback of the normal bundle of $0 \in S^n$ via f. More precisely, the differential of f, $df : TM^k \to TS^n$ restricts to $TM^k|_V$ and factors through the quotient $\nu(V \hookrightarrow M^k)$ to give a map of vector bundles

$$\begin{array}{ccc} \nu(V \hookrightarrow M^k) & \xrightarrow{df} & \nu(0 \hookrightarrow S^n) \\ \downarrow & & \downarrow \\ V & \longrightarrow & 0 \end{array}$$

which is an isomorphism in each fiber.

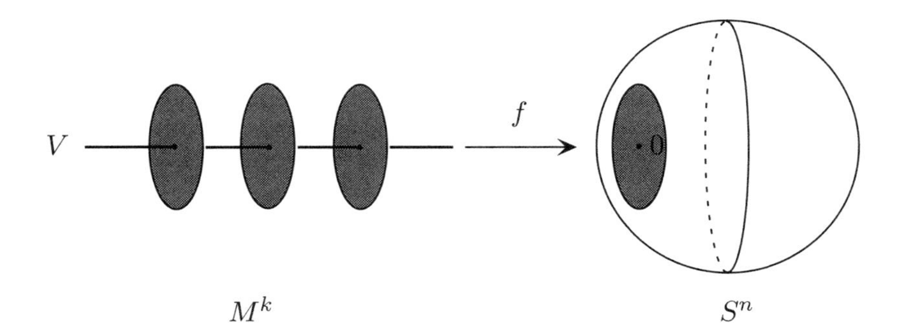

Since the normal bundle of 0 in $\mathbf{R}^n \cup \{\infty\}$ is naturally framed by the standard basis, the second assertion above implies that the normal bundle of V in M^k is also framed; i.e. there is a bundle isomorphism

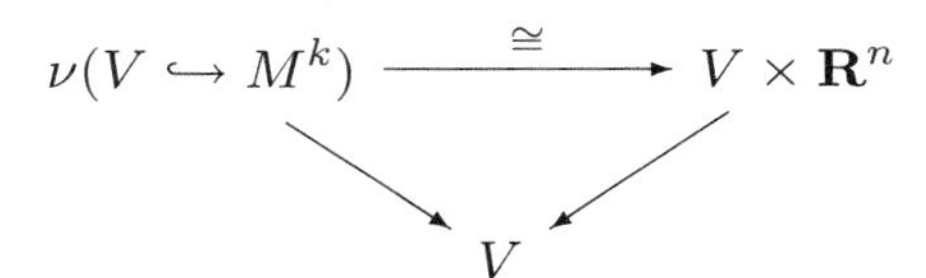

The map d is defined by sending $[f]$ to $f^{-1}(0)$ with the above framing. To see that d is well-defined, consider a homotopy

$$F : M \times I \to S^n$$

where $F|_{M\times\{0,1\}}$ is transverse to $0 \in S^n$. Consider the "trace of F"

$$\begin{aligned} \hat{F} : M \times I &\to S^n \times I \\ (m, t) &\mapsto (F(m, t), t), \end{aligned}$$

which has the advantage that it takes boundary points to boundary points. The (relative) transversality approximation theorem says that $\hat{F}$ is homotopic (rel $M \times \{0, 1\}$) to a map transverse to $0 \times I$. The inverse image of $0 \times I$ equipped with an appropriate normal framing gives a normally framed bordism between $F|^{-1}_{M\times\{0\}}(0)$ and $F|^{-1}_{M\times\{1\}}(0)$.

Our final task is to show that c and d are mutual inverses. It is easy to see that $d \circ c$ is the identity, but to show $c \circ d$ is the identity takes some work. First represent an element of $[M, S^n]$ by a a map f transverse to $0 \in \mathbf{R}^n \cup \{\infty\} = S^n$. It seems plausible that the collapse map associated to $V = f^{-1}(0)$ with the normal framing induced by df is homotopic to f, but there are technical details.

Here goes. Let $\nu = \nu(V \hookrightarrow M)$, let $g : \nu \to M$ be a tubular neighborhood of V, assume ν has a metric, and let $D = g(D(\nu))$ correspond to the disk bundle. Define $\Phi : \nu \to \mathbf{R}^n$ by $\Phi(x) = \lim_{t\to 0} t^{-1} f(g(tx))$. Then $\Phi(x)$

is the velocity vector of a curve, and by the chain rule Φ is the composite of the identification of ν with $\nu(V \hookrightarrow \nu)$ and $df \circ dg$. In particular Φ gives an isomorphism from each fiber of ν to $\mathbf{R}^n$.

There is a homotopy $f_t : D \to \mathbf{R}^n \cup \{\infty\}$ for $-1 \leq t \leq 1$ given by

$$f_t(g(x)) = \begin{cases} \frac{1}{1+t\|x\|}\Phi(x) & \text{if } -1 \leq t \leq 0, \\ t^{-1}f(g(tx)) & \text{if } 0 < t \leq 1. \end{cases}$$

We now have a map

$$\partial D \times [-1,1] \;\cup\; (M - \mathrm{Int}D) \times \{1\} \;\cup\; (M - \mathrm{Int}D) \times \{-1\} \to S^n - \{0\}$$

defined by f_t on the first piece, by f on the second piece, and by the constant map at infinity on the third piece. This extends to a map $(M - \mathrm{Int}D) \times [-1,1] \to S^n - \{0\}$ by the Tietze extension theorem.

We can then paste back in f_t to get a homotopy

$$F : M \times [-1,1] \to S^n$$

from our original f to a map h so that

$$h^{-1}\mathbf{R}^n = \mathrm{Int}D \cong V \times \mathbf{R}^n$$

where the diffeomorphism $\cong$ is defined by mapping to V by using the original tubular neighborhood and by mapping to $\mathbf{R}^n$ by h. Thus $f \simeq h$ where h is in the image of c. It follows that c is surjective and thus that c and d are mutual inverses. □

In reading the above proof you need either a fair amount of technical skill to fill in the details or you need to be credulous. For an alternate approach see [**27**, Chapter 7].

For a real vector bundle over a point, i.e. a vector space, a framing is the same as a choice of orientation of the vector space, since $GL(n, \mathbf{R})$ has two path components. Thus a normal framing of $V \subset S^k$ induces an orientation on the normal bundle $\nu(V \hookrightarrow S^k)$. (See Section 10.7 for more information about orientation.)

Exercise 130. Let V be a normally framed submanifold of a manifold M. Show that an orientation on M induces an orientation on V. (Hint: consider the isomorphism $TV \oplus \nu = TM|_V$.)

Theorem 8.5 (Hopf degree theorem). *Let M^k be a connected, closed, smooth manifold.*

1. *If M^k is orientable, then two maps $M^k \to S^k$ are homotopic if and only if they have the same degree.*
2. *If M^k is nonorientable, then two maps $M^k \to S^k$ are homotopic if and only if they have the same degree mod 2.*

Exercise 131. Prove the Hopf degree theorem in two ways: obstruction theory and framed bordism.

The function $\pi_k S^n \to [S^k, S^n]$ obtained by forgetting base points is a bijection. For $n > 1$ this follows from the fact that S^n is simply connected, and so vacuously the fundamental group acts trivially. For $n = 1$ this is still true because $\pi_k S^1$ is trivial for $k > 1$ and abelian for $k = 1$.

The result that $\pi_n S^n \cong \mathbf{Z}$ is a nontrivial result in algebraic topology; it is cool that this can be proven using differential topology.

Exercise 132. We only showed that the isomorphism of Theorem 8.2 is a bijection of sets. However, since $\pi_k S^n$ is an abelian group, the framed bordism classes inherit an abelian group structure. Prove that this group structure on framed bordism is given by taking the disjoint union

$$[V_0] + [V_1] := V_0 \amalg V_1 \subset S^k \# S^k \cong S^k$$

with negatives given by changing the orientation of the framing (e.g. replacing the first vector field in the framing by its negative)

$$-[V_0] = [-V_0].$$

We will generalize Theorem 8.2 by considering the effect of the suspension map $S : \pi_k S^n \to \pi_{k+1} S^{n+1}$ and eventually passing to the colimit $\operatorname{colim}_{\ell \to \infty} \pi_{k+\ell} S^{n+\ell}$. This has the effect of eliminating the thorny embedding questions of submanifolds in S^k; in the end we will be able to work with abstract framed manifolds V without reference to an embedding of V in some sphere.

Exercise 133. (The J-homomorphism) Let $V^{k-n} \subset M^k$ be a non-empty normally framed manifold. Use twisting to define a function

$$J : [V^{k-n}, O(n)] \to [M^k, S^n].$$

Now let V be the equatorial $S^{k-n} \subset S^k$ with the canonical framing coming from the inclusions $S^{k-n} \subset S^{k-n+1} \subset \cdots \subset S^k$, and show that the function

$$J : \pi_{k-n}(O(n)) \to \pi_k S^n$$

is a homomorphism provided $k > n$. It is called the *J-homomorphism* and can be used to construct interesting elements in $\pi_k S^n$.

Draw an explicit picture of a framed circle in $\mathbf{R}^3 = S^3 - \{\infty\}$ representing $J(\iota)$ where $\iota \in \pi_1 O(2) = \mathbf{Z}$ is the generator.

8.2. Suspension and the Freudenthal theorem

Recall that the (reduced) suspension of a space $X \in \mathcal{K}_*$ with nondegenerate base point is the space

$$SX = X \times I/\sim$$

where the subspace $(x_0 \times I) \cup (X \times \{0,1\})$ is collapsed to a point. This construction is functorial with respect to based maps $f : X \to Y$. In particular, the suspension defines a function

$$S : [X,Y]_0 \to [SX,SY]_0.$$

By Proposition 6.35, $SS^k = S^{k+1}$, so that when $X = S^k$, the suspension defines a function, in fact a homomorphism

$$S : \pi_k(Y) \to \pi_{k+1}(SY)$$

for any space Y. Taking Y to be a sphere one obtains

$$S : \pi_k(S^n) \to \pi_{k+1}(S^{n+1}).$$

We next identify $S^k \subset S^{k+1} = SS^k$ as the equator, and similarly $S^n \subset S^{n+1}$, and interpret the above map in terms of framed bordism.

If $f : S^k \to S^n$ is smooth, then the suspension

$$Sf : S^{k+1} \to S^{n+1}$$

is smooth away from the base points. If $x \in S^n$ is a regular value different from the base point, and $V = f^{-1}(x)$ is the normally framed submanifold of S^k associated to f, then clearly

$$V = (Sf)^{-1}(x) \subset S^k \subset S^{k+1}.$$

Let us compare normal bundles and normal framings.

$$\begin{aligned} \nu(V \hookrightarrow S^{k+1}) &= \nu(V \hookrightarrow S^k) \oplus \nu(S^k \hookrightarrow S^{k+1})|_V \\ &= \nu(V \hookrightarrow S^k) \oplus \varepsilon_V \end{aligned}$$

where $\varepsilon_V = V \times \mathbf{R} =$ trivial line bundle.

Similarly, $\nu(x \hookrightarrow S^{n+1}) = \nu(x \hookrightarrow S^n) \oplus \varepsilon_{\{x\}}$, and the differential of Sf preserves the trivial factor, since, locally (near the equator $S^k \subset S^{k+1}$),

$$Sf \cong f \times Id : S^k \times (-\epsilon, \epsilon) \to S^n \times (-\epsilon, \epsilon).$$

We have shown the following.

Theorem 8.6. *Taking the suspension of a map corresponds, via the Pontrjagin-Thom construction, to the same manifold V, but embedded in the equator $S^k \subset S^{k+1}$, and with normal framing the direct sum of the old normal framing and the trivial 1-dimensional framing.* □

Now consider the effect of multiple suspensions,

$$S^\ell : \pi_k S^n \longrightarrow \pi_{k+\ell} S^{n+\ell}.$$

For each suspension, the effect on the normally framed submanifold V is to replace it by the same manifold embedded in the equator, with the new normal framing $\nu_{\text{new}} = \nu_{\text{old}} \oplus \varepsilon_V$. Thus after ℓ suspensions,

$$\nu_{\text{new}} = \nu_{\text{old}} \oplus \varepsilon_V^\ell.$$

The following fundamental result is the starting point for the investigation of "stable" phenomena in homotopy theory. We will not give a proof at this time, since a spectral sequence proof is the easiest way to go. The proof is given in Section 10.3.

Theorem 8.7 (Freudenthal suspension theorem). *Suppose that X is an $(n-1)$-connected space $(n \geq 2)$. Then the suspension homomorphism*

$$S : \pi_k X \longrightarrow \pi_{k+1} SX$$

is an isomorphism if $k < 2n-1$ and an epimorphism if $k = 2n-1$. □

The most important case is when $X = S^n$, and here the Freudenthal suspension theorem can also be given a differential topology proof using framed bordism and the facts that any j-manifold embeds in S^n for $n \geq 2j+1$, uniquely up to isotopy if $n \geq 2j+2$, and that any embedding of a j-manifold in S^{n+1} is isotopic to an embedding in S^n if $n \geq 2j+1$.

Exercise 134. Show that for any k-dimensional CW-complex X and for any $(n-1)$-connected space Y $(n \geq 2)$ the suspension map

$$[X, Y]_0 \to [SX, SY]_0$$

is bijective if $k < 2n-1$ and surjective if $k = 2n-1$. (Hint: Instead consider the map $[X, Y]_0 \to [X, \Omega SY]_0$. Convert the map $Y \to \Omega SY$ to a fibration and apply cross-section obstruction theory as well as the Freudenthal suspension theorem.)

For a based space X, $\pi^n X = [X, S^n]_0$ is called the *n-th cohomotopy set*. If X is a CW-complex with dim $X < 2n-1$, then Exercise 134 implies that $\pi^n X$ is a group, with group structure given by suspending and using the suspension coordinate in SX. The reader might ponder the geometric meaning (framed bordism) of the cohomotopy group structure when X is a manifold.

Definition 8.8. The k-th *stable homotopy group* of a based space X is the colimit

$$\pi_k^S X = \operatorname*{colim}_{\ell \to \infty} \pi_{k+\ell} S^\ell X.$$

The *stable k-stem* is

$$\pi_k^S = \pi_k^S S^0.$$

The computation of the stable k-stem for all k is the holy grail of the field of homotopy theory.

The Hurewicz theorem implies that if X is $(n-1)$-connected, then SX is n-connected, since $\widetilde{H}_\ell SX = \widetilde{H}_{\ell-1} X = 0$ if $\ell \leq n$ and $\pi_1 SX = 0$ if X is path connected. The following corollary follows from this fact and the Freudenthal theorem.

Corollary 8.9. *If X is path connected,*

$$\pi_k^S X = \pi_{2k}(S^k X) = \pi_{k+\ell}(S^\ell X) \qquad \text{for } \ell \geq k.$$

For the stable k-stem,

$$\pi_k^S = \pi_{2k+2}(S^{k+2}) = \pi_{k+\ell}(S^\ell) \qquad \text{for } \ell \geq k+2.$$

□

Recall from Equation (6.3) that $\pi_k(O(n-1)) \to \pi_k(O(n))$, induced by the inclusion $O(n-1) \hookrightarrow O(n)$, is an isomorphism for $k < n-2$, and therefore letting $O = \operatorname{colim}_{n\to\infty} O(n)$, $\pi_k O = \pi_k(O(n))$ for $k < n-2$. It follows from the definitions that the following diagram commutes:

$$\begin{array}{ccc} \pi_k(O(n-1)) & \xrightarrow{J} & \pi_{k+n-1}(S^{n-1}) \\ \downarrow{\scriptstyle i_*} & & \downarrow{\scriptstyle s} \\ \pi_k(O(n)) & \xrightarrow{J} & \pi_{k+n}(S^n) \end{array}$$

with the horizontal maps the J-homomorphisms, the left vertical map induced by the inclusion, and the right vertical map the suspension homomorphism. If $k < n-2$, then both vertical maps are isomorphisms, and so one obtains the *stable J-homomorphism*

$$J : \pi_k(O) \to \pi_k^S.$$

Corollary 8.10. *The Pontrjagin-Thom construction defines an isomorphism from π_k^S to the normally framed bordism classes of normally framed k-dimensional closed submanifolds of S^n for any $n \geq 2k+2$.* □

8.3. Stable tangential framings

We wish to remove the restriction that our normally framed manifolds be submanifolds of S^n. To this end we need to eliminate the reference to the normal bundle. This turns out to be easy and corresponds to the fact that the normal and tangent bundles of a submanifold of S^n are inverses in a certain stable sense. Since the tangent bundle is an intrinsic invariant of a smooth manifold, and so is defined independently of any embedding in S^k, this will enable us to replace normal framings with tangential framings. On the homotopy level, however, we will need to take suspensions when describing in what way the bundles are inverses. In the end this means that we will obtain an isomorphism between *stably tangentially framed* bordism classes and *stable* homotopy groups.

In what follows, ε^j will denote a *trivialized* j-dimensional real bundle over a space.

Lemma 8.11. *Let $V^k \subset S^n$ be a closed, oriented, normally framed submanifold of S^n. Then*

1. *A normal framing $\gamma : \nu(V \hookrightarrow S^n) \cong \varepsilon^{n-k}$ induces a trivialization*
$$\overline{\gamma} : TV \oplus \varepsilon^{n-k+1} \cong \varepsilon^{n+1}.$$
2. *A trivialization $\overline{\gamma} : TV \oplus \varepsilon \cong \varepsilon^{k+1}$ induces a trivialization*
$$\nu(V \hookrightarrow S^n) \oplus \varepsilon^{k+1} \cong \varepsilon^{n+1}.$$

Proof. The inclusion $S^n \subset \mathbf{R}^{n+1}$ has a trivial 1-dimensional normal bundle which can be framed by choosing the outward unit normal as a basis. This shows that the once stabilized tangent bundle of S^n is *canonically* trivialized
$$TS^n \oplus \varepsilon \cong \varepsilon^{n+1}$$
since the tangent bundle of $\mathbf{R}^{n+1}$ is canonically trivialized.

There is a canonical decomposition
$$(TS^n \oplus \varepsilon)|_V = \nu(V \hookrightarrow S^n) \oplus TV \oplus \varepsilon.$$
Using the trivialization of $TS^n \oplus \varepsilon$, one has a canonical isomorphism
$$\varepsilon^{n+1} \cong \nu(V \hookrightarrow S^n) \oplus TV \oplus \varepsilon.$$
Thus a normal framing $\gamma : \nu(V \hookrightarrow S^k) \cong \varepsilon^{n-k}$ induces an isomorphism
$$\varepsilon^{n+1} \cong \varepsilon^{n-k} \oplus TV \oplus \varepsilon,$$
and, conversely, a trivialization $\overline{\gamma} : TV \oplus \varepsilon \cong \varepsilon^{k+1}$ induces an isomorphism
$$\varepsilon^{n+1} \cong \nu(V \hookrightarrow S^n) \oplus \varepsilon^{k+1}.$$

□

Definition 8.12. A *stable (tangential) framing* of a k-dimensional manifold V is an equivalence class of trivializations of
$$TV \oplus \varepsilon^n$$
where ε^n is the trivial bundle $V \times \mathbf{R}^n$. Two trivializations
$$t_1 : TV \oplus \varepsilon^{n_1} \cong \varepsilon^{k+n_1}, \quad t_2 : TV \oplus \varepsilon^{n_2} \cong \varepsilon^{k+n_2}$$

are considered equivalent if there exists some N greater than n_1 and n_2 such that the direct sum trivializations

$$t_1 \oplus \mathrm{Id} : TV \oplus \varepsilon^{n_1} \oplus \varepsilon^{N-n_1} \cong \varepsilon^{k+n_1} \oplus \varepsilon^{N-n_1} = \varepsilon^{k+N}$$

and

$$t_2 \oplus \mathrm{Id} : TV \oplus \varepsilon^{n_2} \oplus \varepsilon^{N-n_2} \cong \varepsilon^{k+n_2} \oplus \varepsilon^{N-n_2} = \varepsilon^{k+N}$$

are homotopic.

Similarly, a *stable normal framing* of a submanifold V of S^ℓ is an equivalence class of trivializations of $\nu(V \hookrightarrow S^\ell) \oplus \varepsilon^n$, and a *stable framing* of a bundle η is an equivalence class of trivializations of $\eta \oplus \varepsilon^n$.

A tangential framing is easier to work with than a normal framing, since one does not need to refer to an embedding $V \subset S^n$ to define a tangential framing. However, stable normal framings and stable tangential framings are equivalent, essentially because the tangent bundle of S^n is canonically stably framed. Lemma 8.11 generalizes to give the following theorem.

Theorem 8.13. *There is a 1-1 correspondence between stable tangential framings and stable normal framings of a manifold V. More precisely:*

1. *Let $i : V \hookrightarrow S^n$ be an embedding. A stable framing of TV determines stable framing of $\nu(i)$ and conversely.*
2. *Let $i_1 : V \hookrightarrow S^{n_1}$ and $i_2 : V \hookrightarrow S^{n_2}$ be embeddings. For n large enough there exist a canonical (up to homotopy) identification*

$$\nu(i_1) \oplus \varepsilon^{n-n_1} \cong \nu(i_2) \oplus \varepsilon^{n-n_2}.$$

A stable framing of $\nu(i_1)$ determines one of $\nu(i_2)$ and vice versa.

Proof. 1. The proof of Lemma 8.11 gives a canonical identification

$$\nu(V \hookrightarrow S^n) \oplus \varepsilon^\ell \oplus TV \cong \varepsilon^{n+\ell}$$

for all $\ell > 0$. Associativity of $\oplus$ shows stable framings of the normal bundle and tangent bundles coincide.

2. Let $i_1 : V \hookrightarrow S^{n_1}$ and $i_2 : V \hookrightarrow S^{n_2}$ be embeddings. There is a formal proof that stable framings of $\nu(i_1)$ and $\nu(i_2)$ coincide. Namely, a stable framing of $\nu(i_1)$ determines a stable framing of TV by Part 1, which in turn determines a stable framing of $\nu(i_2)$. However, the full statement of Part 2 applies to submanifolds with non-trivial normal bundle, and theorems from differential topology must be used.

Choose n large enough so that any two embeddings of V in S^n are isotopic. (Transversality theorems imply that $n > 2k + 1$ suffices.)

The composite $V \xrightarrow{i_1} S^{n_1} \xrightarrow{j_1} S^n$, with $S^{n_1} \xrightarrow{j_1} S^n$ the equatorial embedding, has normal bundle

$$\nu(j_1 \circ i_1) = \nu(i_1) \oplus \varepsilon^{n-n_1}.$$

Similarly, the composite $V \xrightarrow{i_2} S^{n_2} \xrightarrow{j_2} S^n$ has normal bundle

$$\nu(j_2 \circ i_2) = \nu(i_2) \oplus \varepsilon^{n-n_2}.$$

Then $j_2 \circ i_2$ is isotopic to $j_1 \circ i_1$, and the isotopy induces an isomorphism $\nu(j_2 \circ i_2) \cong \nu(j_1 \circ i_1)$.

If $n > 2(k+1)+1$, then any self-isotopy is isotopic to the constant isotopy, so that the identification $\nu(j_2 \circ i_2) \cong \nu(j_1 \circ i_1)$ is canonical (up to homotopy). □

Definition 8.14. Two real vector bundles E, F over V are called *stably equivalent* if there exists non-negative integers i, j so that $E \oplus \varepsilon^i$ and $F \oplus \varepsilon^j$ are isomorphic.

Since every smooth compact manifold embeds in S^n for some n, the second part of Theorem 8.13 has the consequence that the stable normal bundle (i.e. the stable equivalence class of the normal bundle for some embedding) is a well defined invariant of a smooth manifold, independent of the embedding, just as the tangent bundle is. However, something stronger holds. If $\nu(i_1)$ and $\nu(i_2)$ are normal bundles of two different embeddings of a manifold in a sphere, then not only are $\nu(i_1)$ and $\nu(i_2)$ stably equivalent, but the stable isomorphism is determined up to homotopy.

Returning to bordism, we see that the inclusion $S^n \subset S^{n+1}$ sets up a correspondence between the suspension operation and stabilizing a normal (or equivalently tangential) framing. Consequently Corollary 8.10 can be restated as follows.

Corollary 8.15. *The stable k-stem π_k^S is isomorphic to the stably tangentially framed bordism classes of stably tangentially framed k-dimensional smooth, oriented compact manifolds without boundary.* □

This statement is more appealing since it refers to k-dimensional manifolds intrinsically, without reference to an embedding in some S^n.

Here is a list of some computations of stable homotopy groups of spheres for you to reflect on. (Note: π_k^S has been computed for $k \leq 64$. There is no reasonable conjecture for π_k^S for general k, although there are many results known. For example, in Chapter 10, we will show that the groups are finite for $k > 0$; $\pi_0^S = \mathbf{Z}$ by the Hopf degree theorem.)

k	1	2	3	4	5	6
π_k^S	$\mathbf{Z}/2$	$\mathbf{Z}/2$	$\mathbf{Z}/24$	0	0	$\mathbf{Z}/2$
k	7	8	9	10	11	12
π_k^S	$\mathbf{Z}/240$	$(\mathbf{Z}/2)^2$	$(\mathbf{Z}/2)^3$	$\mathbf{Z}/6$	$\mathbf{Z}/504$	0
k	13	14	15	16	17	18
π_k^S	$\mathbf{Z}/3$	$(\mathbf{Z}/2)^2$	$\mathbf{Z}/480 \oplus \mathbf{Z}/2$	$(\mathbf{Z}/2)^2$	$(\mathbf{Z}/2)^4$	$\mathbf{Z}/8 \oplus \mathbf{Z}/2$
k	19	20	21	22	23	24
π_k^S	$\mathbf{Z}/264 \oplus \mathbf{Z}/2$	$\mathbf{Z}/24$	$(\mathbf{Z}/2)^2$	$(\mathbf{Z}/2)^2$	†	$(\mathbf{Z}/2)^2$

† π_{23}^S is $\mathbf{Z}/65520 \oplus \mathbf{Z}/24 \oplus \mathbf{Z}/2$.

The reference [**32**] is a good source for the tools to compute π_k^S.

We will give stably framed manifolds representing generators of π_k^S for $k < 9$; you may challenge your local homotopy theorist to supply the proofs. In this range there are (basically) two sources of framed manifolds: normal framings on spheres coming from the image of the stable J-homomorphism $J : \pi_k(O) \to \pi_k^S$, and tangential framing coming from Lie groups. There is considerable overlap between these sources.

Bott periodicity (Theorem 6.49) computes $\pi_k(O)$.

k	0	1	2	3	4	5	6	7	8
$\pi_k O$	$\mathbf{Z}/2$	$\mathbf{Z}/2$	0	$\mathbf{Z}$	0	0	0	$\mathbf{Z}$	$\mathbf{Z}/2$

Then $J : \pi_k O \to \pi_k^S$ is an isomorphism for $k = 1$, an epimorphism for $k = 3, 7$, and a monomorphism for $k = 8$.

Another source for framed manifolds are Lie groups. If G is a compact k-dimensional Lie group and $T_eG \cong \mathbf{R}^k$ is an identification of its tangent space at the identity, then one can use the group multiplication to identify $TG \cong G \times \mathbf{R}^k$ and thereby frame the tangent bundle. This is the so-called Lie invariant framing. The generators of the cyclic groups $\pi_0^S, \pi_1^S, \pi_2^S, \pi_3^S, \pi_6^S, \pi_7^S$ are given by $e, S^1, S^1 \times S^1, S^3, S^3 \times S^3, S^7$ with invariant framings. (The unit octonions S^7 fail to be a group because of the lack of associativity, but nonetheless, they do have an invariant framing.)

Finally, the generators of π_8^S are given by S^8 with framing given by the J-homomorphism and the unique exotic sphere in dimension 8. (An exotic sphere is a smooth manifold homeomorphic to a sphere and not diffeomorphic to a sphere.)

We have given a bordism description of the groups π_k^S. If X is any space, $\pi_k^S X$ can be given a bordism description also. In this case one adds the structure of a map from the manifold to X. (A map from a manifold to a space X is sometimes called a *singular manifold in* X.)

Definition 8.16. Let $(V_i, \gamma_i : TV_i \oplus \varepsilon^a \cong \varepsilon^{k+a})$, $i = 0, 1$, be two stably framed k-manifolds and $g_i : V_i \to X$, $i = 0, 1$, two maps.

We say (V_0, γ_0, g_0) is *stably framed bordant to* (V_1, γ_1, g_1) *over* X if there exists a stably framed bordism (W, τ) from (V_0, γ_0) to (V_1, γ_1) and a map

$$G : W \to X$$

extending g_0 and g_1.

We introduce the notation:

1. Let X_+ denote $X \amalg \mathrm{pt}$, the union of X with a disjoint base point.
2. Let $\Omega_k^{\mathrm{fr}}(X)$ denote the stably framed bordism classes of stably framed k-manifolds over X.

Since every space maps uniquely to a point, and since $S^0 = \mathrm{pt}_+$, we can restate Corollary 8.15 in this notation as

$$\Omega_k^{\mathrm{fr}}(\mathrm{pt}) = \pi_k^S(\mathrm{pt}_+)$$

since $\pi_k^S = \pi_k^S(S^0) = \pi_k^S(\mathrm{pt}_+)$.

More generally one can easily prove the following theorem.

Theorem 8.17. $\Omega_k^{\mathrm{fr}}(X) = \pi_k^S(X_+)$.

The proof of this theorem is essentially the same as for $X = \mathrm{pt}$; one just has to carry the map $V \to X$ along for the ride. We give an outline of the argument and indicate a map $\pi_k^S(X_+) \to \Omega_k^{\mathrm{fr}}(X)$.

Sketch of proof. Choose ℓ large so that $\pi_k^S(X_+) = \pi_{k+\ell}(X_+ \wedge S^\ell)$.

The smash product $X_+ \wedge S^\ell = X_+ \times S^\ell / X_+ \vee S^\ell = X \times S^\ell / X \times \mathrm{pt}$ is called the *half smash* of X and S^ℓ and is depicted in the following picture.

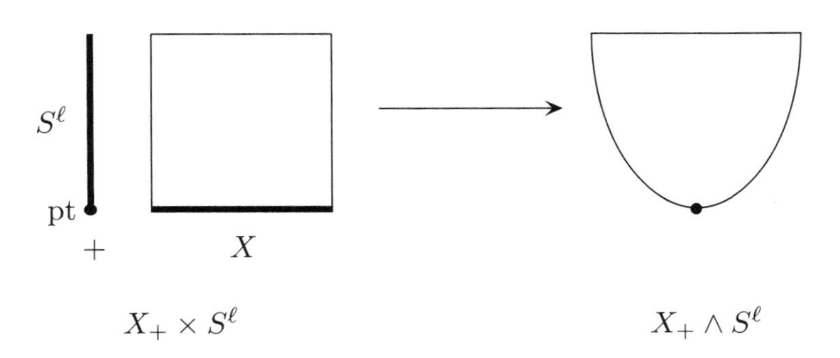

Given $f : S^{k+\ell} \to X_+ \wedge S^\ell$, make f transverse to $X \times \{x\}$, where $x \in S^\ell$ is a point different from the base point. (You should think carefully about what transversality means since X is just a topological space. The point is that smoothness is only needed in the normal directions, since one can project to the sphere.)

Then $f^{-1}(X \times \{x\}) = V$ is a smooth, compact manifold, and since a neighborhood of $X \times \{x\}$ in $X_+ \wedge S^\ell$ is homeomorphic to $X \times \mathbf{R}^\ell$ as indicated in the following figure,

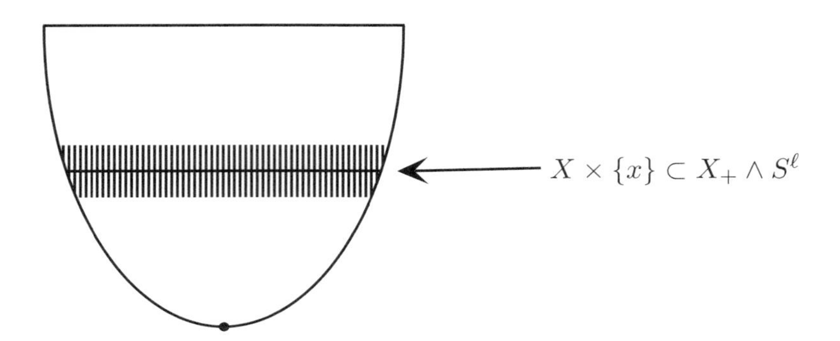

the submanifold V has a framed normal bundle, and $f_{|V} : V \to X \times \{x\} = X$. This procedure shows how to associate a stably framed manifold with a map to X to a (stable) map $f : S^{k+\ell} \to X_+ \wedge S^\ell$. One can show as before, using the Pontrjagin-Thom construction, that the induced map $\pi_{k+\ell}(X_+ \wedge S^\ell) \to \Omega_k^{\mathrm{fr}}(X)$ is an isomorphism. □

Exercise 135. Define the reverse map $\Omega_k^{\mathrm{fr}}(X) \to \pi_k^S(X_+)$.

8.4. Spectra

The collection of spheres, $\{S^n\}_{n=0}^\infty$, together with the maps (in fact homeomorphisms)

$$k_n : SS^n \xrightarrow{\cong} S^{n+1}$$

forms a system of spaces and maps from which one can construct the stable homotopy groups $\pi_n^S(X)$. Another such system is the collection of Eilenberg–MacLane spaces $K(\mathbf{Z}, n)$ from which we can recover the cohomology groups by the identification $H^n(X; \mathbf{Z}) = [X, K(\mathbf{Z}, n)]$ according to the results of Chapter 7.

The notion of a spectrum abstracts from these two examples and introduces a category which measures "stable" phenomena, that is, phenomena which are preserved by suspending. Recall that $\widetilde{H}^n(X) = \widetilde{H}^{n+1}(SX)$ and by definition $\pi_n^S(X) = \pi_{n+1}^S(SX)$. Thus cohomology and stable homotopy groups are measuring stable information about a space X.

Definition 8.18. A *spectrum* is a sequence of pairs $\{K_n, k_n\}$ where the K_n are based spaces and $k_n : SK_n \to K_{n+1}$ are base point preserving maps, where SK_n denotes the suspension.

In Exercise 96 you saw that the n-fold reduced suspension of $S^n X$ of X is homeomorphic to $S^n \wedge X$. Thus we can rewrite the definition of stable homotopy groups as

$$\pi_n^S X = \operatorname*{colim}_{\ell \to \infty} \pi_{n+\ell}(S^\ell \wedge X)$$

where the colimit is taken over the homomorphisms

$$\pi_{n+\ell}(S^\ell \wedge X) \to \pi_{n+\ell+1}(S^{\ell+1} \wedge X).$$

These homomorphisms are composites of the suspension

$$\pi_{n+\ell}(S^\ell \wedge X) \to \pi_{n+\ell+1}(S(S^\ell \wedge X)),$$

the identification $S(S^\ell \wedge X) = S^1 \wedge (S^\ell \wedge X) = S(S^\ell) \wedge X$, and the map $\pi_{n+\ell+1}(S(S^\ell) \wedge X) \to \pi_{n+\ell+1}(S^{\ell+1} \wedge X)$ induced by the map $k_\ell : S(S^\ell) \to S^{\ell+1}$.

Thus we see a natural link between the sphere spectrum

$$\mathbf{S} = \{S^n, k_n : S(S^n) \cong S^{n+1}\}$$

and the stable homotopy groups

$$\pi_n^S(X) = \operatorname*{colim}_{\ell \to \infty} \pi_{n+\ell}(S^\ell \wedge X).$$

Another example is provided by ordinary integral homology. The path space fibration and the long exact sequence in homotopy show that the loop space of the Eilenberg–MacLane space $K(\mathbf{Z}, n+1)$ is homotopy equivalent to $K(\mathbf{Z}, n)$. Fixing a model for $K(\mathbf{Z}, n)$ for each n, there exists a sequence of homotopy equivalences

$$h_n : K(\mathbf{Z}, n) \to \Omega K(\mathbf{Z}, n+1).$$

Then h_n defines, by taking its adjoint, a map

$$k_n : S(K(\mathbf{Z}, n)) \to K(\mathbf{Z}, n+1).$$

In this way we obtain the *Eilenberg–MacLane* spectrum

$$\mathbf{K}(\mathbf{Z}) = \{K(\mathbf{Z}, n), k_n\}.$$

We have seen in Theorem 7.22 that $H^n(X; \mathbf{Z}) = [X, K(\mathbf{Z}, n)]$.

Ordinary homology and cohomology are derived from the Eilenberg–MacLane spectrum, as the next theorem indicates. This point of view generalizes to motivate the definition of homology and cohomology with respect to any spectrum.

Theorem 8.19. *For any space X,*

1. $H_n(X;\mathbf{Z}) = \operatorname{colim}_{\ell\to\infty} \pi_{n+\ell}(X_+ \wedge K(\mathbf{Z},\ell))$,
2. $H^n(X;\mathbf{Z}) = \operatorname{colim}_{\ell\to\infty}[S^\ell(X_+), K(\mathbf{Z}, n+\ell)]_0$.

□

Recall that for $n \geq 0$, $H^n(X) = \widetilde{H}^n(X_+) = \widetilde{H}^{n+1}(SX_+) = H^{n+1}(SX_+)$; in fact the diagram

$$\begin{array}{ccc} [X_+, K(\mathbf{Z},n)]_0 & \xrightarrow{S} & [SX_+, SK(\mathbf{Z},n)]_0 \\ \downarrow{\scriptstyle h_n \cong} & & \downarrow{\scriptstyle k_n} \\ [X_+, \Omega K(\mathbf{Z}, n+1)]_0 & \xrightarrow{\cong} & [SX_+, K(\mathbf{Z}, n+1)]_0 \end{array}$$

commutes. This shows that we could have *defined* the cohomology of a space by

$$H^n(X;\mathbf{Z}) = \operatorname*{colim}_{\ell\to\infty}[S^\ell X_+; K(\mathbf{Z}, n+\ell)]_0$$

and verifies the second part of this theorem. The first part can be proven by starting with this fact and using Spanier-Whitehead duality. See the project on Spanier-Whitehead duality at the end of this chapter.

These two examples and Theorem 8.19 leads to the following definition. Recall that X_+ denotes the space X with a disjoint base point. In particular, if $A \subset X$, then $(X_+/A_+) = X/A$ if A is non-empty and equals X_+ if A is empty.

Definition 8.20. Let $\mathbf{K} = \{K_n, k_n\}$ be a spectrum. Define the *(unreduced) homology and cohomology with coefficients in the spectrum* $\mathbf{K}$ to be the functor taking a space X to the abelian group

$$H_n(X;\mathbf{K}) = \operatorname*{colim}_{\ell\to\infty} \pi_{n+\ell}(X_+ \wedge K_\ell)$$

and

$$H^n(X;\mathbf{K}) = \operatorname*{colim}_{\ell\to\infty}[S^\ell(X_+); K_{n+\ell}]_0,$$

the *reduced homology and cohomology with coefficients in the spectrum* $\mathbf{K}$ to be the functor taking a based space X to the abelian group

$$\widetilde{H}_n(X;\mathbf{K}) = \operatorname*{colim}_{\ell\to\infty} \pi_{n+\ell}(X \wedge K_\ell)$$

and

$$\widetilde{H}^n(X;\mathbf{K}) = \operatorname*{colim}_{\ell\to\infty}[S^\ell X; K_{n+\ell}]_0,$$

and the *homology and cohomology of a pair with coefficients in the spectrum* $\mathbf{K}$ to be the functor taking a pair of spaces (X, A) to the abelian group

$$H_n(X, A; \mathbf{K}) = \operatorname*{colim}_{\ell \to \infty} \pi_{n+\ell}((X_+/A_+) \wedge K_\ell)$$

and

$$H^n(X, A; \mathbf{K}) = \operatorname*{colim}_{\ell \to \infty} [S^\ell(X_+/A_+); K_{n+\ell}]_0.$$

It is a theorem that these are *generalized (co)homology theories*; they satisfy all the Eilenberg–Steenrod axioms except the dimension axiom. We will discuss this in more detail later.

For example, stable homotopy theory $\widetilde{H}_n(X; \mathbf{S}) = \pi_n^S X$ is a reduced homology theory; framed bordism $H_n(X; \mathbf{S}) = \pi_n^S X_+ = \Omega_n^{\text{fr}}(X)$ is an unreduced homology theory.

Note that $H_n(\text{pt}; \mathbf{K})$ can be non-zero for $n \neq 0$, for example $H_n(\text{pt}; \mathbf{S}) = \pi_n^S$. Ordinary homology is characterized by the fact that $H_n(\text{pt}) = 0$ for $n \neq 0$ (see Theorem 1.31). The groups $H_n(\text{pt}; \mathbf{K})$ are called the *coefficients* of the spectrum.

There are many relationships between reduced homology, unreduced homology, suspension, and homology of pairs, some of which are obvious and some of which are not. We list some facts for homology.

- For a based space X, $\widetilde{H}_n(X; \mathbf{K}) = \widetilde{H}_{n+1}(SX; \mathbf{K})$.
- For a space X, $H_n(X; \mathbf{K}) = \widetilde{H}_n(X_+; \mathbf{K})$.
- For a pair of spaces, $H_n(X, A; \mathbf{K}) \cong \widetilde{H}_n(X/A; \mathbf{K})$.
- For a CW-pair, $H_n(X, A; \mathbf{K})$ fits into the long exact sequence of a pair.

8.5. More general bordism theories

(Stably) framed bordism is a special case of a general bordism theory, where one considers bordisms respecting some *specific stable structure* on the normal bundle of a smooth manifold. We will give examples of stable structures now, and then ask you to supply a general definition in Exercise 136. Basically a property of vector bundles is stable if whenever a bundle η has that property, then so does $\eta \oplus \varepsilon^k$ for all k.

8.5.1. Framing. A stable framing on a bundle $[\eta]$ is, as we have seen, a choice of homotopy class of bundle isomorphism

$$\gamma : \eta \oplus \varepsilon^k \cong \varepsilon^{n+k}$$

subject to the equivalence relation generated by the requirement that

$$\gamma \sim \gamma \oplus \mathrm{Id} : \eta \oplus \varepsilon^k \oplus \varepsilon \cong \varepsilon^{n+k+1}.$$

8.5.2. The empty structure. This refers to bundles with no extra structure.

8.5.3. Orientation. This is weaker than requiring a framing. The most succinct way to define an orientation of an n-plane bundle η is to choose a homotopy class of trivialization of the highest exterior power of the bundle,

$$\gamma : \wedge^n(\eta) \cong \varepsilon.$$

Equivalently, an orientation is a reduction of the structure group to $GL_+(n, \mathbf{R})$, the group of n-by-n matrices with positive determinant. An oriented manifold is a manifold with an orientation on its tangent bundle.

Since $\wedge^{a+b}(V \oplus W)$ is canonically isomorphic to $\wedge^a V \otimes \wedge^b W$ if V is an a-dimensional vector space and W is a b-dimensional vector space, it follows that $\wedge^n(\eta)$ is canonically isomorphic to $\wedge^{n+k}(\eta \oplus \varepsilon^k)$ for any $k \geq 0$. Thus an orientation on η induces one on $\eta \oplus \varepsilon$, so an orientation is a well-defined stable property.

8.5.4. Spin structure. Let $\mathrm{Spin}(n) \to SO(n)$ be the double cover where $\mathrm{Spin}(n)$ is connected for $n > 1$. A *spin structure* on an n-plane bundle η over a space M is a reduction of the structure group to $\mathrm{Spin}(n)$. This is equivalent to giving a principal $\mathrm{Spin}(n)$-bundle $P \to M$ and an isomorphism $\eta \cong (P \times_{\mathrm{Spin}(n)} \mathbf{R}^n \to M)$. A *spin manifold* is a manifold whose tangent bundle has a spin structure. Spin structures come up in differential geometry and index theory.

The stabilization map $SO(n) \to SO(n+1)$ induces a map $\mathrm{Spin}(n) \to \mathrm{Spin}(n+1)$. Thus a principal $\mathrm{Spin}(n)$-bundle $P \to M$ induces a principal $\mathrm{Spin}(n+1)$-bundle $P \times_{\mathrm{Spin}(n)} \mathrm{Spin}(n+1) \to M$, and hence a spin structure on η gives a spin structure on $\eta \oplus \varepsilon$. A spin structure is a stable property.

A framing on a bundle gives a spin structure. A spin structure on a bundle gives an orientation. It turns out that a spin structure is equivalent to a framing on the 2-skeleton of M.

8.5.5. Stable complex structure. A *complex structure* on a bundle η is a bundle map $J : \eta \to \eta$ so that $J \circ J = -\mathrm{Id}$. This forces the (real) dimension of η to be even. Equivalently, a complex structure is a reduction of the structure group to $GL(k, \mathbf{C}) \subset GL(2k, \mathbf{R})$. The tangent bundle of a complex manifold admits a complex structure. One calls a manifold with a complex structure on its tangent bundle an *almost complex manifold.* An almost complex manifold may or may not admit the structure of a complex manifold. (It can be shown that S^6 is an almost complex manifold, but whether or not S^6 is a complex manifold is still an open question.)

One way to define a *stable complex structure* on a bundle η is as a section

$$J \in \Gamma(\mathrm{Hom}(\eta \oplus \varepsilon^k,\ \eta \oplus \varepsilon^k))$$

satisfying $J^2 = -\mathrm{Id}$ in each fiber. Given such a J, one can extend it canonically to

$$\widehat{J} = J \oplus i \in \Gamma(\mathrm{Hom}(\eta \oplus \varepsilon^k \oplus \varepsilon^{2\ell},\ \eta \oplus \varepsilon^k \oplus \varepsilon^{2\ell}))$$

by identifying $\varepsilon^{2\ell}$ with $M \times \mathbf{C}^\ell$ and using multiplication by i to define $i \in \Gamma(\mathrm{Hom}(M \times \mathbf{C}^\ell, M \times \mathbf{C}^\ell))$. As usual, two such structures are identified if they are homotopic. Note that odd-dimensional manifolds cannot have almost complex structures but may have stable almost complex structures.

If $\gamma : \eta \oplus \varepsilon^k \cong \varepsilon^\ell$ is a stable framing, up to equivalence we may assume that ℓ is even. Then identifying ε^ℓ with $M \times \mathbf{C}^{\ell/2}$ induces a stable complex structure on $\eta \oplus \varepsilon^k$. Thus stably framed bundles have a stable complex structure.

Similarly, a complex structure determines an orientation, since a complex vector space has a canonical (real) orientation. To see this, notice that if $\{e_1, \dots, e_r\}$ is a complex basis for a complex vector space, then $\{e_1, ie_1, \cdots, e_r, ie_r\}$ is a real basis whose orientation class is independent of the choice of the basis $\{e_1, \dots, e_r\}$.

The orthogonal group $O(n)$ is a strong deformation retract of the general linear group $GL(n, \mathbf{R})$; this can be shown using the Gram-Schmidt process. This leads to a one-to-one correspondence between isomorphism classes of vector bundles and isomorphism classes of $\mathbf{R}^n$-bundles with structure group $O(n)$ over a paracompact base space. An $\mathbf{R}^n$-bundle with a metric has structure group $O(n)$. Conversely an $\mathbf{R}^n$-bundle with structure group $O(n)$ over a connected base space admits a metric, uniquely defined up to scaling. Henceforth in this chapter all bundles will have metrics with orthogonal structure group.

The following exercise indicates how to define a structure on a stable bundle in general.

Exercise 136. Let $\mathbf{G} = \{G_n\}$ be a sequence of *topological groups* with continuous homomorphisms $G_n \to G_{n+1}$ and $G_n \to O(n)$ so that the diagram

$$\begin{array}{ccc} G_n & \longrightarrow & G_{n+1} \\ \downarrow & & \downarrow \\ O(n) & \hookrightarrow & O(n+1) \end{array}$$

commutes for each n, where the injection $O(n) \to O(n+1)$ is defined by

$$A \mapsto \begin{bmatrix} A & 0 \\ 0 & 1 \end{bmatrix}.$$

Use this to define a stable $\mathbf{G}$-structure on a bundle η. (Hint: either use classifying spaces or else consider the overlap functions for the stable bundle.)

Define what a homomorphism $\mathbf{G} \to \mathbf{G}'$ should be in such a way that a bundle with a stable $\mathbf{G}$-structure becomes a bundle with a stable $\mathbf{G}'$-structure.

There are many examples of $\mathbf{G}$-structures. As a perhaps unusual example, one could take G_n to be $O(n)$ or $SO(n)$ with the *discrete* topology. This spectrum arises in the study of flat bundles and algebraic K-theory.

For our previous examples, a framing corresponds to $G_n = 1$, the trivial group for all n. The empty structure corresponds to $G_n = O(n)$. An orientation corresponds to $G_n = SO(n) \subset O(n)$. A spin structure corresponds to $G_n = \text{Spin}(n) \to SO(n)$. A stable complex structure corresponds to $G_n = U([n/2]) \subset O(n)$.

Concepts such as orientation and almost complex structure are more natural on the tangent bundle, while the Pontrjagin-Thom construction and hence bordism naturally deal with the stable normal bundle. The following exercise generalizes Theorem 8.13 and shows that in some cases one can translate back and forth.

Exercise 137.

1. Show that an orientation on the stable tangent bundle of a manifold determines one on the stable normal bundle and conversely.
2. Show that a complex structure on the stable tangent bundle of a manifold determines one on the stable normal bundle and conversely.

(Hint/discussion: The real point is that the tangent bundle and normal bundle are (stably) Whitney sum inverses, so one may as well consider bundles α and β over a finite-dimensional base space with a framing of $\alpha \oplus \beta$. A complex structure on α is classified by a map to $G_n(\mathbf{C}^k)$ and β is equivalent to the pullback of the orthogonal complement of canonical bundle over the complex grassmannian, and hence β is equipped with a complex structure. Part 1 could be done using exterior powers or using the grassmannian of oriented n-planes in $\mathbf{R}^k$.)

Definition 8.21. Given a $\mathbf{G}$-structure, define the *n-th $\mathbf{G}$-bordism group of a space X* to be the $\mathbf{G}$-bordism classes of n-dimensional closed manifolds mapping to X with stable $\mathbf{G}$-structures on the normal bundle of an embedding of the manifold in a sphere. Denote this abelian group (with disjoint union as the group operation) by

$$\Omega_n^G(X).$$

Thus an element of $\Omega_n^G(X)$ is represented by an embedded closed submanifold $M^n \subset S^k$, a continuous map $f : M \to X$, and a stable $\mathbf{G}$-structure γ on the normal bundle $\nu(M \hookrightarrow S^k)$. Bordism is the equivalence relation generated by replacing k by $k+1$, and by

$$(M_0 \subset S^k, f_0, \gamma_0) \sim (M_1 \subset S^k, f_1, \gamma_1)$$

provided that there exist a compact manifold $W \subset S^k \times I$ with boundary $M_0 \times \{0\} \cup M_1 \times \{1\}$ (which we identify with $M_0 \amalg M_1$), a map $F : W \to X$ and a stable $\mathbf{G}$-structure Γ on $\nu(W \hookrightarrow S^k \times I)$ which restricts to $(M_0 \amalg M_1, f_0 \amalg f_1, \gamma_0 \amalg \gamma_1)$.

We previously used the notation Ω^{fr} for framed bordism, i.e. $\Omega^{\text{fr}} = \Omega^{\mathbf{1}}$ where $\mathbf{1} = \mathbf{G} = \{G_n\}$, the trivial group for all n.

We next want to associate spectra to bordism theories based on a stable structure. We have already seen how this works for framed bordism:

$$\Omega_n^{\text{fr}}(X) = \pi_n^S(X_+) = \operatorname*{colim}_{\ell \to \infty} \pi_{n+\ell}(X_+ \wedge S^\ell);$$

i.e. framed bordism corresponds to the sphere spectrum $\mathbf{S} = \{S^n, k_n\}$.

What do the other bordism theories correspond to? Does there exist a spectrum $\mathbf{K}$ for each structure $\mathbf{G}$ so that

$$\Omega_n^G(X) = H_n(X; \mathbf{K}) = \operatorname*{colim}_{\ell \to \infty} \pi_{n+\ell}(X_+ \wedge K_\ell)?$$

The answer is *yes*; the spectra for bordism theories are called *Thom spectra* $\mathbf{MG}$. In particular, one can define $\mathbf{G}$-*cobordism* by taking

$$H^n(X; \mathbf{MG}) = \operatorname*{colim}_{\ell \to \infty} [S^\ell X_+; MG_{n+\ell}]_0.$$

We are using the algebraic topology terminology where *cobordism* is the theory dual (in the Spanier-Whitehead sense) to *bordism.* It is traditional for geometric topologists to call bordant manifolds "cobordant", but we will avoid this terminology in this book.

Thus we know that $\mathbf{M1}$ is the sphere spectrum. We will give a construction for $\mathbf{MG}$ for any structure $\mathbf{G}$.

8.6. Classifying spaces

The construction of Thom spectral is accomplished most easily via the theory of classifying spaces. The basic result about classifying spaces is the following. The construction and the proof of this theorem is one of the student projects for Chapter 4.

Theorem 8.22. *Given any topological group G, there exists a principal G-bundle $EG \to BG$ where EG is a contractible space. The construction is* *functorial, so that any continuous group homomorphism* $\alpha : G \to H$ *induces a bundle map*

$$\begin{array}{ccc} EG & \xrightarrow{E\alpha} & EH \\ \downarrow & & \downarrow \\ BG & \xrightarrow{B\alpha} & BH \end{array}$$

compatible with the actions, so that if $x \in EG, g \in G$,

$$E\alpha(x \cdot g) = (E\alpha(x)) \cdot \alpha(g).$$

The space BG *is called a classifying space for* G.

The function

$$\Phi : Maps(B, BG) \to \{Principal\ G\text{-bundles over } B\}$$

defined by pulling back (so $\Phi(f) = f^*(EG)$*) induces a* bijection *from the homotopy set* $[B, BG]$ *to the set of isomorphism classes of principal* G*-bundles over* B*, when* B *is a CW-complex (or more generally a paracompact space).*

□

The long exact sequence for the fibration $G \to EG \to BG$ shows that $\pi_n BG = \pi_{n-1} G$. In fact, ΩBG is (weakly) homotopy equivalent to G, as one can see by taking the extended fiber sequence $\cdots \to \Omega EG \to \Omega BG \to G \to EG \to BG$, computing with homotopy groups, and observing that EG and ΩEG are contractible. Thus the space BG is a *delooping* of G.

The following lemma is extremely useful.

Lemma 8.23. *Let* $p : E \to B$ *be a principal* G*-bundle, and let* $f : B \to BG$ *be the classifying map. Then the homotopy fiber of* f *is weakly homotopy equivalent to* E.

Proof. Turn $f : B \to BG$ into a fibration $q : B' \to BG$ using Theorem 6.18 and let F' denote the homotopy fiber of $q : B' \to BG$. Thus there is a commutative diagram

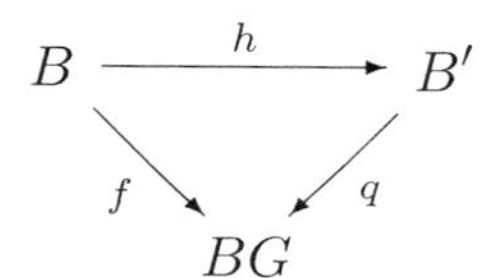

with h a homotopy equivalence. The fact that f is the classifying map for $p : E \to B$ implies that there is a commutative diagram

$$\begin{array}{ccc} E & \xrightarrow{\tilde{f}} & EG \\ {\scriptstyle p}\downarrow & & \downarrow \\ B & \xrightarrow{f} & BG \end{array}$$

and since EG is contractible, $f \circ p = q \circ h \circ p : E \to BG$ is nullhomotopic. By the homotopy lifting property for the fibration $q : B' \to BG$ it follows that $h \circ p : E \to B'$ is homotopic into the fiber F' of $q : B' \to BG$, and so one obtains a homotopy commutative diagram of spaces

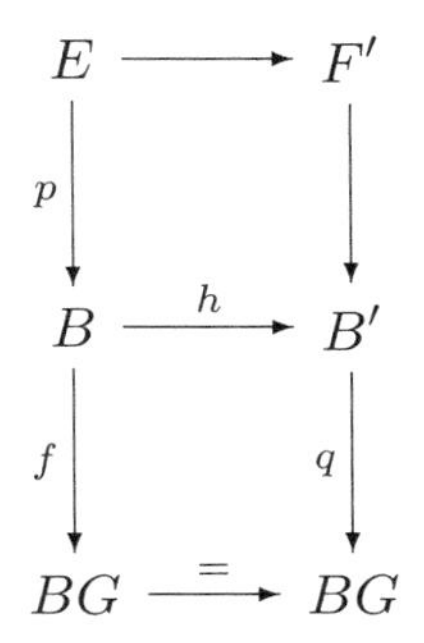

The left edge is a fibration, h is a homotopy equivalence, and by the five lemma the map $\pi_n(E) \to \pi_n(F')$ is an isomorphism for all n. □

In Lemma 8.23 one can usually conclude that the homotopy fiber of $f : B \to BG$ is in fact a homotopy equivalence. This would follow if we know that B' is homotopy equivalent to a CW-complex. This follows for most G by a theorem of Milnor [**24**].

Exercise 138. Show that given a principal G-bundle $E \to B$ there is a fibration

$$\begin{array}{ccc} E & \to & EG \times_G E \\ & & \downarrow \\ & & BG \end{array}$$

where $EG \times_G E$ denotes the Borel construction . How is this fibration related to the fibration of Lemma 8.23?

8.7. Construction of the Thom spectra

We proceed with the construction of the Thom spectra. We begin with a few preliminary notions.

Definition 8.24. If $E \to B$ is any vector bundle over a CW-complex B with metric then *the Thom space of* $E \to B$ is the quotient $D(E)/S(E)$, where $D(E)$ denotes the unit disk bundle of E and $S(E) \subset D(E)$ denotes the unit sphere bundle of E.

Notice that the zero section $B \to E$ defines an embedding of B into the Thom space.

The first part of the following exercise is virtually a tautology, but it is key to understanding why the spectra for bordism are given by Thom spaces.

Exercise 139.

1. If $E \to B$ is a smooth vector bundle over a smooth compact manifold B, then the Thom space of E is a smooth manifold away from one point and the 0-section embedding of B into the Thom space is a smooth embedding with normal bundle isomorphic to the bundle $E \to B$.
2. The Thom space of a vector bundle over a compact base is homeomorphic to the one-point compactification of the total space.

Now let a **G**-structure be given. Recall that this means we have a sequence of continuous groups G_n and homomorphisms $G_n \to O(n)$ and $G_n \to G_{n+1}$ such that the diagram

$$\begin{array}{ccc} G_n & \to & G_{n+1} \\ \downarrow & & \downarrow \\ O(n) & \hookrightarrow & O(n+1) \end{array}$$

commutes.

We will construct the Thom spectrum for this structure from the Thom spaces of vector bundles associated to the principal bundles $G_n \to EG_n \to BG_n$.

Composing the homomorphism $G_n \to O(n)$ with the standard action of $O(n)$ on $\mathbf{R}^n$ defines an action of G_n on $\mathbf{R}^n$. Use this action to form the universal $\mathbf{R}^n$-vector bundle over BG_n

$$\begin{array}{c} EG_n \times_{G_n} \mathbf{R}^n \\ \Big\downarrow \\ BG_n \end{array}$$

Let us denote this vector bundle by $V_n \to BG_n$. Notice that by our assumption that G_n maps to $O(n)$, this vector bundle has a metric, and so the unit sphere and disk bundles are defined.

Functoriality gives vector bundle maps (which are linear injections on fibers).

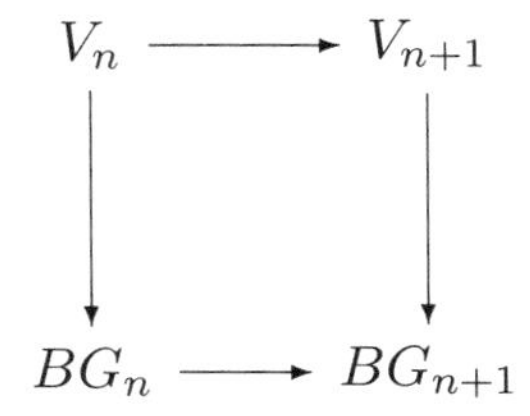

Let MG_n denote the Thom space of $V_n \to BG_n$. Thus MG_n is obtained by collapsing the unit sphere bundle of V_n in the unit disk bundle to a point.

Lemma 8.25.

1. *If $E \to B$ is a vector bundle, then the Thom space of $E \oplus \varepsilon$ is the reduced suspension of the Thom space of E.*
2. *A vector bundle map*

$$\begin{array}{ccc} E & \longrightarrow & E' \\ \downarrow & & \downarrow \\ B & \longrightarrow & B' \end{array}$$

which is an isomorphism preserving the metric on each fiber induces a map of Thom spaces.

Proof. To see why the first statement is true, note that an $O(n)$-equivariant homeomorphism $D^{n+1} \to D^n \times I$ determines a homeomorphism of $D(E \oplus \varepsilon)$ with $D(E) \times I$ which induces a homeomorphism $D(E \oplus \varepsilon)/S(E \oplus \varepsilon)$ with

$$(D(E) \times I)/(S(E) \times I \ \cup \ D(E) \times \{0, 1\}).$$

But it is easy to see that this identification space is the same as the (reduced) suspension of $D(E)/S(E)$.

The second statement is clear. □

The following theorem states that the collection $\mathbf{MG} = \{MG_n\}$ forms a spectrum and that the corresponding homology theory is the bordism theory defined by the corresponding structure.

Theorem 8.26. *The fiberwise injection $V_n \to V_{n+1}$ extends to a (metric preserving) bundle map $V_n \oplus \varepsilon \to V_{n+1}$ which is an isomorphism on each fiber and hence defines a map*

$$k_n : SMG_n \to MG_{n+1}.$$

Thus $\{MG_n, k_n\} = \mathbf{MG}$ is a spectrum, called the Thom spectrum.

Moreover, the bordism groups $\Omega_n^G(X)$ *are isomorphic to* $H_n(X;\mathbf{MG})$.

Proof. Since the diagram

$$\begin{array}{ccc} G_n & \rightarrow & G_{n+1} \\ \downarrow & & \downarrow \\ O(n) & \hookrightarrow & O(n+1) \end{array}$$

commutes, where $O(n) \hookrightarrow O(n+1)$ is the homomorphism

$$A \mapsto \begin{bmatrix} A & 0 \\ 0 & 1 \end{bmatrix},$$

it follows by the construction of V_n that the pullback of V_{n+1} by the map $\gamma_n : BG_n \to BG_{n+1}$ splits canonically into a direct sum $\gamma_n^*(V_{n+1}) = V_n \oplus \varepsilon$. Thus the diagram

$$\begin{array}{ccc} V_n & \longrightarrow & V_{n+1} \\ \downarrow & & \downarrow \\ BG_n & \longrightarrow & BG_{n+1} \end{array}$$

extends to a diagram

$$\begin{array}{ccc} V_n \oplus \varepsilon & \longrightarrow & V_{n+1} \\ \downarrow & & \downarrow \\ BG_n & \longrightarrow & BG_{n+1} \end{array}$$

which is an isomorphism on each fiber; this isomorphism preserves the metrics since the actions are orthogonal.

By Lemma 8.25, the above bundle map defines a map

$$k_n : SMG_n \to MG_{n+1},$$

establishing the first part of the theorem.

We now outline how to establish the isomorphism

$$\Omega_n^G(X) = \operatorname*{colim}_{\ell\to\infty} \pi_{n+\ell}(X_+ \wedge MG_\ell).$$

This is a slightly more complicated version of the Pontrjagin-Thom construction we described before, using the basic property of classifying spaces.

We will first define the collapse map

$$c : \Omega_n^G(X) \to \operatorname*{colim}_{\ell\to\infty} \pi_{n+\ell}(X_+ \wedge MG_\ell).$$

Suppose $[W, f, \gamma] \in \Omega_n^G(X)$. So W is an n-manifold with a $\mathbf{G}$-structure on its stable normal bundle, and $f : W \to X$ is a continuous map. Embed W in $S^{n+\ell}$ for some large ℓ so that the normal bundle $\nu(W)$ has a G_ℓ-structure.

Let $F \to W$ be the principal $O(\ell)$-bundle of orthonormal frames in $\nu(W)$. The statement that $\nu(W)$ has a G_ℓ-structure is equivalent to saying that there is a principal G_ℓ-bundle $P \to W$ and a bundle map

$$\begin{array}{ccc} P & \longrightarrow & F \\ & \searrow \quad \swarrow & \\ & W & \end{array}$$

which is equivariant with respect to the homomorphism

$$G_\ell \to O(\ell).$$

Let $c_1 : W \to BG_\ell$ classify the principal bundle $P \to W$. Then by definition $\nu(W)$ is isomorphic to the pullback $c_1^*(V_\ell)$. Let U be a tubular neighborhood of W in $S^{n+\ell}$ and $D \subset U \subset S^{n+\ell}$ correspond to the disk bundle. Define a map

$$h : S^{n+\ell} \to MG_\ell$$

by taking everything outside of D to the base point, and on D, take the composite

$$D \cong D(\nu(W)) \to D(V_\ell) \to MG_\ell.$$

The product

$$f \times h : S^{n+\ell} \to X \times MG_\ell$$

composes with the collapse

$$X \times MG_\ell \to X_+ \wedge MG_\ell$$

to give a map

$$\alpha = f \wedge h : S^{n+\ell} \to X_+ \wedge MG_\ell.$$

We have thus defined the collapse map

$$c : \Omega_n^G(X) \to \operatorname*{colim}_{\ell \to \infty} \pi_{n+\ell}(X_+ \wedge MG_\ell) = H_n(X; \mathbf{MG}).$$

To motivate the definition of the inverse of c, we will make a few comments on the above construction. The figure below illustrates that the composite of the zero section $z : BG_\ell \to D(V_\ell)$ and the quotient map $D(V_\ell) \to MG_\ell$ is an embedding.

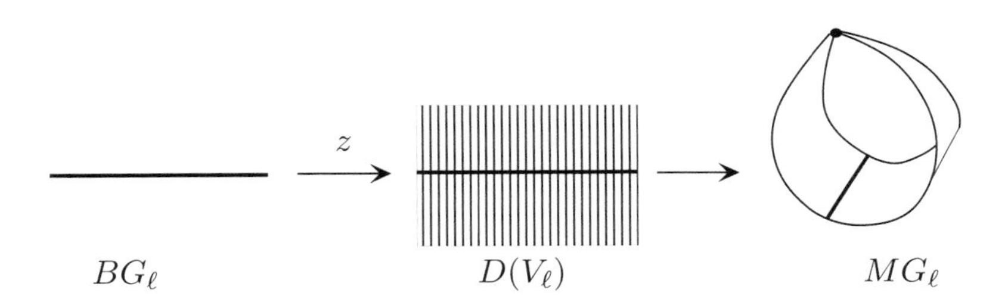

We thus will consider BG_ℓ to be a subset of MG_ℓ. Then in the above construction of the collapse map c, $W = \alpha^{-1}(X \times BG_\ell)$.

Next we use transversality to define the inverse of the collapse map c. Represent $\hat{\alpha} \in H_n(X; \mathbf{MG})$ by

$$\alpha : S^{n+\ell} \to X_+ \wedge MG_\ell.$$

Observe that the composite

$$X \times BG_\ell \hookrightarrow X_+ \times MG_\ell \to X_+ \wedge MG_\ell$$

is an embedding, since:

1. BG_ℓ misses the base point of MG_ℓ, and
2. the base point of X_+ misses X.

(The following figure gives an analogue by illustrating the embedding of $X \times B$ in $X_+ \wedge M$ if B is a point, M is a D^2-bundle over B, and X is an interval.)

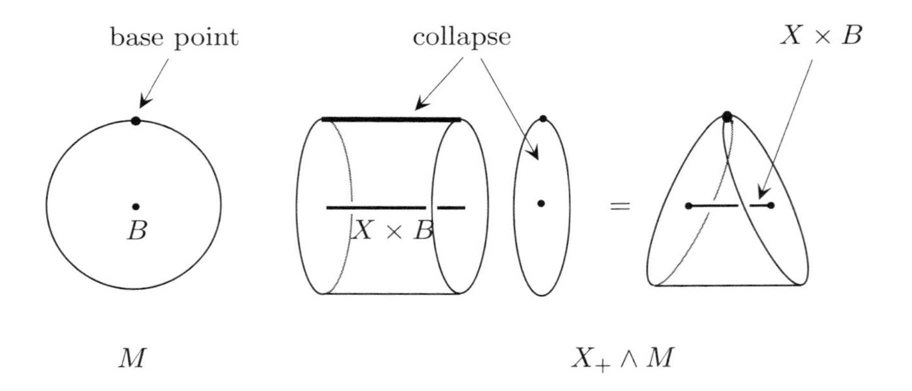

Furthermore,

$$X \times BG_\ell \subset X_+ \wedge MG_\ell$$

has a neighborhood which is isomorphic to the pullback $\pi_2^* V_\ell$ where $\pi_2 : X \times BG_\ell \to BG_\ell$ is the projection on the second factor. Transversality, adapted to this setting, says that $\alpha : S^{n+\ell} \to X_+ \wedge MG_\ell$ can be homotoped slightly to a map β so that $W = \beta^{-1}(X \times BG_\ell)$ is a smooth manifold whose tubular neighborhood, i.e. the normal bundle of W, has a $\mathbf{G}$-structure. The composite of $\beta : W \to X \times BG_\ell$ and $\mathrm{pr}_1 : X \times BG_\ell \to X$ gives the desired element $(W \to X) \in \Omega_n^G(X)$.

We sort of rushed through the construction of the inverse map to c, so we will backtrack and discuss some details. For every point in BG_ℓ, there is a neighborhood $U \subset BG_\ell$ over which the bundle $V_\ell \to BG_\ell$ is trivial, and so there is a map

$$\alpha^{-1}(X \times U) \to D^\ell/S^\ell$$

defined by composing α with projection on the fiber. Transversality then applies to this map between manifolds and one can patch together to get β using partitions of unity. Furthermore, transversality gives a diagram of bundle maps, isomorphisms in each fiber,

$$\begin{array}{ccccc} \nu(W \hookrightarrow S^{n+\ell}) & \longrightarrow & X \times V_\ell & \longrightarrow & V_\ell \\ \downarrow & & \downarrow & & \downarrow \\ W & \xrightarrow{\beta} & X \times BG_\ell & \xrightarrow{\mathrm{pr}_2} & BG_\ell \end{array}$$

so that the normal bundle of W inherits a **G**-structure.

Next note that replacing ℓ by $\ell+1$ leads to the same bordism element. Stabilizing the normal bundle

$$\nu(W \hookrightarrow S^{n+\ell}) \longrightarrow \nu(W \hookrightarrow S^{n+\ell}) \oplus \varepsilon = \nu(W \hookrightarrow S^{n+\ell+1})$$

corresponds to including $W \subset S^{n+\ell} \subset S^{n+\ell+1}$. Since the composite

$$SS^{n+\ell} \xrightarrow{Sf} S(X_+ \wedge MG_\ell) \xrightarrow{k_\ell \wedge Id} X_+ \wedge MG_{\ell+1}$$

replaces the tubular neighborhood of $X \times BG_\ell$, i.e. $X \times V_\ell$ by $X \times (V_\ell \oplus \varepsilon)$, the construction gives a well-defined stable **G**-structure on the stable normal bundle of W.

The full proof that the indicated map $H_n(X; \mathbf{MG}) \to \Omega_n^G(X)$ is well-defined and is the inverse of c is a careful but routine check of details involving bordisms, homotopies, and stabilization. □

Taking X to be a point, we see that the groups (called the coefficients) $\Omega_n^G = \Omega_n^G(\mathrm{pt})$ are isomorphic to the homotopy groups $\operatorname{colim}_{\ell\to\infty} \pi_{n+\ell}(MG_\ell)$, since $\mathrm{pt}_+ \wedge M = M$.

As an example of how these coefficients can be understood geometrically, consider oriented bordism, corresponding to $G_n = SO(n)$. The coefficients Ω_n^{SO} equal $\pi_{n+\ell}(MSO_\ell)$ for ℓ large enough. Here are some basic computations:

1. An oriented closed 0-manifold is just a signed finite number of points. This bounds a 1-manifold if and only if the sum of the signs is zero. Hence $\Omega_0^{SO} \cong \mathbf{Z}$. Also, $\pi_\ell MSO_\ell = \mathbf{Z}$ for $\ell \geq 2$.
2. Every oriented closed 1-manifold bounds an oriented 2-manifold, since $S^1 = \partial D^2$. Therefore $\Omega_1^{SO} = 0$.

3. Every oriented 2-manifold bounds an oriented 3-manifold since any oriented 2-manifold embeds in $\mathbf{R}^3$ with one of the two complementary components compact. Thus $\Omega_2^{SO} = 0$.
4. A theorem of Rohlin states that every oriented 3-manifold bounds a 4-manifold. Thus $\Omega_3^{SO} = 0$.
5. An oriented 4-manifold has a signature in $\mathbf{Z}$, i.e. the signature of its intersection form. A good exercise using Poincaré duality (see the projects for Chapter 3) shows that this is an oriented bordism invariant and hence defines a homomorphism $\Omega_4^{SO} \to \mathbf{Z}$. This turns out to be an isomorphism. More generally the signature defines a map $\Omega_{4k}^{SO} \to \mathbf{Z}$ for all k. This is a surjection since the signature of $\mathbf{C}P^{2k}$ is 1.
6. It is a fact that away from multiples of 4, the oriented bordism groups are torsion; i.e. $\Omega_n^{SO} \otimes \mathbf{Q} = 0$ if $n \neq 4k$.
7. For all n, Ω_n^{SO} is finitely generated, in fact, a finite direct sum of $\mathbf{Z}$'s and $\mathbf{Z}/2$'s.

Statements 5, 6, and 7 can be proven by computing $\pi_{n+\ell}(MSO_\ell)$. How does one do this? A starting point is the *Thom isomorphism theorem*, which says that for all k,

$$H_n(BSO(\ell)) \cong \tilde{H}_{n+\ell}(MSO_\ell)$$

(where $\tilde{H}$ denotes reduced cohomology). The cohomology of $BSO(n)$ can be studied in several ways, and so one can obtain information about the cohomology of MSO_ℓ by this theorem. Combining this with the Hurewicz theorem and other methods leads ultimately to a complete computation of oriented bordism (due to C.T.C. Wall), and this technique was generalized by Adams to a machine called the Adams spectral sequence. We will return to the Thom isomorphism theorem in Chapter 10.

Once the coefficients are understood, one can use the fact that bordism is a homology theory to compute $\Omega_n^{SO}(X)$. For now we just remark that there is a map $\Omega_n^{SO}(X) \to H_n(X)$ defined by taking $f : M \to X$ to the image of the fundamental class $f_*[M]$. Thus, for example, the identity map on a closed, oriented manifold M^n is non-zero in $\Omega_n^{SO}(M)$.

We can also make an elementary remark about unoriented bordism, which corresponds to $G_n = O(n)$. Notice first that for any $\alpha \in \Omega_n^O(X)$, $2\alpha = 0$. Indeed, if $f : V^n \to X$ represents α, take $F : V \times I \to X$ to be $F(x,t) = f(x)$; then $\partial(V \times I, F) = 2(V, f)$. Thus $\Omega_n^O(X)$ consists only of elements of order 2. The full computation of unoriented bordism is due to Thom. We will discuss this more in Section 10.10.

Exercise 140. Show that $\Omega_0^O = \mathbf{Z}/2$, $\Omega_1^O = 0$, and $\Omega_2^O = \mathbf{Z}/2$. (Hint: for Ω_2^O use the classification theorem for closed surfaces; then show that if a surface F is a boundary of a 3-manifold, $\dim H^1(F;\mathbf{Z}/2)$ is even.)

There are several conventions regarding notation for bordism groups; each has its advantages. Given a structure defined by a sequence $\mathbf{G} = \{G_n\}$, one can use the notation

$$\Omega_*^G(X), \quad H_*(X;\mathbf{MG}) \text{ or } MG_*(X).$$

There is a generalization of a $\mathbf{G}$-structure called a $\mathbf{B}$-structure. It is given by a sequence of commutative diagrams

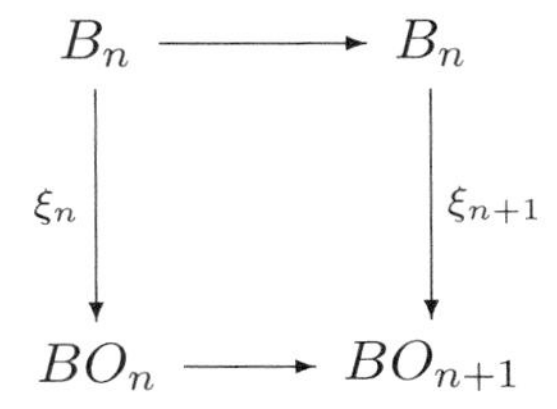

where the vertical maps are fibrations. A $\mathbf{G}$-structure in the old sense gives a $\mathbf{BG} = \{BG_n\}$-structure. A $\mathbf{B}$-structure has a Thom spectrum $\mathbf{TB} = \{T(\xi_n)\}$, where ξ_n here denotes the vector bundle pulled back from the canonical bundle over BO_n. There is a notion of a stable $\mathbf{B}$-structure on a normal bundle of an embedded M, which implies that there is a map from the (stablized) normal bundle to ξ_k. There is a Pontrjagin-Thom isomorphism

$$\Omega_n^B(X) \cong H_n(X;\mathbf{TB}).$$

For a precise discussion of $\mathbf{B}$-bordism and for further information on bordism in general, see [**30**], [**39**] and the references therein.

8.8. Generalized homology theories

We have several functors from (based) spaces to graded abelian groups: stable homotopy $\pi_n^S(X)$, bordism $\Omega_n^G(X)$, or, more generally, homology of a space with coefficients in a spectrum $H_n(X;\mathbf{K})$. These are examples of *generalized homology theories.* Generalized homology theories come in two (equivalent) flavors, *reduced* and *unreduced.* Unreduced theories apply to unbased spaces and pairs. Reduced theories are functors on based spaces. The equivalence between the two points of view is obtained by passing from (X,A) to X/A and from X to X_+.

There are three high points to look out for in our discussion of homology theories.

- The axioms of a (co)homology theory are designed for computations. One first computes the coefficients of the theory (perhaps using the

Adams spectral sequence) and then computes the homology of a CW-complex X, using excision, Mayer–Vietoris, or a generalization of cellular homology discussed in the next chapter, the Atiyah–Hirzebruch spectral sequence.

- There is a uniqueness theorem. A natural transformation of (co)homology theories inducing an isomorphism on coefficients induces an isomorphism for all CW-complexes X.
- A (co)homology theory is given by (co)homology with coefficients in a spectrum $\mathbf{K}$.

8.8.1. Reduced homology theories.

Let $\mathcal{K}_*$ be the category of compactly generated spaces with non-degenerate base points.

Definition 8.27. A *reduced homology theory* is

1. A family of covariant functors
$$h_n : \mathcal{K}_* \to \mathcal{A} \text{ for } n \in \mathbf{Z}$$
where $\mathcal{A}$ denotes the category of abelian groups. (Remark: We *do not* assume h_n is zero for $n < 0$.)
2. A family of natural transformations
$$e_n : h_n \to h_{n+1} \circ \mathcal{S}$$
where $S : \mathcal{K}_* \to \mathcal{K}_*$ is the (reduced) suspension functor.

These must satisfy the three following axioms:

A1. *(Homotopy)* If $f_0, f_1 : X \to Y$ are homotopic, then
$$h_n(f_0) = h_n(f_1) : h_n(X) \to h_n(Y).$$

A2. *(Exactness)* For $f : X \to Y$, let C_f be the mapping cone of f, and $j : Y \hookrightarrow C_f$ the inclusion. Then
$$h_n(X) \xrightarrow{h_n(f)} h_n(Y) \xrightarrow{h_n(j)} h_n(C_f)$$
is exact for all $n \in \mathbf{Z}$.

A3. *(Suspension)* The homomorphism
$$e_n(X) : h_n(X) \to h_{n+1}(SX)$$
given by the natural transformation e_n is an isomorphism for all $n \in \mathbf{Z}$.

Exercise 141. Show that ordinary singular homology defines a homology theory in this sense by taking $h_n(X)$ to be the reduced homology of X.

There are two other "nondegeneracy" axioms which a given generalized homology theory may or may not satisfy.

A4. *(Additivity)* If X is a wedge product $X = \bigvee_{j \in J} X_j$, then

$$\bigoplus_{j \in J} h_n(X_j) \to h_n(X)$$

is an isomorphism for all $n \in \mathbf{Z}$.

A5. *(Isotropy)* If $f : X \to Y$ is a weak homotopy equivalence, then $h_n(f)$ is an isomorphism for all $n \in \mathbf{Z}$.

If we work in the category of based CW-complexes instead of $\mathcal{K}_*$, then Axiom A5 follows from Axiom A1 by the Whitehead theorem. Given a reduced homology theory on based CW-complexes, it extends uniquely to an isotropic theory on $\mathcal{K}_*$.

For any reduced homology theory, $h_n(\mathrm{pt}) = 0$ for all n, since

$$h_n(\mathrm{pt}) \to h_n(\mathrm{pt}) \to h_n(\mathrm{pt}/\mathrm{pt}) = h_n(\mathrm{pt})$$

is exact, but also each arrow is an isomorphism. Thus the reduced homology of a point says nothing about the theory; instead one makes the following definition.

Definition 8.28. The *coefficients* of a reduced homology theory are the groups $\{h_n(S^0)\}$.

A homology theory is called *ordinary* (or *proper*) if it satisfies

$$h_n(S^0) = 0 \text{ for } n \neq 0.$$

(This is the dimension axiom of Eilenberg–Steenrod.) Singular homology with coefficients in an abelian group A is an example of an ordinary theory. It follows from a simple argument using the Atiyah-Hirzebruch spectral sequence that any ordinary reduced homology theory is isomorphic to reduced singular homology with coefficients in $A = h_0(S^0)$.

If (X, A) is an NDR pair, then we saw in Chapter 6 that the mapping cone C_f is homotopy equivalent to X/A. Thus $h_n(A) \to h_n(X) \to h_n(X/A)$ is exact. Also in Chapter 6 we proved that the sequence

$$A \to X \to X/A \to SA \to SX \to S(X/A) \to \cdots$$

has each three term sequence a (homotopy) cofibration. Thus

$$h_n(A) \to h_n(X) \to h_n(X/A) \to h_n(SA) \to h_n(SX) \to \cdots$$

is exact. Applying the transformations e_n and using Axiom A3, we conclude that

$$\to h_n(A) \to h_n(X) \to h_n(X/A) \to h_{n-1}(A) \to h_{n-1}(X) \to \cdots$$

is exact. Thus to any reduced homology theory one obtains a long exact sequence associated to a cofibration.

Exercise 142. Let X be a based CW-complex with subcomplexes A and B, both of which contain the base point. Show that for any reduced homology theory h_* there is a Mayer–Vietoris long exact sequence

$$\cdots \to h_n(A\cap B) \to h_n(A)\oplus h_n(B) \to h_n(X) \to h_{n-1}(A\cap B) \to \cdots .$$

8.8.2. Unreduced homology theories. We will derive unreduced theories from reduced theories to emphasize that these are the same concept presented slightly differently.

Let $\mathcal{K}^2$ denote the category of NDR pairs (X, A), allowing the case when A is empty. Given a reduced homology theory $\{h_n, e_n\}$ define functors H_n on $\mathcal{K}^2$ as follows (for this discussion, H_n *does not* denote ordinary singular homology!).

1. Let
$$H_n(X,A) = h_n(X_+/A_+) = \begin{cases} h_n(X/A) & \text{if } A \neq \phi, \\ h_n(X_+) & \text{if } A = \phi \end{cases}$$

2. Let $\partial_n : H_n(X,A) \to H_{n-1}(A)$ be the composite:
$$H_n(X,A) = h_n(X_+/A_+) \xrightarrow{\cong} h_n(C_i) \longrightarrow h_n(SA_+) \xrightarrow{\cong} h_{n-1}(A_+) = H_{n-1}(A)$$
where C_i is the mapping cone of the inclusion $i : A_+ \hookrightarrow X_+$, and $C_i \to SA_+$ is the quotient
$$C_i \to C_i/X_+ = SA_+.$$

Then $\{H_n, \partial_n\}$ satisfy the Eilenberg–Steenrod axioms:

A1. *(Homotopy)* If $f_0, f_1 : (X,A) \to (Y,B)$ are (freely) homotopic, then
$$H_n(f_0) = H_n(f_1) : H_n(X,A) \to H_n(Y,B).$$

A2. *(Exactness)* For a cofibration $i : A \hookrightarrow X$, let $j : (X,\phi) \hookrightarrow (X,A)$; then
$$\cdots \to H_{n+1}(X,A) \xrightarrow{\partial_{n+1}} H_n(A) \xrightarrow{H_n(i)} H_n(X) \xrightarrow{H_n(j)} H_n(X,A) \to \cdots$$
is exact.

A3. *(Excision)* Suppose that $X = A\cup B$, with A, B closed, and suppose that $(A, A\cap B)$ is an NDR pair. Then
$$H_n(A, A\cap B) \to H_n(X,B)$$
is an isomorphism for all $n \in \mathbf{Z}$.

Exercise 143. Prove that these three properties hold using the axioms of a reduced theory.

If a reduced theory is additive and/or isotropic, the functors H_n likewise satisfy

A4. *(Additivity)* Let $X = \amalg_{j\in J} X_j,\ A \subset X,\ A_j = X_j \cap A$. Then

$$\bigoplus_{j\in J} H_n(X_j, A_j) \to H_n(X, A)$$

is an isomorphism for all $n \in \mathbf{Z}$.

A5. *(Isotropy)* If $f : X \to Y$ is a weak homotopy equivalence, then $H_n(f) : H_n(X) \to H_n(Y)$ is an isomorphism for all $n \in \mathbf{Z}$.

Notice that if the reduced theory is ordinary, then $H_n(\mathrm{pt}) = 0$ for $n \neq 0$.

One uses these properties to define an unreduced homology theory.

Definition 8.29. A collection of functors $\{H_n, \partial_n\}$ on $\mathcal{K}^2$ is called an *(unreduced) homology theory* if it satisfies the three axioms A1, A2, and A3. It is called *additive* and/or *isotropic* if Axiom A4 and/or Axiom A5 hold. It is called *ordinary* or *proper* if $H_n(\mathrm{pt}) = 0$ for $n \neq 0$.

The *coefficients* of the unreduced homology theory are $\{H_n(\mathrm{pt})\}$.

One can go back and forth: an unreduced homology theory $\{H_n, \partial_n\}$ defines a reduced one by taking $h_n(X) = H_n(X, \{*\})$. The following theorem is proved in [**43**, Section XII.6].

Theorem 8.30. *These constructions set up a $1-1$ correspondence (up to natural isomorphism) between reduced homology theories on $\mathcal{K}_*$ and (unreduced) homology theories on $\mathcal{K}^2$. Moreover the reduced theory is additive, isotropic, or ordinary if and only if the corresponding unreduced theory is.*

□

The uniqueness theorem below has an easy inductive cell-by-cell proof in the case of finite CW-complexes, but requires a more delicate limiting argument for infinite CW-complexes.

Theorem 8.31 (Eilenberg–Steenrod uniqueness theorem)**.**

1. *Let $T : (H_n, \partial_n) \to (H'_n, \partial'_n)$ be a natural transformation of homology theories defined on the category of finite CW-pairs such that $T : H_*(\mathrm{pt}) \to H'_*(\mathrm{pt})$ is an isomorphism. Then $T : H_*(X, A) \to H'_*(X, A)$ is an isomorphism for all finite CW-pairs.*
2. *Let $T : (H_n, \partial_n) \to (H'_n, \partial'_n)$ be a natural transformation of additive homology theories defined on the category of CW-pairs where $T : H_*(\mathrm{pt}) \to H'_*(\mathrm{pt})$ is an isomorphism. Then $T : H_*(X, A) \to H'_*(X, A)$ is an isomorphism for all CW-pairs.*

8.8.3. Homology theories and spectra.

Theorem 8.32. *(Reduced) homology with coefficients in a spectrum* $\mathbf{K}$

$$\widetilde{H}_n(-;\mathbf{K}) : X \mapsto \operatorname*{colim}_{\ell\to\infty} \pi_{n+\ell}(X \wedge K_\ell)$$
$$H_n(-;\mathbf{K}) : (X, A) \mapsto \operatorname*{colim}_{\ell\to\infty} \pi_{n+\ell}((X_+/A_+) \wedge K_\ell)$$

is a (reduced) homology theory satisfying the additivity axiom.

One needs to prove the axioms A1, A2, A3, and A5. The homotopy axiom is of course obvious. The axiom A2 follows from the facts about the Puppe sequences we proved in Chapter 6 by passing to the limit. The suspension axiom holds almost effortlessly from the fact that the theory is defined by taking the direct limit (i.e. colimit) over suspension maps. The additivity axiom follows from the fact that the image of a sphere is compact and that a compact subspace of an infinite wedge is contained in a finite wedge.

A famous theorem of E. Brown (the Brown representation theorem) gives a converse of the above theorem. It leads to a shift in perspective on the functors of algebraic topology by prominently placing spectra as the source of homology theories. Here is a precise statement.

Theorem 8.33.

1. *Let* $\{H_n, \partial_n\}$ *be a homology theory. There there exists a spectrum* $\mathbf{K}$ *and a natural isomorphism* $H_n(X, A) \cong H_n(X, A; \mathbf{K})$ *for all finite CW-pairs.*
2. *Let* $\{H_n, \partial_n\}$ *be an additive homology theory. There there exists a spectrum* $\mathbf{K}$ *and a natural isomorphism* $H_n(X, A) \cong H_n(X, A; \mathbf{K})$ *for all CW-pairs.*

We have seen several examples: an ordinary homology theory corresponds to the Eilenberg–MacLane spectrum $\mathbf{K}(A)$, stable homotopy corresponds to the sphere spectrum $\mathbf{S}$, and the bordism theories correspond to Thom spectra. Note that the Brown representation theorem shows that for any homology theory there is a spectrum and hence an associated generalized cohomology theory.

Exercise 144. Give a definition of a map of spectra. Define maps of spectra $\mathbf{S} \to \mathbf{K}(\mathbf{Z})$ and $\mathbf{S} \to \mathbf{MG}$ inducing the Hurewicz map $\pi_n^S(X) \to \widetilde{H}_n(X)$ and the map $\Omega_n^{\mathrm{fr}}(X) \to \Omega_n^G(X)$ from framed to G-bordism.

8.8.4. Generalized cohomology theories. The development of cohomology theories parallels that of homology theories following the principle of reversing arrows.

Exercise 145. Define reduced and unreduced *cohomology* theories.

There is one surprise however. In order for $H^n(\ ;\mathbf{K})$ to be an additive theory (which means the cohomology of a disjoint union is a direct *product*), one must require that $\mathbf{K}$ is an *Ω-spectrum*, a spectrum so that the adjoints

$$K_n \to \Omega K_{n+1}$$

of the structure maps k_n are homotopy equivalences. Conversely, the Brown representation theorem applied to an additive cohomology theory produces an Ω-spectrum. The Eilenberg–MacLane spectrum is an Ω-spectrum while the sphere spectrum or more generally bordism spectra are not.

An important example of a generalized cohomology theory is topological K-theory. It is the subject of one of the projects at the end of this chapter. Complex topological K-theory has a definition in terms of stable equivalence classes of complex vector bundles, but we instead indicate the definition in terms of a spectrum. Most proofs of the Bott periodicity theorem (Theorem 6.51, which states that $\pi_n U \cong \mathbf{Z}$ for n odd and $\pi_n U = 0$ for n even) actually prove a stronger result, that there is a homotopy equivalence

$$\mathbf{Z} \times BU \simeq \Omega^2(\mathbf{Z} \times BU).$$

This allows the definition of the complex K-theory spectrum with

$$K_n = \begin{cases} \mathbf{Z} \times BU & \text{if } n \text{ is even,} \\ \Omega(\mathbf{Z} \times BU) & \text{if } n \text{ is odd.} \end{cases} \tag{8.1}$$

The structure maps k_n

$$S(\mathbf{Z} \times BU) \to \Omega(\mathbf{Z} \times BU)$$
$$S\Omega(\mathbf{Z} \times BU) \to \mathbf{Z} \times BU$$

are given by the adjoints of the Bott periodicity homotopy equivalence and the identity map

$$\mathbf{Z} \times BU \to \Omega^2(\mathbf{Z} \times BU)$$
$$\Omega(\mathbf{Z} \times BU) \to \Omega(\mathbf{Z} \times BU).$$

Thus the complex K-theory spectrum is an Ω-spectrum. The corresponding cohomology theory is called complex K-theory and satisfies

$$K^n(X) = K^{n+2}(X) \quad \text{for all } n \in \mathbf{Z}.$$

In particular this is a non-connective cohomology theory, where a *connective* cohomology theory is one that satisfies $H^n(X) = 0$ for all $n < n_0$. Ordinary homology, as well as bordism theories, are connective, since a manifold of negative dimension is empty.

A good reference for the basic results in the study of spectra (stable homotopy theory) is Adams' book [**2**].

8.9. Projects for Chapter 8

8.9.1. Basic notions from differential topology.

Define a smooth manifold and submanifold, the tangent bundle of a smooth manifold, a smooth map between manifolds and its differential, an isotopy, the Sard theorem, transversality, the tubular neighborhood theorem, and the decomposition

$$TM|_P = TP \oplus \nu(P \hookrightarrow M),$$

where $P \subset M$ is a smooth submanifold, and show that if $f : M \to N$ is a smooth map transverse to a submanifold $Q \subset N$, with $P = f^{-1}(Q)$, then the differential of f induces a bundle map $df : \nu(P \hookrightarrow M) \to \nu(Q \hookrightarrow n)$ which is an isomorphism in each fiber. A good reference is Hirsch's book [**17**].

8.9.2. Definition of K-theory.

Define complex (topological) K-theory of a space in terms of vector bundles. Indicate why the spectrum for this theory is $\{K_n\}$ given in Equation (8.1). State the Bott periodicity theorem. Discuss vector bundles over spheres. Discuss real K-theory. References for this material are the books by Atiyah [**3**] and Husemoller [**18**].

8.9.3. Spanier-Whitehead duality.

Spanier-Whitehead duality is a generalization of Alexander duality which gives a geometric method of going back and forth between a generalized homology theory and a generalized cohomology theory. Suppose that $X \subset S^{n+1}$ is a finite simplicial complex, and let $Y = S^{n+1} - X$, or better, $Y = S^{n+1} - U$ where U is some open simplicial neighborhood of X which deformation retracts to X. Recall that Alexander duality implies that

$$\widetilde{H}^p(X) \cong \widetilde{H}_{n-p}(Y).$$

(See Theorem 3.26.) What this means is that the cohomology of X determines the homology of Y and vice versa.

The strategy is to make this work for generalized cohomology theories and any space X and to remove the dependence on the embedding. The best way to do this is to do it carefully using spectra. Look at Spanier's article [**35**]. There is a good sequence of exercises developing this material in [**36**, pages 462-463]. Another reference using the language of spectra is [**39**, page 321].

Here is a slightly low-tech outline. You should lecture on the following, providing details.

Given based spaces X and Y, let

$$\{X, Y\} = \operatorname*{colim}_{k\to\infty}[S^k \wedge X, S^k \wedge Y]_0.$$

Given a finite simplicial subcomplex $X \subset S^{n+1}$, let $D_n X \subset S^{n+1}$ be a finite simplicial subcomplex which is a deformation retract of $S^{n+1} - X$. Then $SD_n X$ is homotopy equivalent to $S^{n+2} - X$.

For k large enough, the homotopy type of the suspension $S^k D_n X$ depends only on X and $k + n$, and not on the choice of embedding into S^{n+1}. Moreover, for any spaces Y and Z

$$\{S^q Y, D_n X \wedge Z\} = \{S^{q-n} Y \wedge X, Z\}. \tag{8.2}$$

As an example, taking $Y = S^0$ and $Z = K(\mathbf{Z}, p + q - n)$, Equation (8.2) says that

$$\{S^q, D_n X \wedge K(\mathbf{Z}, p + q - n)\} = \{S^{q-n} \wedge X, K(\mathbf{Z}, p + q - n)\}. \tag{8.3}$$

Definition 8.20 says that the left side of Equation (8.3) is $\widetilde{H}_{n-p}(D_n X; \mathbf{K}(\mathbf{Z}))$. The right side is $\widetilde{H}^p(X; \mathbf{Z})$, using the fact that

$$[SA, K(\mathbf{Z}, k)] = [A, \Omega K(\mathbf{Z}, k)] = [A, K(\mathbf{Z}, k - 1)].$$

What this means is that by combining Alexander duality, the result $H^q(X) = [X, K(\mathbf{Z}, q)]$ of obstruction theory, and Spanier-Whitehead duality (i.e. Equation (8.2)), the definition of homology with coefficients in the Eilenberg–MacLane spectrum given in Definition 8.20 coincides with the usual definition of (ordinary) homology (at least for finite simplicial complexes, but this works more generally).

This justifies Definition 8.20 of homology with coefficients in an arbitrary spectrum $\mathbf{K}$. It also gives a duality $\widetilde{H}_{n-p}(D_n X; \mathbf{K}) = \widetilde{H}^p(X; \mathbf{K})$, which could be either considered as a generalization of Alexander duality or as a further justification of the definition of (co)homology with coefficients in a spectrum.

Chapter 9

Spectral Sequences

Spectral sequences are powerful computational tools in topology. They also can give quick proofs of important theoretical results such as the Hurewicz theorem and the Freudenthal suspension theorem. Computing with spectral sequences is somewhat like computing integrals in calculus; it is helpful to have ingenuity and a supply of tricks, and even so, you may not arrive at the final solution to your problem. There are many spectral sequences which give different kinds of information. We will focus on one important spectral sequence, the *Leray-Serre-Atiyah-Hirzebruch spectral sequence* which takes as input a fibration over a CW-complex and a generalized homology or cohomology theory. This spectral sequence exhibits a complicated relationship between the generalized (co)homology of the total space and fiber and the ordinary (co)homology of the base. Many other spectral sequences can be derived from this one by judicious choice of fibration and generalized (co)homology theory.

Carefully setting up and proving the basic result requires very careful bookkeeping; the emphasis in these notes will be on applications and how to calculate. The project for this chapter is to outline the proof of the main theorem, Theorem 9.6.

9.1. Definition of a spectral sequence

Definition 9.1. A *spectral sequence* is a homological object of the following type:

One is given a sequence of chain complexes

$$(E^r, d^r) \text{ for } r = 1.2. \dots$$

and isomorphisms

$$E^{r+1} \cong H(E^r, d^r) = \frac{\ker d^r : E^r \to E^r}{\operatorname{im} d^r : E^r \to E^r}.$$

The isomorphisms are fixed as part of the structure of the spectral sequence, so henceforth we will fudge the distinction between "$\cong$" and "$=$" in the above context.

In this definition the term "chain complex" just means an abelian group (or R-module) with an endomorphism whose square is zero. In many important contexts, the spectral sequence has more structure: namely the chain complexes E^r are graded or even *bigraded*; that is, E^r decomposes as a direct sum of terms $E^r_{p,q}$ for $(p,q) \in \mathbf{Z} \oplus \mathbf{Z}$. Moreover the differentials d^r have a well-defined *bidegree*. For example, in a homology spectral sequence, usually d^r has bidegree $(-r, r-1)$. In other words $d^r(E^r_{p,q}) \subset E^r_{p-r,q+r-1}$.

A student first exposed to this plethora of notation may be intimidated; the important fact to keep in mind is that a bigrading decomposes a big object (E^r) into bite-sized pieces ($E^r_{p,q}$). Information about the $E^r_{p,q}$ for some pairs (p,q) gives information about $E^{r+1}_{p,q}$ for (probably fewer) pairs (p,q). But with luck one can derive valuable information. For example, from what has been said so far you should easily be able to see that if $E^r_{p,q} = 0$ for some fixed pair (p,q), then $E^{r+k}_{p,q} = 0$ for all $k \geq 0$. This simple observation can sometimes be used to derive highly non-trivial information. When computing with spectral sequences it is very useful to draw diagrams like the following:

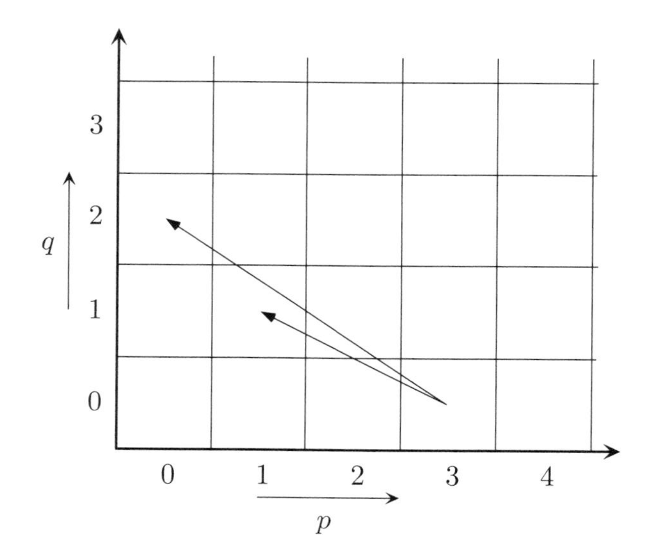

In this picture the short arrow depicts the differential $d^2 : E^2_{3,0} \to E^2_{1,1}$, and the long arrow corresponds to the differential $d^3 : E^3_{3,0} \to E^3_{0,2}$.

One usually computes with a spectral sequence in the following way. A theorem will state that there exists a spectral sequence so that:

1. the modules E^2 (or E^1) can be identified with something known, and
2. the colimit
$$E^\infty = \operatorname{colim}_{r\to\infty} E^r$$
is related to something one wishes to compute.

It can also work the opposite way: E^∞ can be related to something known, and E^2 can be related to something we wish to compute. In either case, this gives a complicated relationship between two things. The relationship usually involves exact sequences. In favorable circumstances information can be derived by carefully analyzing this relationship.

As an example to see how this may be used, the Leray–Serre spectral sequence of a fibration implies that if $F \hookrightarrow E \to B$ is a fibration with B simply connected, then there is a spectral sequence with

$$E^2_{p,q} \cong H_p(B;\mathbf{Q}) \otimes H_q(F;\mathbf{Q})$$

and with

$$H_n(E;\mathbf{Q}) \cong \oplus_p E^\infty_{p,n-p}.$$

This establishes a relationship between the homology of the base, total space, and fiber of a fibration. Of course, the hard work when computing with this spectral sequence is in getting from E^2 to E^∞. But partial computations and results are often accessible. For example, we will show later (and the reader may wish to show as an exercise now) that if $\oplus_p H_p(B;\mathbf{Q})$ and $\oplus_q H_q(F;\mathbf{Q})$ are finite-dimensional, then so is $\oplus_n H_n(E;\mathbf{Q})$ and

$$\chi(B) \cdot \chi(F) = \chi(E).$$

Another example: if $H_p(B;\mathbf{Q}) \otimes H_{n-p}(F;\mathbf{Q}) = 0$ for all p, then $H_n(E;\mathbf{Q}) = 0$. This generalizes a similar fact which can be proven for the trivial fibration $B \times F \to B$ using the Künneth theorem.

The next few definitions will provide us with a language to describe the way that the parts of the spectral sequence fit together.

Definition 9.2. A *filtration of an R-module A* is an increasing union

$$0 \subset \cdots \subset F_{-1} \subset F_0 \subset F_1 \subset \cdots \subset F_p \subset \cdots \subset A$$

of submodules. A filtration is *convergent* if the union of the F_p's is A and their intersection is 0.

If A itself is graded, then the filtration is assumed to preserve the grading, i.e. $F_p \cap A_n \subset F_{p+1} \cap A_n$. If A is graded, then we bigrade the filtration by setting

$$F_{p,q} = F_p \cap A_{p+q}.$$

We will mostly deal with filtrations that are *bounded below*, i.e. $F_s = 0$ for some s; or *bounded above*, i.e. $F_t = A$ for some t; or *bounded*, i.e. bounded above and bounded below. In this book, we will always have $F_{-1} = 0$.

Definition 9.3. Given a filtration $F = \{F_n\}$ of an R-module A, the *associated graded module* is the graded R-module denoted by $\mathrm{Gr}(A, F)$ and defined by

$$\mathrm{Gr}(A, F)_p = \frac{F_p}{F_{p-1}}.$$

We will usually just write $\mathrm{Gr}(A)$ when the filtration is clear from context.

In general, one is interested in the algebraic structure of A rather than $\mathrm{Gr}(A)$. Notice that $\mathrm{Gr}(A)$ contains some (but not necessarily all) information about A. For example, for a convergent filtration:

1. If $\mathrm{Gr}(A) = 0$, then $A = 0$.
2. If R is a field and A is a finite dimensional vector space, then each F_i is a subspace and $\mathrm{Gr}(A)$ and A have the same dimension. Thus in this case $\mathrm{Gr}(A)$ determines A up to isomorphism. This holds for more general R if each $\mathrm{Gr}(A)_n$ is free and the filtration is bounded above.
3. If $R = \mathbf{Z}$, then given a prime b, information about the b-primary part of $\mathrm{Gr}(A)$ gives information about the b-primary part of A; e.g. if $\mathrm{Gr}(A)_p$ has no b-torsion for all p, then A has no b-torsion for all p. However, the b-primary part of $\mathrm{Gr}(A)$ does not determine the b-primary part of A; e.g. if $\mathrm{Gr}(A)_0 = \mathbf{Z}$, $\mathrm{Gr}(A)_1 = \mathbf{Z}/2$, and $\mathrm{Gr}(A)_n = 0$ for $n \neq 0, 1$, it is impossible to determine whether $A \cong \mathbf{Z}$ or $A \cong \mathbf{Z} \oplus \mathbf{Z}/2$.

In short, knowing the quotients $\mathrm{Gr}(A)_p = F_p/F_{p-1}$ determines A up to "extension questions", at least when the filtration is bounded.

Definition 9.4. A bigraded spectral sequence $(E^r_{p,q}, d^r)$ is called a *homology spectral sequence* if the differential d^r has bidegree $(-r, r-1)$.

Definition 9.5. Given a bigraded homology spectral sequence $(E^r_{p,q}, d^r)$ and a graded R-module A_*, we say *the spectral sequence converges to* A_* and write

$$E^2_{p,q} \Rightarrow A_{p+q}$$

if:

1. for each p, q, there exists an r_0 so that $d^r_{p,q}$ is zero for each $r \geq r_0$ (by Exercise 146 below this implies $E^r_{p,q}$ surjects to $E^{r+1}_{p,q}$ for $r \geq r_0$), and
2. there is a convergent filtration of A_*, so that for each n, the colimit $E^\infty_{p,n-p} = \operatorname{colim}_{r\to\infty} E^r_{p,n-p}$ is isomorphic to the associated graded module $\mathrm{Gr}(A_*)_p$.

In many favorable situations (e.g. *first-quadrant spectral sequences* where $E^2_{p,q} = 0$ if $p < 0$ or $q < 0$) the convergence is stronger; namely for each pair (p, q) there exists an r_0 so that $E^r_{p,q} = E^\infty_{p,q}$ for all $r \geq r_0$.

An even stronger notion of convergence is the following. Suppose that there exists an r_0 so that for each (p, q) and all $r \geq r_0$, $E^r_{p,q} = E^\infty_{p,q}$. When this happens we say *the spectral sequence collapses at E^{r_0}.*

Exercise 146. Fix $p, q \in \mathbf{Z} \oplus \mathbf{Z}$.

1. Show that if there exists $r_0(p, q)$ so that $d^r_{p,q} = 0$ for all $r \geq r_0(p, q)$, then there exists a surjection $E^r_{p,q} \to E^{r+1}_{p,q}$ for all $r \geq r_0(p, q)$.
2. Show that if $E^2_{p,q} = 0$ whenever $p < 0$, then there exists a number $r_0 = r_0(p, q)$ as above.

Theorems on spectral sequences usually take the form: "There exists a spectral sequence with $E^2_{p,q}$ some known object converging to A_*." This is an abbreviated way of saying that the E^∞-terms are, on the one hand, the colimits of the E^r-terms and, on the other, the graded pieces in the associated graded module $\mathrm{Gr}(A_*)$ to A_*.

9.2. The Leray-Serre-Atiyah-Hirzebruch spectral sequence

Serre, based on earlier work of Leray, constructed a spectral sequence converging to $H_*(E)$, given a fibration

$$F \hookrightarrow E \xrightarrow{f} B.$$

Atiyah and Hirzebruch, based on earlier work of G. Whitehead, constructed a spectral sequence converging to $G_*(B)$ where G_* is an additive generalized homology theory and B is a CW-complex . The spectral sequence we present here is a combination of these spectral sequences and converges to $G_*(E)$ when G_* is an additive homology theory. The spectral sequence is carefully constructed in [**43**], and we refer you there for a proof.

We may assume B is path connected by restricting to path components, but we do not wish to assume B is simply connected. In order to deal with this case we will have to use local coefficients derived from the fibration. Theorem 6.12 shows that the homotopy lifting property gives rise to a homomorphism

$$\pi_1 B \to \{\text{Homotopy classes of homotopy equivalences } F \to F\}.$$

Applying the (homotopy) functor G_n, one obtains a representation

$$\pi_1 B \to \operatorname{Aut}(G_n F)$$

for each integer n. Thus for each n, $G_n(F)$ has the structure of a $\mathbf{Z}[\pi_1 B]$ module, or, equivalently, one has a system of local coefficients over B with fiber $G_n(F)$. (Of course, if $\pi_1 B = 1$, then this is a trivial local coefficient system.) Taking (ordinary) homology with local coefficients, we can associate the group $H_p(B; G_q F)$ to each pair of integers p, q. Notice that $H_p(B; G_q F)$ is zero if $p < 0$.

Theorem 9.6. *Let $F \hookrightarrow E \xrightarrow{f} B$ be a fibration, with B a path connected CW-complex. Let G_* be an additive homology theory. Then there exists a spectral sequence*

$$H_p(B; G_q F) \cong E^2_{p,q} \Rightarrow G_{p+q}(E).$$

□

Exercise 147. If G_* is an additive, isotropic homology theory, then the hypothesis that B is a CW-complex can be omitted. (Hint: for any space B there is a weak homotopy equivalence from a CW-complex to B.)

As a service to the reader, we will explicitly unravel the statement of the above theorem. There exists

1. A (bounded below) filtration
$$0 = F_{-1,n+1} \subset F_{0,n} \subset F_{1,n-1} \subset \cdots \subset F_{p,n-p} \subset \cdots \subset G_n(E)$$
of $G_n(E) = \cup_p F_{p,n-p}$ for each integer n.
2. A bigraded spectral sequence $(E^r_{*,*}, d^r)$ such that the differential d^r has bidegree $(-r, r-1)$ (i.e. $d^r(E^r_{p,q}) \subset E^r_{p-r,q+r-1}$), and so
$$E^{r+1}_{p,q} = \frac{\ker\, d^r : E^r_{p,q} \to E^r_{p-r,q+r-1}}{\operatorname{im}\, d^r : E^r_{p+r,q-r+1} \to E^r_{p,q}}.$$
3. Isomorphisms $E^2_{p,q} \cong H_p(B; G_q F)$.

This spectral sequence converges to $G_*(E)$. That is, for each fixed p, q, there exists an r_0 so that

$$d^r : E^r_{p,q} \to E^r_{p-r,q+r-1}$$

is zero for all $r \geq r_0$ and so

$$E^{r+1}_{p,q} = E^r_{p,q} / d^r(E^r_{p+r,q-r+1})$$

for all $r \geq r_0$.

Define $E^\infty_{p,q} = \operatorname{colim}_{r\to\infty} E^r_{p,q}$. There is an isomorphism

$$F_{p,q}/F_{p-1,q+1} \cong E^\infty_{p,q},$$

i.e.

$$\mathrm{Gr}(G_n E)_p \cong E^\infty_{p,n-p}$$

with respect to the filtration of $G_n(E)$.

In this spectral sequence, some filtrations of the groups $G_n(E)$ are given, with the associated graded groups made up of the pieces $E^\infty_{p,n-p}$. So, for example, if $G_n(E) = 0$, then $E^\infty_{p,n-p} = 0$ for each $p \in \mathbf{Z}$.

The filtration is given by

$$F_{p,n-p} = \operatorname{im}(G_n(f^{-1}(B^p)) \to G_n E)$$

where $f : E \to B$ is the fibration and B^p denotes the p-skeleton of B.

As a first non-trivial example of computing with spectral sequences we consider the problem of computing the homology of the loop space of a sphere. Given $k > 1$, let $P = P_{x_0} S^k$ be the space of paths in S^k which start at $x_0 \in S^k$. As we saw in Chapter 6 evaluation at the end point defines a fibration $P \to S^k$ with fiber the loop space ΩS^k. Moreover, the path space P is contractible.

The spectral sequence for this fibration (using homology with integer coefficients for G_*) has $E^2_{p,q} = H_p(S^k; H_q(\Omega S^k))$. The coefficients are untwisted since $\pi_1(S^k) = 0$. Therefore

$$E^2_{p,q} = \begin{cases} H_q(\Omega S^k) & \text{if } p = 0 \text{ or } p = k, \\ 0 & \text{otherwise.} \end{cases} \tag{9.1}$$

In particular this is a first-quadrant spectral sequence.

Since $H_n(P) = 0$ for all $n \neq 0$, the filtration of $H_n(P)$ is trivial for $n > 0$ and so $E^\infty_{p,q} = 0$ if $p + q > 0$. Since this is a first-quadrant spectral sequence, $E^\infty_{p,q} = 0$ for all $(p,q) \neq (0,0)$, and, furthermore, given any $(p,q) \neq (0,0)$, $E^r_{p,q} = 0$ for some r large enough.

Now here's the cool part. Looking at the figure and keeping in mind the fact that the bidegree of d^r is $(-r, r-1)$, we see that all differentials $d^r : E^r_{p,q} \to E^r_{p-r,q+r-1}$ either:

1. start or end at a zero group, or
2. $r = k$ and $(p,q) = (k,q)$ with $q \geq 0$, so that

$$d^k : E^k_{k,q} \to E^k_{0,q+k-1}.$$

The following picture shows the E^k-stage and the differential $d^k : E^k_{k,0} \to E^k_{0,k-1}$. The shaded columns contain the only possible non-zero entries, since $E^2_{p,q} = 0$ if $p \neq 0$ or k.

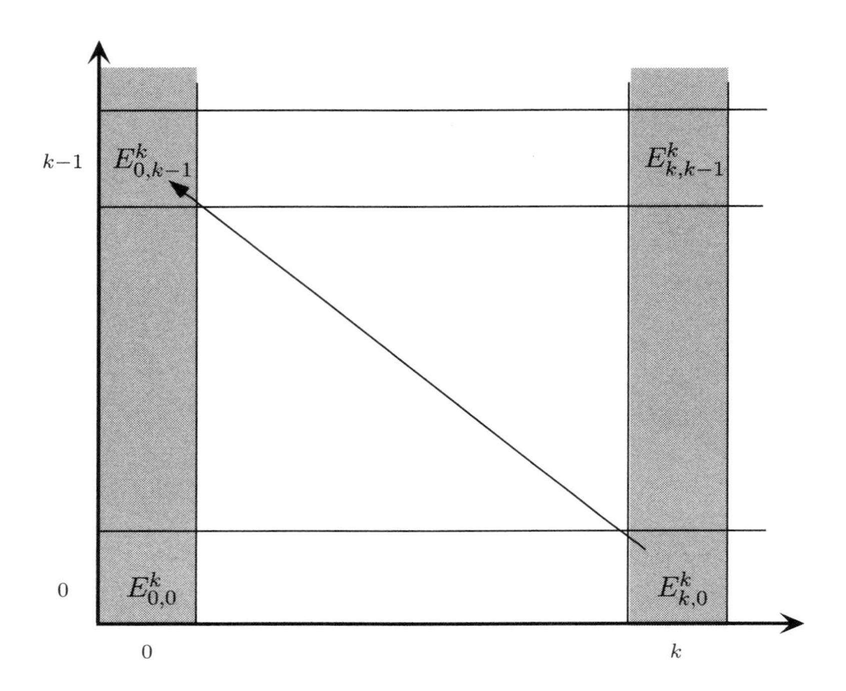

Hence

$$E^2_{p,q} = E^3_{p,q} = \cdots = E^k_{p,q}. \tag{9.2}$$

Thus if $(p,q) \neq (0,0)$,

$$0 = E^\infty_{p,q} = E^{k+1}_{p,q} = \begin{cases} \ker d^k : E^k_{k,q} \to E^k_{0,q+k-1} & \text{if } (p,q) = (k,q), \\ \operatorname{Coker} d^k : E^k_{k,q} \to E^k_{0,q+k-1} & \text{if } (p,q) = (0, q+k-1), \\ 0 & \text{otherwise.} \end{cases}$$

Therefore, the spectral sequence collapses at E^{k+1}.

It follows that d^k is an isomorphism, i.e. $E^k_{k,q} \cong E^k_{0,q+k-1}$ whenever $(k,q) \neq (0,0)$ or $q \neq 1-k$. Using Equations (9.2) and (9.1) we can restate this as

$$H_q(\Omega S^k) \cong H_{q+k-1}(\Omega S^k).$$

Using induction, starting with $H_0(\Omega S^k) = 0$, we conclude that

$$H_q(\Omega S^k) = \begin{cases} \mathbf{Z} & \text{if } q = a(k-1), a \geq 0 \\ 0 & \text{otherwise.} \end{cases} \tag{9.3}$$

Exercise 148. If $S^k \hookrightarrow S^\ell \xrightarrow{f} S^m$ is a fibration, then $\ell = 2m-1$ and $k = m-1$. (In fact, it is a result of Adams that there are only such fibrations for $m = 1, 2, 4$ and 8.)

Returning to our general discussion, notice that $E^{r+1}_{p,q}$ and $E^\infty_{p,q}$ are *subquotients* of $E^r_{p,q}$; in particular, since $E^2_{p,q} \cong H_p(B; G_qF)$ we conclude the following fundamental fact.

Theorem 9.7. *The associated graded module to the filtration of $G_n(E)$ has graded summands which are subquotients of $H_p(B; G_{n-p}F)$.* □

This fact is the starting point for many spectral sequence calculations. For example,

Theorem 9.8. *If $H_p(B; G_{n-p}F) = 0$ for all p, then $G_n(E) = 0$.*

Proof. Since $E^2_{p,n-p} = 0$ for each p, it follows that $E^\infty_{p,n-p} = 0$ for each p and so $G_n(E) = 0$. □

9.3. The edge homomorphisms and the transgression

Before we turn to more involved applications, it is useful to know several facts about the Leray-Serre-Atiyah-Hirzebruch spectral sequence. These facts serve to identify certain homomorphisms which arise in the guts of the spectral sequence with natural maps induced by the inclusion of the fiber or the projection to the base in the fibration.

Lemma 9.9. *In the Leray-Serre-Atiyah-Hirzebruch spectral sequence there is a surjection*

$$E^2_{0,n} \to E^\infty_{0,n}$$

for all n.

Proof. Notice that

$$E^{r+1}_{0,n} = \frac{\ker d^r : E^r_{0,n} \to E^r_{-r,n+r-1}}{\text{im } d^r : E^r_{r,n-r+1} \to E^r_{0,n}} \quad \text{for } r > 1.$$

But, since $E^2_{p,q} = 0$ for $p < 0$, we must have $E^2_{-r,q} = 0$ for all q and so also its subquotient $E^r_{-r,q} = 0$ for all q.

Hence $(\ker d^r : E^r_{0,n} \to E^r_{-r,n+r-1}) = E^r_{0,n}$ and so

$$E^{r+1}_{0,n} = \frac{E^r_{0,n}}{\operatorname{im} d^r}.$$

Thus each $E^r_{0,n}$ surjects to $E^{r+1}_{0,n}$ and hence also to the colimit $E^\infty_{0,n}$. □

Proposition 5.14 says that if V is any local coefficient system over a path connected space B, then

$$H_0(B;V) = V/\langle v - \alpha \cdot v \mid v \in V,\ \alpha \in \pi_1 B\rangle.$$

Applying this to $V = G_n(F)$, it follows that there is a surjection

$$G_n(F) \to H_0(B; G_nF). \tag{9.4}$$

We can now use the spectral sequence to construct a homomorphism $G_*(F) \to G_*(E)$. Theorem 9.10 below asserts that this homomorphism is just the homomorphism induced by the inclusion of the fiber into the total space.

Since $F_{-1,n-1} = 0$,

$$E^\infty_{0,n} \cong F_{0,n}/F_{-1,n+1} = F_{0,n} \subset G_n(E).$$

This inclusion can be precomposed with the surjections of Lemma 9.9 and Equation (9.4) to obtain a homomorphism (called an *edge homomorphism*)

$$G_n(F) \to H_0(B; G_nF) \cong E^2_{0,n} \to E^\infty_{0,n} \subset G_n(E). \tag{9.5}$$

Theorem 9.10. *The edge homomorphism given by (9.5) equals the map $i_* : G_n(F) \to G_n(E)$ induced by the inclusion $i : F \hookrightarrow E$ by the homology theory G_*.* □

Another simple application of the spectral sequence is to compute oriented bordism groups of a space in low dimensions. We apply the Leray-Serre-Atiyah-Hirzebruch spectral sequence to the fibration $\mathrm{pt} \hookrightarrow X \xrightarrow{\mathrm{Id}} X$, and take $G_* = \Omega^{SO}_*$, oriented bordism.

In this case the Leray-Serre-Atiyah-Hirzebruch spectral sequence says

$$H_p(X; \Omega^{SO}_q(\mathrm{pt})) \Rightarrow \Omega^{SO}_{p+q}(X).$$

Notice that the coefficients are untwisted; this is because the fibration is trivial. Write $\Omega^{SO}_n = \Omega^{SO}_n(\mathrm{pt})$. Note that $\mathrm{pt} \hookrightarrow X$ is split by the constant map; hence the edge homomorphism $\Omega^{SO}_n \to \Omega^{SO}_n(X)$ is a split injection, so by Theorem 9.10, the differentials $d^r : E^r_{r,n-r+1} \to E^r_{0,n}$ whose targets are on the vertical edge of the first quadrant must be zero; i.e. every element of $E^2_{0,n}$ *survives* to $E^\infty_{0,n} = \Omega^{SO}_n$.

Recall from Section 8.7 that $\Omega^{SO}_q = 0$ for $q = 1, 2, 3$, and $\Omega^{SO}_q = \mathbf{Z}$ for $q = 0$ and 4. Of course $\Omega^{SO}_q = 0$ for $q < 0$. Thus for $n = p + q \leq 4$, the only (possibly) non-zero terms are $E^2_{n,0} \cong H_n(X)$ and $E^2_{0,4} = \Omega_4^{SO}$. Hence $E^2_{p,n-p} = E^\infty_{p,n-p}$ for $n = 0, 1, 2, 3$, and 4. From the spectral sequence one concludes

$$\Omega_n^{SO}(X) \cong H_n(X) \qquad \text{for } n = 0, 1, 2, 3$$
$$\Omega_4^{SO}(X) \cong \mathbf{Z} \oplus H_4(X).$$

It can be shown that the map $\Omega_n^{SO}(X) \to H_n(X)$ is a Hurewicz map which takes $f : M \to X$ to $f_*([M])$. In particular this implies that any homology class in $H_n(X)$ for $n = 0, 1, 2, 3$, and 4 is *represented* by a map from an oriented manifold to X. The map $\Omega_4^{SO}(X) \to \mathbf{Z}$ is the map taking $f : M \to X$ to the signature of M.

We next identify another edge homomorphism which can be constructed in the same manner as (9.5). The analysis will be slightly more involved, and we will state it only in the case when G_* is ordinary homology with coefficients in an R-module (we suppress the coefficients).

In this context $E^2_{p,q} = H_p(B; H_qF) = 0$ for $q < 0$ or $p < 0$. So $E^*_{*,*}$ is a first-quadrant spectral sequence; i.e. $E^r_{p,q} = E^\infty_{p,q} = 0$ for $q < 0$ or $p < 0$.

This implies that the filtration of $H_n(E)$ has finite length

$$0 = F_{-1,n+1} \subset F_{0,n} \subset F_{1,n-1} \subset \cdots \subset F_{n,0} = H_n(E)$$

since

$$0 = E^\infty_{p,n-p} = F_{p,n-p}/F_{p-1,n-p+1}$$

if $p < 0$ or $n - p < 0$.

The second map in the short exact sequence

$$0 \to F_{n-1,1} \to F_{n,0} \to E^\infty_{n,0} \to 0$$

can thus be thought of as a homomorphism

$$H_n(E) \to E^\infty_{n,0}. \tag{9.6}$$

Lemma 9.11. *There is an inclusion*

$$E^\infty_{n,0} \subset E^2_{n,0}$$

for all n.

Proof. Since $E^r_{n+r,1-r} = 0$ for $r > 1$,

$$E^{r+1}_{n,0} = \frac{\ker\ d^r : E^r_{n,0} \to E^r_{n-r,r-1}}{\text{im}\ d^r : E^r_{n+r,1-r} \to E^r_{n,0}} = \ker d^r : E^r_{n,0} \to E^r_{n-r,r-1}.$$

Thus

$$\cdots \subset E^{r+1}_{n,0} \subset E^r_{n,0} \subset E^{r-1}_{n,0} \subset \cdots$$

and hence

$$E^\infty_{n,0} = \bigcap_r E^r_{n,0} \subset E^2_{n,0}.$$

□

Note that the constant map from the fiber F to a point induces a homomorphism $H_n(B; H_0F) \to H_nB$. If F is path connected, then the local coefficient system H_0F is trivial and $H_n(B; H_0(F)) = H_n(B)$ for all n.

Theorem 9.12. *The composite map (also called an edge homomorphism)*

$$H_n(E) = F_{n,0} \to E^\infty_{n,0} \subset E^2_{n,0} \cong H_n(B; H_0F) \to H_n(B)$$

is just the map induced on homology by the projection $f : E \to B$ of the fibration. □

The long differential $d^k : E^k_{k,0} \to E^k_{0,k-1}$ in the spectral sequence for a fibration (for ordinary homology) has an alternate geometric interpretation called the *transgression*. It is defined as follows. Suppose $f : E \to B$ is a fibration with fiber F. Fix $k > 0$. We assemble the homomorphism $f_* : H_k(E,F) \to H_k(B,b_0)$, the isomorphism $H_k(B) \cong H_k(B,b_0)$, and the connecting homomorphism $\delta : H_k(E,F) \to H_{k-1}(F)$ for the long exact sequence of the pair (E,F) to define a (not well-defined, multi–valued) function $\tau : H_k(B)$"$\to$"$H_{k-1}(F)$ as the "composite"

$$\tau : H_k(B) \cong H_k(B,b_0) \xleftarrow{f_*} H_k(E,F) \xrightarrow{\delta} H_{k-1}(F).$$

To make this more precise, we take as the domain of τ the image of $f_* : H_k(E,F) \to H_k(B,b_0) \cong H_k(B)$, and as the range of τ the quotient of $H_{k-1}(F)$ by $\delta(\ker f_* : H_k(E,F) \to H_k(B,b_0))$. A simple diagram chase shows τ is well-defined with this choice of domain and range. Thus the transgression τ is an honest homomorphism from a subgroup of $H_k(B)$ to a quotient group of $H_{k-1}(F)$. Intuitively, the transgression is trying his/her best to imitate the boundary map in the long exact homotopy sequence for a fibration (see Theorem 9.15 below).

Assume for simplicity that F is path connected, and consider the differential

$$d^k : E^k_{k,0} \to E^k_{0,k-1}$$

in the spectral sequence for this fibration (taking $G_* = H_* =$ ordinary homology). Its domain, $E^k_{k,0}$, is a subgroup of $E^2_{k,0} = H_k(B; H_0(F)) = H_k(B)$ because all differentials d^r into $E^k_{k,0}$ are zero for $r < k$ (this is a first-quadrant spectral sequence), and hence $E^k_{k,0}$ is just the intersection of the kernels of $d^r : E^r_{k,0} \to E^r_{k-r,r-1}$ for $r < k$.

Similarly the range $E^k_{0,k-1}$ of $d^k : E^k_{k,0} \to E^k_{0,k-1}$ is a quotient of $E^2_{0,k-1} = H_0(B; H_{k-1}(F))$, which by Proposition 5.14 is just the quotient of $H_{k-1}(F)$ by the action of $\pi_1(B)$.

We have shown that like the transgression, the differential $d^k : E^k_{k,0} \to E^k_{0,k-1}$ has domain identified with a subgroup of $H_k(B)$ and range a quotient of $H_{k-1}(F)$. The following theorem identifies the transgression and this differential.

Theorem 9.13 (transgression theorem). *The differential $d^k : E^k_{k,0} \to E^k_{0,k-1}$ in the spectral sequence of the fibration $F \hookrightarrow E \to B$ coincides with the transgression*

$$H_k(B) \supset \text{domain}(\tau) \xrightarrow{\tau} \text{range}(\tau) = H_{k-1}(F)/\delta(\ker f_*).$$

□

The proofs of Theorems 9.10, 9.12, and 9.13 are not hard, but require an examination of the construction which gives the spectral sequence. We omit the proofs, but you should look them up when working through the project for this chapter.

9.4. Applications of the homology spectral sequence

9.4.1. The five-term and Serre exact sequences.

Corollary 9.14 (five-term exact sequence). *Suppose that $F \hookrightarrow E \xrightarrow{f} B$ is a fibration with B and F path connected. Then there exists an exact sequence*

$$H_2(E) \xrightarrow{f_*} H_2(B) \xrightarrow{\tau} H_0(B; H_1(F)) \to H_1(E) \xrightarrow{f_*} H_1(B) \to 0.$$

The composite of the surjection $H_1(F) \to H_0(B; H_1(F))$ with the map $H_0(B; H_1(F)) \to H_1(E)$ in this exact sequence is the homomorphism induced by the inclusion $F \hookrightarrow E$, and τ is the transgression.

Proof. Take $G_* = H_*(-)$, ordinary homology, perhaps with coefficients. The corresponding first quadrant spectral sequence has

$$E^2_{p,q} \cong H_p(B; H_qF)$$

and converges to $H_*(E)$.

The local coefficient system $\pi_1 B \to \text{Aut}(H_0(F))$ is trivial since F is path connected. Thus $E^2_{p,0} = H_p(B; H_0(F)) = H_p(B)$.

The following facts either follow immediately from the statement of Theorem 9.6 or are easy to verify, using the bigrading of the differentials and the fact that the spectral sequence is a first–quadrant spectral sequence.

1. $H_1(B) \cong E^2_{1,0} = E^r_{1,0} = E^\infty_{1,0}$ for all $r \geq 2$.
2. $H_2(B) \cong E^2_{2,0}$.
3. $H_0(B; H_1F) = E^2_{0,1}$.
4. $E^\infty_{2,0} = E^r_{2,0} = E^3_{2,0} = \ker d^2 : E^2_{2,0} \to E^2_{0,1}$ for all $r \geq 3$.
5. $E^\infty_{0,1} = E^r_{0,1} = E^3_{0,1} = \operatorname{coker} d^2 : E^2_{2,0} \to E^2_{0,1}$ for all $r \geq 3$.

Exercise 149. Prove these five facts.

The last two facts give an exact sequence

$$0 \to E^\infty_{2,0} \to E^2_{2,0} \xrightarrow{d^2} E^2_{0,1} \to E^\infty_{0,1} \to 0$$

or, making the appropriate substitutions, the exact sequence

$$0 \to E^\infty_{2,0} \to H_2(B) \to H_0(B; H_1(F)) \to E^\infty_{0,1} \to 0. \tag{9.7}$$

Since the spectral sequence converges to $H_*(E)$, and the $E^\infty_{p,n-p}$ form the associated graded groups for $H_n(E)$, the two sequences

$$0 \to E^\infty_{0,1} \to H_1(E) \to E^\infty_{1,0} \to 0 \tag{9.8}$$

and

$$E^\infty_{1,1} \to H_2(E) \to E^\infty_{2,0} \to 0 \tag{9.9}$$

are exact.

Splicing the sequences (9.7), (9.8), and (9.9) together and using the first fact above, one obtains the exact sequence

$$E^\infty_{1,1} \to H_2(E) \to H_2(B) \to H_0(B; H_1(F)) \to H_1(E) \to H_1(B) \to 0.$$

In this sequence the homomorphism $H_i(E) \to H_i(B)$ is the edge homomorphism and hence is induced by the fibration $f : E \to B$. The map $H_0(B; H_1(F)) \to H_2(E)$ composes with $H_1(F) \to H_0(B; H_1(F))$ to give the other edge homomorphism, induced by the inclusion of the fiber. The map $H_2(B) \to H_0(B; H_1(F))$ is the transgression. These assertions follow by chasing definitions and using Theorems 9.10, 9.12 and 9.13. □

We have seen, beginning with our study of the Puppe sequences, that cofibrations give exact sequences in homology and fibrations give exact sequences in homotopy. One might say that a map is a "fibration or cofibration in some range" if there are partial long exact sequences. Corollary 9.14 implies that if $\pi_1 B$ acts trivially on $H_1(F)$, then *the fibration is a cofibration in a certain range.* A more general result whose proof is essentially identical to that of Corollary 9.14 is given in the following important theorem.

Theorem 9.15 (Serre exact sequence). *Let $F \xrightarrow{i} E \xrightarrow{f} B$ be a fibration with B and F path connected and with $\pi_1 B$ acting trivially on H_*F. Suppose $H_pB = 0$ for $0 < p < m$ and $H_qF = 0$ for $0 < q < n$. Then the sequence:*

$$H_{m+n-1}F \xrightarrow{i_*} H_{m+n-1}E \xrightarrow{f_*} H_{m+n-1}B \xrightarrow{\tau} H_{m+n-2}F \xrightarrow{i_*} \cdots$$
$$\cdots \xrightarrow{f_*} H_1 B \to 0.$$

is exact. □

Exercise 150. Prove Theorem 9.15.

To understand this result, suppose B is $(m-1)$-connected and F is $(n-1)$-connected. The long exact sequence for a fibration shows that E is $(\min(m,n)-1)$-connected, so that by the Hurewicz theorem, $H_qE = 0$ for $q < \min(m,n)$. So trivially the low-dimensional part of the Serre exact sequence is exact; indeed all groups are zero for $q < \min(m,n)$. The remarkable fact is that the sequence remains exact for all $\min\{m,n\} \leq q < m+n$.

9.4.2. Euler characteristics and fibrations. Let k be a field. Recall that the Euler characteristic of a space Z is defined to be the alternating sum $\chi(Z) = \sum_n (-1)^n \beta_n(Z;k)$ of the Betti numbers $\beta_n(Z;k) = \dim_k(H_n(Z;k))$ whenever this sum is a finite sum of finite ranks. For finite CW-complexes it is equal to the alternating sum of the number of n-cells by the following standard exercise applied to the cellular chain complex.

Exercise 151. Let (C_*, ∂) be a chain complex over a field with $\oplus_i C_i$ finite-dimensional. Show that the alternating sum of the ranks of the C_i equals the alternating sum of the ranks of the cohomology groups $H_i(C_*, \partial)$.

Given a product space $E = B \times F$ with B and F finite CW-complexes, the Künneth theorem implies that the homology with field coefficients is a tensor product

$$H_*(E;k) \cong H_*(B;k) \otimes H_*(F;k)$$

from which it follows that the Euler characteristic is multiplicative

$$\chi(E) = \chi(B) \cdot \chi(F).$$

The following theorem extends this formula to the case when E is only a product locally, i.e. fiber bundles, and even to fibrations.

Notice that the homology itself need not be multiplicative for a non-trivial fibration. For example, consider the Hopf fibration $S^3 \hookrightarrow S^7 \to S^4$. The graded groups $H_*(S^7;k)$ and $H_*(S^3;k) \otimes H_*(S^4;k)$ are not isomorphic, even though the Euler characteristics multiply $(0 = 0 \cdot 2)$.

Theorem 9.16. *Let $p : E \to B$ be a fibration with fiber F, let k be a field, and suppose the action of $\pi_1 B$ on $H_*(F;k)$ is trivial. Assume that the Euler characteristics $\chi(B), \chi(F)$ are defined (e.g. if B, F are finite cell complexes). Then $\chi(E)$ is defined and*

$$\chi(E) = \chi(B) \cdot \chi(F).$$

Proof. Since k is a field and the action of $\pi_1 B$ on $H_*(F;k)$ is trivial,

$$H_p(B; H_q(F;k)) \cong H_p(B;k) \otimes_k H_q(F;k)$$

by the universal coefficient theorem. Theorem 9.6 with $G_* = H_*(-;k)$ implies that there exists a spectral sequence with

$$E^2_{p,q} \cong H_p(B;k) \otimes H_q(F;k).$$

By hypothesis, $E^2_{p,q}$ is finite-dimensional over k and is zero for all but finitely many pairs (p,q). This implies that the spectral sequence collapses at some stage and so $E^\infty_{p,q} = E^r_{p,q}$ for r large enough.

Define

$$E^r_n = \oplus_p E^r_{p,n-p}$$

for each n and $r \geq 2$ including $r = \infty$.

Then since the Euler characteristic of the tensor product of two graded vector spaces is the product of the Euler characteristics,

$$\chi(E^2_*) = \chi(B)\chi(F).$$

Notice that (E^r_*, d^r) is a (singly) graded chain complex with homology E^{r+1}_*. Exercise 151 shows that for any $r \geq 2$,

$$\chi(E^r_*) = \chi(H_*(E^r_*, d^r)) = \chi(E^{r+1}_*).$$

Since the spectral sequence collapses $\chi(E^2_*) = \chi(E^\infty_*)$.

Since we are working over a field, $H_n(E;k)$ is isomorphic to its associated graded vector space $\oplus_p E^\infty_{p,n-p} = E^\infty_n$. In particular $H_n(E;k)$ is finite-dimensional and $\dim H_n(E;k) = \dim E^\infty_n$.

Therefore,

$$\chi(B)\chi(F) = \chi(E^2_*) = \chi(E^\infty_*) = \chi(H_*(E;k)) = \chi(E).$$

□

9.4.3. The homology Gysin sequence.

Theorem 9.17. *Let R be a commutative ring. Suppose $F \hookrightarrow E \xrightarrow{f} B$ is a fibration, and suppose F is an R-homology n-sphere; i.e.*

$$H_i(F;R) \cong \begin{cases} R & \text{if } i = 0 \text{ or } n, \\ 0 & \text{otherwise.} \end{cases}$$

Assume that $\pi_1 B$ acts trivially on $H_n(F;R)$. Then there exists an exact sequence (R-coefficients):

$$\cdots \to H_r E \xrightarrow{f_*} H_r B \to H_{r-n-1} B \to H_{r-1} E \xrightarrow{f_*} H_{r-1} B \to \cdots .$$

Proof. The spectral sequence for the fibration (using ordinary homology with R-coefficients) has

$$E^2_{p,q} \cong H_p(B; H_q F) = \begin{cases} H_p(B;R) & \text{if } q = 0 \text{ or } n, \\ 0 & \text{otherwise.} \end{cases}$$

The following diagram shows the E^2-stage. The two shaded rows ($q = 0$ and $q = n$) are the only rows that might contain a non-zero $E^2_{p,q}$.

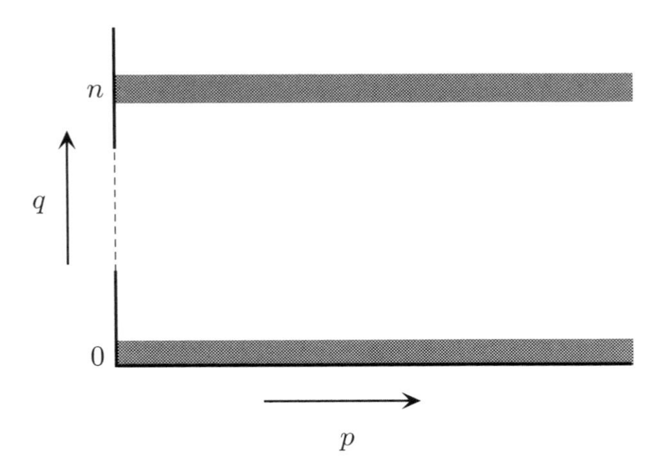

Thus the only possibly non-zero differentials are

$$d^{n+1} : E^{n+1}_{p,0} \to E^{n+1}_{p-n-1,n}.$$

It follows that

$$E^{n+1}_{p,q} \cong E^2_{p,q} \cong H_p(B; H_q F) = \begin{cases} H_p B & \text{if } q = 0 \text{ or } n, \\ 0 & \text{otherwise,} \end{cases}$$

and

$$(9.10)\qquad E^{\infty}_{p,q} \cong \begin{cases} 0 & \text{if } q \neq 0 \text{ or } n, \\ \ker\ d^{n+1} : E^{n+1}_{p,0} \to E^{n+1}_{p-n-1,n} & \text{if } q = 0, \\ \operatorname{coker}\ d^{n+1} : E^{n+1}_{p+n+1,0} \to E^{n+1}_{p,q} & \text{if } q = n. \end{cases}$$

The filtration of $H_r(E)$ reduces to

$$0 \subset E^{\infty}_{r-n,n} \cong F_{r-n,n} \subset F_{r,0} = H_r E,$$

and so the sequences

$$0 \to E^{\infty}_{r-n,n} \to H_r E \to E^{\infty}_{r,0} \to 0$$

are exact for each r. Splicing these with the exact sequences

$$0 \to E^{\infty}_{p,0} \to E^{n+1}_{p,0} \xrightarrow{d^{n+1}} E^{n+1}_{p-n-1,n} \to E^{\infty}_{p-n-1,n} \to 0$$

(obtained from Equation (9.10)) gives the desired exact sequence

$$\cdots \to H_r E \xrightarrow{f_*} H_r B \to H_{r-n-1} B \to H_{r-1} E \to H_{r-1} B \to \cdots$$

with the map labelled f_* induced by $f : E \to B$ by Theorem 9.12. $\square$

Exercise 152. Derive the *Wang sequence*. If $F \hookrightarrow E \to S^n$ is a fibration over S^n, then there is an exact sequence

$$\cdots \to H_r F \to H_r E \to H_{r-n} F \to H_{r-1} F \to \cdots .$$

9.5. The cohomology spectral sequence

The examples in the previous section show that spectral sequences are a useful tool for establishing relationships between the homology groups of the three spaces forming a fibration. Much better information can often be obtained by using the ring structure on cohomology. We next introduce the cohomology spectral sequence and relate the ring structures on cohomology and the spectral sequence. The ring structure makes the cohomology spectral sequence a much more powerful computational tool than the homology spectral sequence.

Definition 9.18. A bigraded spectral sequence $(E_r^{p,q}, d_r)$ is called a *cohomology spectral sequence* if the differential d_r has bidegree $(r, 1-r)$.

Notice the change in placement of the indices in the cohomology spectral sequence. The contravariance of cohomology makes it necessary to change the notion of a filtration. There is a formal way to do this, namely by "lowering indices"; for example rewrite $H^p(X)$ as $H_{-p}(X)$, rewrite F^p as F_{-p}, replace $E_r^{p,q}$ by $E^r_{-p,-q}$ and so forth. Unfortunately for this to work the notion of convergence of a spectral sequence has to be modified; with the definition we gave above the cohomology spectral sequence of a fibration will not converge. Rather than extending the formalism and making the notion of convergence technically more complicated, we will instead just make new definitions which apply in the cohomology setting.

Definition 9.19. A *(cohomology) filtration of an R-module A* is an increasing union

$$0 \subset \cdots \subset F^p \subset \cdots \subset F^2 \subset F^1 \subset F^0 \subset F^{-1} \subset \cdots \subset A$$

of submodules. A filtration is *convergent* if the union of the F_p's is A and their intersection is 0.

If A itself is graded, then the filtration is assumed to preserve the grading i.e. $F^p \cap A^n \subset F^{p-1} \cap A^n$. If A is graded, then we bigrade the filtration by setting

$$F^{p,q} = F^p \cap A^{p+q}.$$

Definition 9.20. Given a cohomology filtration $F = \{F^n\}$ of an R-module A, the *associated graded module* is the graded R-module denoted by $\mathrm{Gr}(A, F)$ and defined by

$$\mathrm{Gr}(A, F)^p = \frac{F^p}{F^{p+1}}.$$

Definition 9.21. Given a bigraded cohomology spectral sequence $(E_r^{p,q}, d_r)$ and a graded R-module A^*, we say *the spectral sequence converges to A^** and write

$$E_2^{p,q} \Rightarrow A^{p+q}$$

if:

1. for each (p,q) there exists an r_0 so that $d_r : E_r^{p-r,q+r-1} \to E_r^{p,q}$ is zero for all $r \geq r_0$; in particular there is an injection $E_{r+1}^{p,q} \hookrightarrow E_r^{p,q}$ for all $r \geq r_0$, and
2. there is a convergent filtration of A^* so that for each n, the limit $E_\infty^{p,q} = \cap_{r \geq r_0} E_r^{p,q}$ is isomorphic to the associated graded $\mathrm{Gr}(A^*)^p$.

Theorem 9.22. *Let $F \hookrightarrow E \xrightarrow{f} B$ be a fibration, with B a path connected CW-complex. Let G^* be an additive cohomology theory. Assume either that B is a finite-dimensional CW-complex or else that there exists an N so that $G^q(F) = 0$ for all $q < N$. Notice that $\pi_1(B)$ acts on $G^q(F)$ determining a local coefficient system.*

Then there exists a (cohomology) spectral sequence

$$H^p(B; G^q F) \cong E_2^{p,q} \Rightarrow G^{p+q}(E).$$

□

There is a version of this theorem which applies to infinite CW-complexes; see [**43**].

Exercise 153. State and prove the cohomology versions of the Serre, Gysin, and Wang sequences. Construct the cohomology edge homomorphisms and the cohomology transgression and state the analogues of Theorems 9.10, 9.12, and 9.13.

As an example we show how to compute the complex K-theory of complex projective space $\mathbf{C}P^k$ (see Section 8.8.4 and the project for Chapter 8). The computation of complex K-theory was the original motivation for Atiyah-Hirzebruch to set up their spectral sequence. Complex K-theory is a cohomology theory satisfying $K^n(X) = K^{n+2}(X)$, and its coefficients are given by

$$K^{2n}(\text{pt}) = \pi_0(\mathbf{Z} \times BU) = \mathbf{Z}$$

and

$$K^{2n+1}(\text{pt}) = \pi_1(\mathbf{Z} \times BU) = 0.$$

Theorem 9.22, applied to the trivial fibration

$$\text{pt} \hookrightarrow \mathbf{C}P^k \xrightarrow{\text{Id}} \mathbf{C}P^k,$$

says there exists a cohomology spectral sequence $E_r^{p,q}$ satisfying

$$H^p(\mathbf{C}P^k; K^q(\text{pt})) \cong E_2^{p,q} \Rightarrow K^{p+q}(\mathbf{C}P^k).$$

The coefficients are untwisted since the fibration is trivial. Since

$$H^p(\mathbf{C}P^k) = \begin{cases} \mathbf{Z} & \text{if } p \text{ is even, } 0 \leq p \leq 2k \\ 0 & \text{otherwise,} \end{cases}$$

it follows that

$$E_2^{p,q} = \begin{cases} \mathbf{Z} & \text{if } p \text{ and } q \text{ are even, } 0 \leq p \leq 2k \\ 0 & \text{otherwise.} \end{cases}$$

This checkerboard pattern forces every differential to be zero, since one of the integers $(r, 1-r)$ must be odd! Notice, by the way, that this is not a first-quadrant spectral sequence since the K-theory of a point is non-zero in positive and negative dimensions.

Therefore $E_2^{p,q} = E_\infty^{p,q}$ and the associated graded group to $K^n(\mathbf{C}P^k)$, $\oplus_p E_\infty^{p,n-p}$, is a direct sum of $k+1$ copies of $\mathbf{Z}$, one for each pair (p,q) so that $p+q=n$, both p and q are even, and $0 \leq p \leq 2k$. Inducting down the filtration we see that $K^n(\mathbf{C}P^k)$ has no torsion and hence is isomorphic to its associated graded group. Therefore

$$K^n(\mathbf{C}P^k) = \begin{cases} \mathbf{Z}^{k+1} & \text{if } n \text{ is even,} \\ 0 & \text{otherwise.} \end{cases}$$

To study the multiplicative properties of the cohomology spectral sequence, take G^* to be ordinary cohomology with coefficients in a commutative ring $R: \ G^*(E) = H^*(E;R)$. Let $F \hookrightarrow E \to B$ be a fibration. To avoid working with cup products with local coefficients, we assume that $\pi_1 B$ acts trivially on $H^*(F)$.

Lemma 9.23. *$H^p(B; H^qF) \cong E_2^{p,q}$ is a bigraded algebra over R.*

Proof. The cup product on H^*B induces a bilinear map

$$H^p(B; H^qF) \times H^r(B; H^sF) \to H^{p+r}(B; H^qF \otimes H^sF).$$

Composing with the coefficient homomorphism induced by the cup product on $H^*(F)$

$$H^q(F) \otimes H^s(F) \to H^{q+s}(F)$$

gives the desired multiplication

$$E_2^{p,q} \otimes E_2^{r,s} = H^p(B; H^qF) \otimes H^r(B; H^sF) \to H^{p+r}(B; H^{q+s}F) = E_2^{p+r,q+s}.$$

□

In many contexts the map $E_2^{*,0} \otimes E_2^{0,*} \to E_2^{*,*}$ is an isomorphism. Theorem 2.33 can be quite useful in this regard, For example if R is a field and B and F are simply connected finite CW-complexes, then the map is an isomorphism.

Theorem 9.24. *The (Leray-Serre) cohomology spectral sequence of the fibration is a spectral sequence of R-algebras. More precisely:*

1. $E_t^{*,*}$ *is a* bigraded R-algebra; *i.e. there are products*

$$E_t^{p,q} \times E_t^{r,s} \to E_t^{p+r,q+s}.$$

2. $d_t : E_t \to E_t$ *is a* derivation. *This means that if $a \in E_t^{p,q}$ and $b \in E_t^{r,s}$,*

$$d_t(a \cdot b) = (d_t a) \cdot b + (-1)^{p+q} a \cdot d_t b. \tag{9.11}$$

3. *The product on E_{t+1} is induced from the one on E_t (see Exercise 154 below) starting with the product on E_2 given by cup products, as in Lemma 9.23.*

4. *The following two ring structures on E_∞ coincide. (This assertion is a compatibility condition which relates the cup products on B, F, and E.)*
   (a) *Make $E_\infty^{*,*}$ a bigraded R-algebra by using the fact that each $(a,b) \in E_\infty^{p,q} \times E_\infty^{r,s}$ is represented by an element of $E_t^{p,q} \times E_t^{r,s}$ for t large enough.*

(b) *The (usual) cup product*

$$\cup : H^*(E) \times H^*(E) \to H^*(E)$$

is "filtration preserving"; i.e. the diagram

$$\begin{array}{ccc} F^{p,q} \times F^{r,s} & \longrightarrow & F^{p+r,q+s} \\ \downarrow & & \downarrow \\ H^{p+q}E \times H^{r+s}E & \xrightarrow{\cup} & H^{p+q+r+s}E \end{array}$$

commutes (this comes from the construction of the filtration), and so this cup product induces a product on the associated graded module, i.e. on E_∞ (see Exercise 155).

□

Exercise 154. Suppose that E is a graded ring and $d : E \to E$ is a differential ($d^2 = 0$) and a derivation (Equation 9.11). Show that the cohomology $H^*(E, d)$ inherits a graded ring structure.

Exercise 155. Show that a filtration-preserving multiplication on a filtered algebra induces a multiplication on the associated graded algebra.

Proposition 9.25. *The rational cohomology ring of $K(\mathbf{Z}, n)$ is a polynomial ring on one generator if n is even and a truncated polynomial ring on one generator (in fact an exterior algebra on one generator) if n is odd:*

$$H^*(K(\mathbf{Z}, n); \mathbf{Q}) = \begin{cases} \mathbf{Q}[\iota_n] & \text{if } n \text{ is even,} \\ \mathbf{Q}[\iota_n]/\iota_n^2 & \text{if } n \text{ is odd} \end{cases}$$

where $\deg(\iota_n) = n$.

Proof. We induct on n. For $n = 1$, $K(\mathbf{Z}, 1) = S^1$ which has cohomology ring $\mathbf{Z}[\iota_1]/\iota_1^2$.

Suppose the theorem is true for $k < n$. Consider the Leray–Serre spectral sequence for path space fibration $K(\mathbf{Z}, n-1) \hookrightarrow P \to K(\mathbf{Z}, n)$ for cohomology with rational coefficients. Then

$$E_2^{p,q} = H^p(K(\mathbf{Z}, n); \mathbf{Q}) \otimes_{\mathbf{Q}} H^q(K(\mathbf{Z}, n-1); \mathbf{Q}) \Rightarrow H^{p+q}(P; \mathbf{Q}).$$

Since $H^{p+q}(P; \mathbf{Q}) = 0$ for $(p, q) \neq (0, 0)$, the differential

$$d_n : E_n^{0,n-1} \to E_n^{n,0}$$

must be an isomorphism. Since $E_n^{0,n-1} = H^{n-1}(K(\mathbf{Z}, n-1); \mathbf{Q}) \cong \mathbf{Q}$, generated by ι_{n-1}, and $E_n^{n,0} = E_2^{n,0} = H^n(K(\mathbf{Z}, n); \mathbf{Q}) \cong \mathbf{Q}$, generated by ι_n, it follows that $d_n(\iota_{n-1})$ is a non-zero multiple of ι_n. By rescaling the generator ι_n by a rational number, assume inductively that $d_n(\iota_{n-1}) = \iota_n$.

Consider the cases n even and n odd separately. If n is even, then since $H^q(K(\mathbf{Z}, n-1); \mathbf{Q}) = 0$ unless $q = 0$ or $n-1$, $E_2^{p,q} = 0$ unless $q = 0$ or $n-1$. This implies that $0 = E_\infty^{p,q} = E_{n+1}^{p,q}$ for $(p,q) \neq (0,0)$ and the derivation property of d_n says that $d_n(\iota_{n-1}\iota_n^r) = \iota_n^{r+1}$ which, by induction on r, is non-zero. It follows easily from $0 = E_\infty^{p,q} = E_{n+1}^{p,q}$ for $(p,q) \neq (0,0)$ that $H^p(K(\mathbf{Z}, n); \mathbf{Q}) = 0$ if p is not a multiple of n and is isomorphic to $\mathbf{Q}$ for $p = nr$. Since ι_n^r is non-zero it generates $H^{nr}(K(\mathbf{Z}, n); \mathbf{Q}) \cong \mathbf{Q}$, and so $H^*(K(\mathbf{Z}, n); \mathbf{Q})$ is a polynomial ring on ι_n as required.

If n is odd, the derivation property of d_n implies that

$$d_n(\iota_{n-1}^2) = d_n(\iota_{n-1})\iota_{n-1} + (-1)^{n-1}\iota_{n-1}d_n(\iota_{n-1}) = 2\iota_{n-1}\iota_n.$$

Hence $d_n : E_n^{0,2n-2} \to E_2^{n,n-1}$ is an isomorphism. More generally by induction one sees that $d_n(\iota_{n-1}^r) = r\iota_n\iota_{n-1}^{r-1}$, so that $d_n : E_n^{0,r(n-1)} \to E_2^{n,(r-1)(n-1)}$ is an isomorphism. It is then easy to see that the spectral sequence collapses at E_{n+1}, and hence $H^p(K(\mathbf{Z}, n); \mathbf{Q}) = \mathbf{Q}$ for $p = 0$ or n and zero otherwise. □

We will show how to use Theorem 9.24 to compute $\pi_4 S^3$. This famous theorem was first proven by G.W. Whitehead and Rohlin (independently). The argument is effortless using spectral sequences.

Theorem 9.26. $\pi_4 S^3 = \mathbf{Z}/2$.

Proof. Since $\mathbf{Z} = H^3(S^3) = [S^3, K(\mathbf{Z}, 3)]$, choose a map $f : S^3 \to K(\mathbf{Z}, 3)$ representing the generator. For example, $K(\mathbf{Z}, 3)$ can be obtained by adding 5 cells, 6 cells, etc., to S^3 inductively to kill all the higher homotopy groups of S^3, and then f can be taken to be the inclusion. The Hurewicz theorem implies that $f_* : \pi_3 S^3 \to \pi_3(K(\mathbf{Z}, 3))$ is an isomorphism.

Pull back the fibration

$$K(\mathbf{Z}, 2) \to * \to K(\mathbf{Z}, 3)$$

(this is shorthand for $\Omega K(\mathbf{Z}, 3) \hookrightarrow P \to K(\mathbf{Z}, 3)$ where P is the contractible path space) via f to get a fibration

$$K(\mathbf{Z}, 2) \to X \to S^3. \tag{9.12}$$

Alternatively, let X be the homotopy fiber of f; i.e. $X \to S^3 \to K(\mathbf{Z}, 3)$ is a fibration up to homotopy. Then $\Omega K(\mathbf{Z}, 3) \sim K(\mathbf{Z}, 2)$ is the homotopy fiber of $X \to S^3$ by Theorem 6.40. (We will use the fibration (9.12) again in Chapter 10.)

From the long exact sequence of homotopy groups for a fibration we see that $\partial : \pi_3 S^3 \to \pi_2(K(\mathbf{Z},2))$ is an isomorphism. Hence

$$\pi_k X = \begin{cases} 0 & \text{if } k \leq 3, \\ \pi_k S^3 & \text{if } k > 3. \end{cases}$$

In particular, $H_4 X = \pi_4 X = \pi_4 S^3$. We will try to compute $H_4 X$ using a spectral sequence.

Consider the cohomology spectral sequence for the fibration (9.12). Then $E_2^{p,q} = H^p(S^3; H^q K(\mathbf{Z},2))$. Recall that $K(\mathbf{Z},2)$ is the infinite complex projective space $\mathbf{C}P^\infty$ whose cohomology algebra is the 1-variable polynomial ring $H^*(K(\mathbf{Z},2)) = \mathbf{Z}[c]$ where $\deg(c) = 2$.

Exercise 156. Give another proof of the fact that $H^*(K(\mathbf{Z},2)) = \mathbf{Z}[c]$ using the spectral sequence for the path space fibration

$$K(\mathbf{Z},1) \to * \to K(\mathbf{Z},2)$$

and the identification of $K(\mathbf{Z},1)$ with S^1. (Hint: the argument is contained in the proof of Proposition 9.25.)

Let $i \in H^3(S^3)$ denote the generator. Then the E_2-stage in the spectral sequence is indicated in the following diagram. The labels mean that the groups in question are infinite cyclic with the indicated generators. The empty entries are zero. The entries in this table are computed using Lemma 9.23.

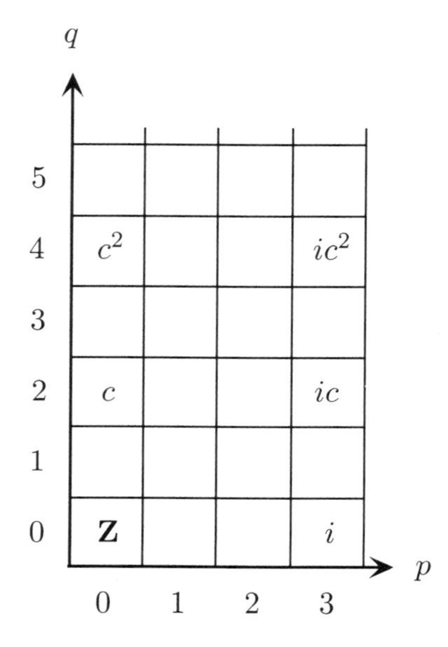

Since $H^2X = 0 = H^3X$ it follows that $d^3c = i$. Therefore,

$$d^3c^2 = ic + ci = 2ci.$$

This implies that $\mathbf{Z}/2 \cong E_4^{3,2} = E_\infty^{3,2} \cong H^5X$ and $0 = E_4^{0,4} = E_\infty^{0,4} = H^4X$. The universal coefficient theorem implies that $H_4X = \mathbf{Z}/2$. We conclude that $\mathbf{Z}/2 \cong \pi_4X = \pi_4S^3$, as desired. □

Corollary 9.27. *$\pi_{n+1}S^n = \mathbf{Z}/2$ for all $n \geq 3$. In particular, $\pi_1^S = \mathbf{Z}/2$.*

Proof. This is an immediate consequence of the Freudenthal suspension theorem (Theorem 8.7). □

Corollary 9.28. *$\pi_4S^2 = \mathbf{Z}/2$.*

Proof. Apply the long exact sequence of homotopy groups to the Hopf fibration $S^1 \hookrightarrow S^3 \to S^2$. □

The reader should think about the strategy used to make these computations. On the one hand fibrations were used to relate homotopy groups of various spaces; on the other spectral sequences are used to compute homology groups. The Hurewicz theorem is then used to conclude that a homology group computation in fact gives a homotopy group computation.

9.6. Homology of groups

Definition 9.29. Let G be a group. Define the *cohomology of G with $\mathbf{Z}$ coefficients* by

$$H^k(G; \mathbf{Z}) = H^k(K(G, 1); \mathbf{Z}).$$

Similarly define the *homology of G*

$$H_k(G; \mathbf{Z}) = H_k(K(G, 1); \mathbf{Z}).$$

More generally define the homology and cohomology of G with coefficients in any R-module A to be the corresponding homology or cohomology of $K(G, 1)$.

Corollary 7.27 implies that the homology and cohomology of a group are well-defined. Moreover, the assignment $G \mapsto K(G, 1)$ is functorial and takes short exact sequences to fibrations. (The functoriality can be interpreted in two different ways. For every group one associates a homotopy type of spaces, and a group homomorphism leads to a homotopy class of maps between the spaces. Alternatively, one can construct an honest functor from the category of groups to the category of spaces by giving a specific model of $K(G, 1)$ related to the bar resolution in homological algebra.)

Groups are very mysterious non-abelian things and thus are hard to study. The homology of groups gives abelian invariants and has been very useful in group theory as well as topology.

It follows that to understand the homology of groups related by exact sequences amounts to understanding the homology of a fibration, for which, as we have seen, spectral sequences are a good tool.

It is easy to see that $K(A \times B, 1) = K(A, 1) \times K(B, 1)$, and so the Künneth theorem can be used to compute the cohomology of products of groups. Therefore the following result is all that is needed to obtain a complete computation of the cohomology of finitely generated abelian groups.

Theorem 9.30. *The cohomology of* $\mathbf{Z}/n$ *is given by*

$$H^q(\mathbf{Z}/n; \mathbf{Z}) = \begin{cases} \mathbf{Z} & \text{if } q = 0, \\ 0 & \text{if } q \text{ is odd, and} \\ \mathbf{Z}/n & \text{if } q > 0 \text{ is even.} \end{cases}$$

Proof. The exact sequence $0 \to \mathbf{Z} \xrightarrow{\times n} \mathbf{Z} \to \mathbf{Z}/n \to 0$ induces a fibration sequence

$$K(\mathbf{Z}, 2) \to K(\mathbf{Z}, 2) \to K(\mathbf{Z}/n, 2)$$

(see Proposition 7.28). By looping this fibration twice (i.e. taking iterated homotopy fibers twice; see Theorem 6.40) we obtain the fibration

$$K(\mathbf{Z}, 1) \to K(\mathbf{Z}/n, 1) \to K(\mathbf{Z}, 2).$$

The fiber $K(\mathbf{Z}, 1)$ is a circle.

Consider the spectral sequence for this fibration. The base is simply connected so there is no twisting in the coefficients. Notice that

$$E_2^{p,q} = H^p(K(\mathbf{Z}, 2); H^q S^1) = \begin{cases} 0 & \text{if } q > 1, \text{ and} \\ H^p(K(\mathbf{Z}, 2); \mathbf{Z}) & \text{if } q = 0 \text{ or } 1. \end{cases}$$

Using Lemma 9.23, the E_2-stage is given by the following table, with the empty entries equal to 0 and the others infinite cyclic with the indicated generators (where i is the generator of $H^1(S^1)$).

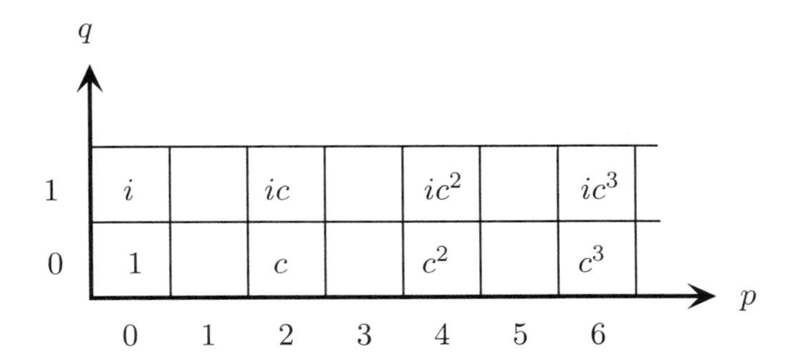

Of course $d^2(i) = kc$ for some integer k, and the question is: what might k be? We can find out by "peeking at the answer". Since $E_\infty^{0,2} = 0 = E_\infty^{1,1}$, we see that $H^2(K(\mathbf{Z}/n,1)) = E_\infty^{2,0} \cong \mathbf{Z}/k$. Since $\pi_1(K(\mathbf{Z}/n,1)) = \mathbf{Z}/n$, by the universal coefficient theorem, we see that H^2 must be $\mathbf{Z}/n$ and hence $k = \pm n$. (Neat, huh?)

Let $\bar{c}$ be the image of c in $E_3^{2,0}$. Here is a picture of the E^3-stage.

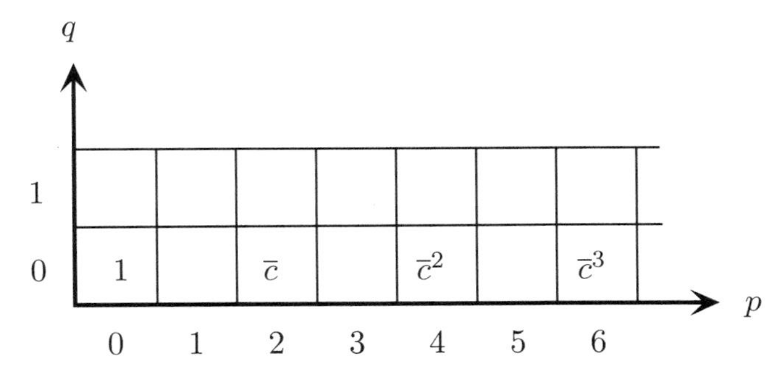

From this we see that the spectral sequence collapses at E^3, and that as graded rings $E_\infty^{*,0} \cong H^*(K(\mathbf{Z}/n,1))$. This not only completes the proof of the theorem, but also computes the cohomology ring

$$H^*(K(\mathbf{Z}/n,1)) = \mathbf{Z}[\bar{c}] \;/\langle n\bar{c}\rangle.$$

□

Also, we can get the homology from the cohomology by using the universal coefficient theorem:

$$H_q(\mathbf{Z}/n) = \begin{cases} \mathbf{Z} & \text{if } q = 0, \\ \mathbf{Z}/n & \text{if } q \text{ is odd, and} \\ 0 & \text{if } q > 0 \text{ is even.} \end{cases}$$

In applications, it is important to know the mod p-cohomology ring (which is the mod p-cohomology ring on an infinite-dimensional lens space).

By the Künneth theorem (which implies that, if we use field coefficients, $H^*(X \times Y) \cong H^*(X) \otimes H^*(Y)$), it suffices to consider the case where n is a prime power. Let $\mathbf{F}_p$ denote the field $\mathbf{Z}/p\mathbf{Z}$ for a prime p.

Exercise 157. Show that $H^*(\mathbf{Z}/2; \mathbf{F}_2) \cong \mathbf{F}_2[a]$ where a has degree one, and if $p^k \neq 2$, $H^*(\mathbf{Z}/p^k; \mathbf{F}_p) \cong \Lambda(a) \otimes \mathbf{F}_p[b]$, where a has degree one and b has degree 2. Here $\Lambda(a)$ is the 2-dimensional graded algebra over $\mathbf{F}_p$ with $\Lambda(a)^0 \cong \mathbf{F}_p$ with generator 1, and $\Lambda(a)^1 \cong \mathbf{F}_p$ with generator a. (Hint: use $\mathbf{R}P^\infty = K(\mathbf{Z}/2, 1)$ and $a \cdot a = -a \cdot a$ for $a \in H^1$.)

Exercise 158. Compute $H^p(K(\mathbf{Z}/2, n); \mathbf{Z}/2)$ for as many p and n as you can. (Hint: try induction on n, using the fibration

$$K(\mathbf{Z}/2, n) \to * \to K(\mathbf{Z}/2, n+1).)$$

9.7. Homology of covering spaces

Suppose that $f : \tilde{X} \to X$ is a regular cover of a path connected space X. Letting $G = \pi_1(X)/f_*(\pi_1(\tilde{X}))$, $f : \tilde{X} \to X$ is a principal G-bundle (with G discrete). Thus $G \hookrightarrow \tilde{X} \to X$ is pulled back from the universal G-bundle $G \hookrightarrow EG \to BG$ (see Theorem 8.22). In other words, there is a diagram

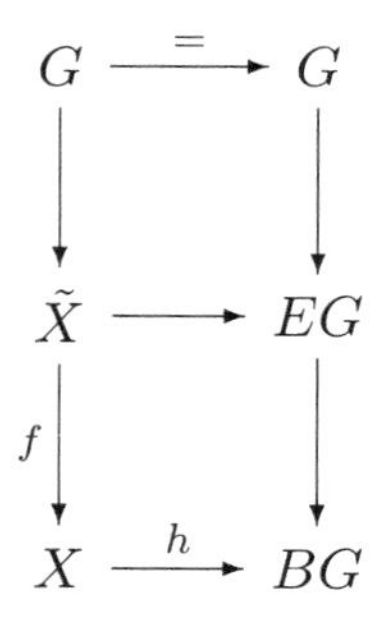

It follows that the sequence

$$\tilde{X} \to X \to BG$$

is a fibration (up to homotopy). (One way to see this is to consider the Borel fibration $\tilde{X} \hookrightarrow \tilde{X} \times_G EG \to X$. Since G acts freely on $\tilde{X}$, there is another fibration $EG \hookrightarrow \tilde{X} \times_G EG \to \tilde{X}/G$. Since EG is contractible we see that the total space of the Borel fibration is homotopy equivalent to X.) Since G is discrete, $BG = K(G, 1)$. Applying the homology (or cohomology) spectral sequence to this fibration immediately gives the following *spectral sequence of a covering space* (we use the notation $H_*(G) = H_*(K(G, 1))$).

Theorem 9.31. *Given a regular cover $f : \tilde{X} \to X$ with group of covering automorphisms $G = \pi_1(X)/f_*(\pi_1(\tilde{X}))$, there is a homology spectral sequence*

$$H_p(G; H_q(\tilde{X})) \cong E^2_{p,q} \Rightarrow H_{p+q}(X)$$

and a cohomology spectral sequence

$$H^p(G; H^q(\tilde{X})) \cong E_2^{p,q} \Rightarrow H^{p+q}(X).$$

The twisting of the coefficients is just the one induced by the action of G on $\tilde{X}$ by covering transformations. □

Applying the five-term exact sequence (Corollary 9.14) in this context gives the very useful exact sequence

$$H_2(X) \to H_2(G) \to H_0(G; H_1(\tilde{X})) \to H_1(X) \to H_1(G) \to 0.$$

Exercise 159. Use the spectral sequence of the universal cover to show that for a path connected space X the sequence

$$\pi_2(X) \xrightarrow{\rho} H_2(X) \to H_2(\pi_1(X)) \to 0$$

is exact, where ρ denotes the Hurewicz map.

As an application we examine the problem of determining which finite groups G can act freely on S^k. Equivalently, what are the fundamental groups of manifolds covered by the k-sphere? First note that if $g : S^k \to S^k$ is a fixed-point free map, then g is homotopic to the antipodal map (can you remember how to prove this?), and so is orientation-preserving if k is odd and orientation-reversing if k is even. Thus if k is even, the composite of any two non-trivial elements of G must be trivial, from which it follows that G has 1 or 2 elements. We shall henceforth assume k is odd, and hence that G acts by orientation-preserving fixed-point free homeomorphisms.

Thus the cohomology spectral sequence for the cover has

$$E_2^{p,q} = \begin{cases} H^p(G; H^q(S^k)) = H^p(G) & \text{if } q = 0 \text{ or } q = k, \\ 0 & \text{otherwise} \end{cases}$$

and converges to $H^{p+q}(S^k/G)$. This implies that the only possible non-zero differentials are

$$d_k : E_{k+1}^{p,k} \to E_{k+1}^{p-k-1,0}$$

and that the spectral sequence collapses at E_{k+2}.

Notice that S^k/G is a compact manifold of dimension k, and in particular $H^n(S^k/G) = 0$ for $n > k$. This forces $E_\infty^{p,q} = 0$ whenever $p + q > k$. Hence the differentials $d_k : E_{k+1}^{p,k} \to E_{k+1}^{p+k+1,0}$ are isomorphisms for $p \geq 1$, and since these are the only possible non-zero differentials we have

$$E_{k+1}^{p,k} = E_2^{p,k} \cong H^p(G) \text{ and } E_{k+1}^{p+k+1,0} = E_2^{p+k+1,0} \cong H^{p+k+1}(G)$$

so that $H^p(G) \cong H^{p+k+1}(G)$ for $p \geq 1$.

Thus G has *periodic cohomology* with *period* $k + 1$. Any subgroup of G also acts freely on S^k by restricting the action. This implies the following theorem.

Theorem 9.32. *If the finite group G acts freely on an odd-dimensional sphere S^k, then every subgroup of G has periodic cohomology of period $k+1$.*
□

As an application, first note the group $\mathbf{Z}/p \times \mathbf{Z}/p$ does not have periodic cohomology; this can be checked using the Künneth theorem. We conclude that any finite group acting freely on a sphere cannot contain a subgroup isomorphic to $\mathbf{Z}/p \times \mathbf{Z}/p$.

9.8. Relative spectral sequences

In studying maps of fibrations, it is useful to have relative versions of the homology and cohomology spectral sequence theorems. There are two relative versions, one involving a subspace of the base and one involving a subspace of the fiber.

Theorem 9.33. *Let $F \hookrightarrow E \xrightarrow{f} B$ be a fibration with B a CW-complex. Let $A \subset B$ a subcomplex. Let $D = p^{-1}(A)$.*

1. *There is a homology spectral sequence with*
$$H_p(B, A; G_qF) \cong E^2_{p,q} \Rightarrow G_{p+q}(E, D).$$
2. *If B is finite-dimensional or if there exists an N so that $G^q(F) = 0$ for all $q < N$, there is a cohomology spectral sequence with*
$$H^p(B, A; G^qF) \cong E_2^{p,q} \Rightarrow G^{p+q}(E, D).$$

□

Theorem 9.34. *Let $F \hookrightarrow E \xrightarrow{f} B$ be a fibration with B a CW-complex. Let $E_0 \subset E$ so that $f|_{E_0} : E_0 \to B$ is a fibration with fiber F_0.*

1. *There is a homology spectral sequence with*
$$H_p(B; G_q(F, F_0)) \cong E^2_{p,q} \Rightarrow G_{p+q}(E, E_0).$$
2. *If B is finite-dimensional or if there exists an N so that $G^q(F, F_0) = 0$ for all $q < N$, there is a cohomology spectral sequence with*
$$H^p(B; G^q(F, F_0)) \cong E_2^{p,q} \Rightarrow G^{p+q}(E, E_0).$$

□

9.9. Projects for Chapter 9

9.9.1. Construction of the spectral sequence. Give (or outline) the construction of the Leray-Serre-Atiyah-Hirzebruch spectral sequence and prove the main theorem, Theorem 9.6. References include [**43**, Sect. XIII.5] and [**36**, Ch. 9] (only for ordinary homology).

Chapter 10

Further Applications of Spectral Sequences

10.1. Serre classes of abelian groups

Definition 10.1. A *Serre class of abelian groups* is a non-empty collection $\mathcal{C}$ of abelian groups satisfying:

1. If $0 \to A \to B \to C \to 0$ is a short exact sequence, then $B \in \mathcal{C}$ if and only if $A, C \in \mathcal{C}$.

Moreover, there are additional axioms which can be useful:

2A. If $A, B \in \mathcal{C}$, then $A \otimes B \in \mathcal{C}$ and $\mathrm{Tor}(A, B) \in \mathcal{C}$.

2B. If $A \in \mathcal{C}$, then $A \otimes B \in \mathcal{C}$ for *any* abelian group B.

3. If $A \in \mathcal{C}$, then $H_n(A; \mathbf{Z}) = H_n(K(A, 1); \mathbf{Z})$ is in $\mathcal{C}$ for every $n > 0$.

Exercise 160. Prove that Axiom 2B implies Axiom 2A. (Hint: show that $\mathrm{Tor}(A, B) \subset A \otimes F$ for some F.)

There are many examples of Serre classes, including the trivial class, the class of all abelian groups, the class of torsion abelian groups, torsion abelian groups such that no element is p^r-torsion for a fixed prime p, the class of finite abelian groups, and the class of abelian p-groups. You should think about which of the axioms these classes satisfy.

It suffices for our exposition to consider the following two examples.

1. The class $\mathcal{C}_{FG}$ of finitely generated abelian groups. Axioms 1 and 2A clearly hold (see Exercise 28 and the remark preceding it). Note, however, that $\mathbf{Z} \in \mathcal{C}_{FG}$, but $\mathbf{Z} \otimes \mathbf{Q}$ is not in $\mathcal{C}_{FG}$, so 2B does not hold.

Axiom 3 follows from Theorem 9.30, the Künneth theorem, and the fact that $K(\mathbf{Z}, 1) = S^1$.

2. Let P denote a subset of the set of all prime numbers. Let $\mathcal{C}_P$ denote the class of torsion abelian groups A so that no element of A has order a positive power of p for $p \in P$. Thus for example, if P is empty, then $\mathcal{C}_P$ is the class of all torsion abelian groups. If P denotes all primes, then $\mathcal{C}_P$ is the class containing only the trivial group. If P consists of the single prime p, then we use the notation $\mathcal{C}_p$ for $\mathcal{C}_P$.

We will show that the class $\mathcal{C}_P$ satisfies Axioms 1, 2B, and 3. First some terminology: given a prime p, the *p-primary subgroup* of an abelian group consists of the subgroup of those elements whose order is a power of p. Thus $\mathcal{C}_P$ consists of those torsion abelian groups whose p-primary subgroup is trivial for any $p \in P$.

Lemma 10.2. *The class $\mathcal{C}_P$ satisfies Axioms 1, 2B, and 3.*

Proof. Say that an integer $n \neq 0$ is *prime to P* if p does not divide n for all $p \in P$. Then an abelian group A is in $\mathcal{C}_P$ if and only if for all $a \in A$, there is an n prime to P so that $na = 0$.

We first prove Axiom 1. Let $0 \to A \xrightarrow{\alpha} B \xrightarrow{\beta} C \to 0$ be an exact sequence of abelian groups. If $B \in \mathcal{C}_P$, then for $a \in A$, there is an n prime to P so that $n\alpha(a) = 0$. Hence $na = 0$, and so $A \in \mathcal{C}_P$. If $B \in \mathcal{C}_P$, then for $c \in C$, choose $b \in \beta^{-1}(c)$ and n prime to P so that $nb = 0$. Then $nc = n\beta(b) = 0$ and hence $C \in \mathcal{C}_P$. Conversely assume $A, C \in \mathcal{C}_P$. Then for $b \in B$, there exists an n prime to P so that $n\beta(b) = 0$. By exactness $nb = \alpha(a)$ for some a. Choose m prime to P so that $ma = 0$. Then $mnb = m\alpha(a) = 0$, so $B \in \mathcal{C}_P$.

Next comes Axiom 2B. Suppose that $A \in \mathcal{C}_P$ and let B be an arbitrary abelian group. Pick an element $t = \sum_i a_i \otimes b_i \in A \otimes B$. Since $A \in \mathcal{C}_P$, we can find integers n_i prime to P so that $n_i a_i = 0$. Let $n = \prod n_i$; this is prime to P and $nt = 0$. Thus $A \otimes B \in \mathcal{C}_P$.

We turn to the proof of Axiom 3. Let $A \in \mathcal{C}_P$. Suppose first that A is finitely generated. Then A is isomorphic to the finite direct sum of cyclic groups $A \cong \oplus_i \mathbf{Z}/p_i^{r_i}$ where $p_i \notin P$. Using Theorem 9.30, the Künneth theorem, and induction, it follows that $H_n(K(A, 1); \mathbf{Z})$ is a finitely generated torsion abelian group with trivial p-primary subgroup for any $p \in P$.

Next let $A \in \mathcal{C}_P$ be arbitrary and pick an $\alpha \in H_n(K(A, 1); \mathbf{Z})$. Choose a cycle z representing the homology class α. Since z is a finite sum of singular simplices there is a finite subcomplex $X \subset K(A, 1)$ containing the image of every singular simplex in z. Therefore α lies in the image of $H_n(X) \to H_n(K(A, 1))$. Let $A' \subset A$ denote the finitely generated subgroup $\operatorname{im}(\pi_1(X) \to \pi_1(K(A, 1)))$. Since $\pi_1(K(A, 1)) = A \in \mathcal{C}_P$, the subgroup A' is also in $\mathcal{C}_P$. The space $K(A',1)$ can be constructed by adding k-cells to X for $k \geq 2$, and since $A' \to A$ is injective the inclusion $X \subset K(A,1)$ can be extended to give the commutative diagram

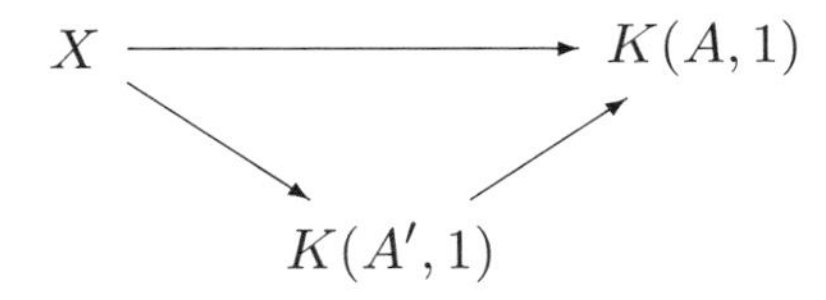

Thus $\alpha \in \operatorname{im}(H_n(K(A',1)) \to H_n(K(A,1)))$, and since A' is finitely generated, α is torsion with order relatively prime to p for $p \in P$. Thus $H_n(K(A,1)) \in \mathcal{C}_P$. $\square$

Definition 10.3. Given a Serre class $\mathcal{C}$, a homomorphism $\varphi : A \to B$ between two abelian groups is called:

1. a $\mathcal{C}$-*monomorphism* if $\ker \varphi \in \mathcal{C}$,
2. a $\mathcal{C}$-*epimorphism* if $\operatorname{coker} \varphi \in \mathcal{C}$, and
3. a $\mathcal{C}$-*isomorphism* if $\ker \varphi \in \mathcal{C}$ and $\operatorname{coker} \varphi \in \mathcal{C}$.

Two abelian groups A and B are called $\mathcal{C}$-*isomorphic* if there exists an abelian group C and two $\mathcal{C}$-isomorphisms $f : C \to A$ and $g : C \to B$.

Lemma 10.4. *Let $\alpha : A \to B$ and $\beta : B \to C$ be homomorphisms of abelian groups. If two of the three maps α, β, and $\beta \circ \alpha$ are $\mathcal{C}$-isomorphisms, then so is the third.*

Proof. This follows from the exact sequence

$$0 \to \ker \alpha \to \ker \beta \circ \alpha \xrightarrow{\alpha} \ker \beta \to \operatorname{coker} \alpha \xrightarrow{\beta} \operatorname{coker} \beta \circ \alpha \to \operatorname{coker} \beta \to 0.$$

$\square$

We will sometimes write $A \cong B \bmod \mathcal{C}$ to indicate that A and B are $\mathcal{C}$-isomorphic.

Exercise 161. Let $\mathcal{C}_\phi$ be the class of torsion abelian groups. Show that $A \cong B \bmod \mathcal{C}_\phi$ if and only if $A \otimes \mathbf{Q} \cong B \otimes \mathbf{Q}$.

Exercise 162. Prove the five-lemma "mod $\mathcal{C}$".

The Hurewicz theorem has the following extremely useful generalization.

Theorem 10.5.

1. (mod $\mathcal{C}$ Hurewicz theorem) *Let X be 1-connected, and suppose $\mathcal{C}$ satisfies Axioms 1, 2A, and 3.*
   (a) *If $\pi_i X \in \mathcal{C}$ for all $i < n$, then $H_i X \in \mathcal{C}$ for all $0 < i < n$ and the Hurewicz map $\pi_n X \to H_n X$ is a $\mathcal{C}$-isomorphism.*

   (b) *If* $H_iX \in \mathcal{C}$ *for all* $0 < i < n$*, then* $\pi_iX \in \mathcal{C}$ *for all* $i < n$ *and the Hurewicz map* $\pi_nX \to H_nX$ *is a* $\mathcal{C}$*-isomorphism.*

2. (mod $\mathcal{C}$ relative Hurewicz theorem) *Suppose* $A \subset X$*,* A *and* X *are 1-connected, and* $\pi_2(X, A) = 0$*. Suppose* $\mathcal{C}$ *satisfies Axioms 1, 2B, and 3.*
   (a) *If* $\pi_i(X, A) \in \mathcal{C}$ *for all* $i < n$*, then* $H_i(X, A) \in \mathcal{C}$ *for all* $i < n$ *and the Hurewicz map* $\pi_n(X, A) \to H_n(X, A)$ *is a* $\mathcal{C}$*-isomorphism.*
   (b) *If* $H_i(X, A) \in \mathcal{C}$ *for all* $i < n$*, then* $\pi_i(X, A) \in \mathcal{C}$ *for all* $i < n$ *and the Hurewicz map* $\pi_n(X, A) \to H_n(X, A)$ *is a* $\mathcal{C}$*-isomorphism.*

Actually, as you can easily check, the part (b)'s above follow from the part (a)'s. We will give a proof of the theorem using spectral sequences and the fact that $\pi_1Y \to H_1Y$ is an isomorphism when the fundamental group is abelian. By taking $\mathcal{C}$ to be the class consisting of the trivial group, we obtain proofs of the classical Hurewicz and relative Hurewicz theorems. The proof we give simplifies a bit in the classical case. A proof of the classical case without the use of spectral sequences was a project in Chapter 6.

The mod $\mathcal{C}$ relative Hurewicz theorem implies the mod $\mathcal{C}$ Whitehead theorem.

Theorem 10.6 (mod $\mathcal{C}$ Whitehead theorem). *Let* $f : A \to X$*, where* A, X *are 1-connected, and suppose* $f : \pi_2A \to \pi_2X$ *is an epimorphism. Let* $\mathcal{C}$ *satisfy Axioms 1, 2B, and 3. Then the following two statements are equivalent.*

1. $f_* : \pi_iA \to \pi_iX$ *is an* $\mathcal{C}$*-isomorphism for* $i < n$ *and a* $\mathcal{C}$*-epimorphism for* $i = n$*.*
2. $f_* : H_iA \to H_iX$ *is a* $\mathcal{C}$*-isomorphism for* $i < n$ *and a* $\mathcal{C}$*-epimorphism for* $i = n$*.*

Exercise 163. Show that Theorem 10.6 follows from Theorem 10.5.

Since the homology groups of a finite CW-complex are all in $\mathcal{C}_{FG}$, the mod $\mathcal{C}$ Hurewicz theorem has the following important consequence.

Corollary 10.7. *If* X *is a simply connected finite CW-complex, then all the homotopy groups of* X *are finitely generated. More generally a simply connected space has finitely generated homology groups in every dimension if and only if it has finitely generated homotopy groups in each dimension.*

This sounds great, but we warn you that the only simply connected finite CW-complexes for which all homotopy groups have been computed are contractible.

The hypothesis in the corollary that X be simply connected is necessary, as the following exercise shows.

Exercise 164. Prove that $\pi_2(S^1 \vee S^2)$ is not finitely generated.

Exercise 165. Show that Corollary 10.7 holds more generally when $\pi_1 X$ is finite.

We turn now to the proof of the Hurewicz theorem.

Proof of Theorem 10.5. Here is the idea of the proof. For a space X, consider the path fibration

$$\Omega X \to PX \xrightarrow{f} X.$$

There is a commutative diagram

$$\begin{array}{ccccc} \pi_n(X, x_0) & \xleftarrow[\cong]{f_*} & \pi_n(PX, \Omega X) & \xrightarrow[\cong]{\partial} & \pi_{n-1}(\Omega X) \\ \downarrow{\scriptstyle\rho} & & \downarrow{\scriptstyle\rho} & & \downarrow{\scriptstyle\rho} \\ H_n(X, x_0) & \xleftarrow{f_*} & H_n(PX, \Omega X) & \xrightarrow[\cong]{\partial} & H_{n-1}(\Omega X) \end{array} \tag{10.1}$$

where the vertical maps are Hurewicz maps. The boundary maps ∂ are isomorphisms since PX is contractible. The top f_* is an isomorphism since f is a fibration (see Lemma 6.54). With the mod $\mathcal{C}$-connectivity hypothesis, a spectral sequence argument given below shows that the bottom f_* is a $\mathcal{C}$-isomorphism. Inductively, the right-hand ρ is a $\mathcal{C}$-isomorphism, so thereby the left-hand ρ is a $\mathcal{C}$-isomorphism. There are three difficulties with this outline. We have to get the induction started; we have to make the mod $\mathcal{C}$ spectral sequence argument; and we have to deal with the fact that if $\pi_2 X \neq 0$, then ΩX is not simply connected, so, strictly speaking, the inductive hypothesis does not apply.

We will use the following lemma, which shows why Serre classes are tailor-made to be used with spectral sequences.

Lemma 10.8. *Let $(E^r_{*,*}, d^r)$ be a first quadrant spectral sequence. Let $\mathcal{C}$ denote a class of abelian groups.*

1. *For any bigraded spectral sequence, if $\mathcal{C}$ satisfies Axiom 1, $E^n_{p,q} \in \mathcal{C}$ for some p, q implies that $E^r_{p,q} \in \mathcal{C}$ for all $r \geq n$.*

2. *Let $F \hookrightarrow E \xrightarrow{f} B$ be a fibration over a simply connected base space. If $\mathcal{C}$ satisfies Axioms 1 and 2A, if $H_p B \in \mathcal{C}$ for $0 < p < n$, and if $H_q F \in \mathcal{C}$ for $0 < q < n-1$, then $f_* : H_i(E, F) \to H_i(B, b_0)$ is a $\mathcal{C}$-isomorphism for $i \leq n$.*

Proof. 1. A subgroup or a quotient group of a group in $\mathcal{C}$ is in $\mathcal{C}$ by Axiom 1. Thus a subquotient of a group in $\mathcal{C}$ is in $\mathcal{C}$. Since $E^r_{p,q}$ is a subquotient of $E^n_{p,q}$, the first statement follows.

2. Consider the spectral sequence of the relative fibration (Theorem 9.33)

$$F \to (E, F) \to (B, b_0).$$

The E^2-term is

$$E^2_{p,q} \cong H_p(B, b_0; H_qF) \cong (H_p(B, b_0) \otimes H_qF) \oplus \operatorname{Tor}(H_{p-1}(B, b_0), H_qF),$$

and the spectral sequence converges to $H_*(E, F)$. Thus $E^2_{p,q} \in \mathcal{C}$ and hence $E^\infty_{p,q} \in \mathcal{C}$ when $p = 0, 1$ or when $1 < p < n$ and $0 < q < n - 1$ (see the shaded area in the picture below).

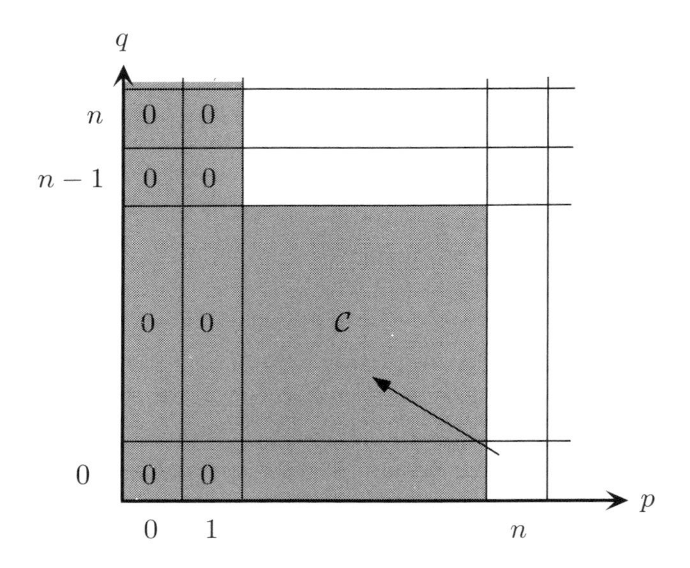

The picture gives a convincing argument that

$$H_i(E, F) \cong H_i(B, b_0) \mod \mathcal{C} \quad \text{for } i \leq n,$$

but here is a precise one.

The spectral sequence gives a filtration

$$0 = F_{-1,i+1} \subset F_{0,i} \subset F_{1,i-1} \subset \cdots \subset F_{i-1,1} \subset F_{i,0} = H_i(E, F)$$

with

$$E^\infty_{p,i-p} = F_{p,i-p}/F_{p-1,i-p+1}.$$

It follows by induction on p that $F_{p,i-p} \in \mathcal{C}$ for $p < i$, and hence that

$$H_i(E, F) \to E^\infty_{i,0} \quad \text{is a } \mathcal{C}\text{-isomorphism.} \tag{10.2}$$

On the other hand, for $r \geq 2$, the exact sequence

$$0 \to E^{r+1}_{i,0} \to E^r_{i,0} \xrightarrow{d^r} E^r_{i-r,r-1}$$

with the range of d^r in $\mathcal{C}$ shows that $E^{r+1}_{i,0} \to E^r_{i,0}$ is a $\mathcal{C}$-isomorphism. Since the composite of $\mathcal{C}$-isomorphisms is a $\mathcal{C}$-isomorphism, it follows by induction that

$$(10.3) \qquad E^{\infty}_{i,0} \to E^2_{i,0} \qquad \text{is a } \mathcal{C}\text{-isomorphism.}$$

Therefore, the composite of (10.2) and (10.3) is a $\mathcal{C}$-isomorphism. But this composite is identified with the edge homomorphism (see Theorem 9.12)

$$f_* : H_i(E, F) \to H_i(B, b_0).$$

□

We return to the proof of Part 1(a). The proof is by induction on n. For $n = 1$, $\pi_1 X = 0 = H_1 X$. For $n = 2$, $\pi_1 X = 0 = H_1 X$, and so Lemma 10.8 shows that $f_* : H_2(PX, \Omega X) \to H_2(X, x_0)$ is an isomorphism (use the class consisting only of the trivial group). Hence diagram (10.1), the fact that $\pi_1(\Omega X)$ is abelian since ΩX is an H-space, and Theorem 1.6 (the π_1-version of the Hurewicz theorem) show that $\rho : \pi_2 X \to H_2 X$ is an isomorphism for any simply connected X.

Now suppose $n > 2$ and inductively assume for simply connected spaces Y with $\pi_i Y \in \mathcal{C}$ for $i < n - 1$ that $H_i Y \in \mathcal{C}$ for $0 < i < n - 1$ and that the Hurewicz map $\rho : \pi_{n-1} Y \to H_{n-1} Y$ is a $\mathcal{C}$-isomorphism. Let X be a simply connected space so that $\pi_i X \in \mathcal{C}$ for $i < n$. There will be two cases: where $\pi_2 X = 0$ and where $\pi_2 X \neq 0$.

In the first case $\pi_1(\Omega X) = \pi_2 X = 0$, and $\pi_i(\Omega X) = \pi_{i+1} X$, so we can apply the inductive hypothesis to ΩX and conclude that the right-hand $\rho : \pi_{n-1}(\Omega X) \to H_{n-1}(\Omega X)$ in diagram (10.1) is a $\mathcal{C}$-isomorphism and that $H_i(\Omega X) \in \mathcal{C}$ for $i < n - 1$. Then Lemma 10.8 applied to the path fibration

$$\Omega X \to PX \xrightarrow{f} X$$

shows that the lower f_* in diagram (10.1) is a $\mathcal{C}$-isomorphism. Then Lemma 10.4 applied repeatedly to diagram (10.1) shows that $\pi_n X \to H_{n-1} X$ is a $\mathcal{C}$-isomorphism as desired.

Now suppose we are in the case where $\pi_2 X \neq 0$. By hypothesis, $\pi_2 X \in \mathcal{C}$. There is a map $f : X \to K(\pi_2 X, 2)$ inducing the identity on π_2. Let $X_2 \to X$ be the homotopy fiber of this map. Now turn this map into a fibration (see Theorem 6.40) to obtain the fibration

$$K(\pi_2 X, 1) \to X_2 \to X.$$

Note by Axiom 3, $H_i(K(\pi_2 X, 1)) \in \mathcal{C}$ for $i > 0$. This has two consequences, first that

$$H_i(X_2) \to H_i(X_2, K(\pi_2 X, 1))$$

is a $\mathcal{C}$-isomorphism for $i > 0$, and second that Lemma 10.8 applies and so

$$H_i(X_2, K(\pi_2 X, 1)) \to H_i(X, x_0)$$

is a $\mathcal{C}$-isomorphism for $0 < i < n$. Thus the composite of the last two maps $H_i(X_2) \to H_i X$ is a $\mathcal{C}$-isomorphism for $0 < i < n$. Summarizing,

1. $H_i X_2 \in \mathcal{C}$ for all $0 < i < n$,
2. $H_n X_2 \cong H_n X \bmod \mathcal{C}$,
3. $\pi_i X_2 = \pi_i X$ for all $i > 2$,
4. $\pi_1 X_2 = 0 = \pi_2 X_2$.

Thus $\rho : \pi_n(X_2) \to H_n(X_2)$ is a $\mathcal{C}$-isomorphism and $H_i(X_2) \in \mathcal{C}$ for $0 < i < n$ by the $\pi_2 = 0$ case; hence the same is true for X. This completes the proof of Part 1(a) of the mod $\mathcal{C}$ Hurewicz theorem.

Part 1(b) follows formally from Part 1(a). Indeed, let X be simply-connected and suppose $H_i X \in \mathcal{C}$ for $0 < i < n$. Then use induction on i to show $\rho : \pi_i X \to H_i X$ is a $\mathcal{C}$-isomorphism for $1 < i \leq n$.

We now show how to deduce the relative Hurewicz theorem from the absolute theorem. We assume that X and A are simply connected, nonempty, and that $\pi_2(X, A) = 0$. The diagram

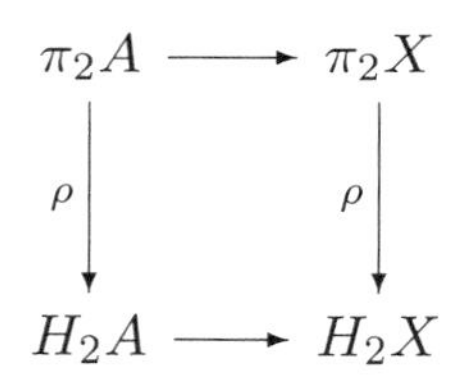

commutes, with the vertical maps Hurewicz isomorphisms and the horizontal maps induced by inclusion. Since the top horizontal map is surjective, so is the bottom one, and it follows from the long exact sequence in homology that $H_2(X, A) = 0$.

Now suppose $n > 2$ and inductively assume that for any simply connected pair $B \subset Y$ with $\pi_2(Y, B) = 0$ and $\pi_i(Y, B) \in \mathcal{C}$ for $i < n - 1$, that $H_i(Y, B) \in \mathcal{C}$ and that the Hurewicz map $\rho : \pi_i(Y, B) \to H_i(Y, B)$ is a $\mathcal{C}$-isomorphism for $i < n - 1$. Let (X, A) be a pair of simply connected spaces with $H_i(X, A) \in \mathcal{C}$ for $i < n$.

Now let $n > 2$ and assume that $\pi_k(X, A) \in \mathcal{C}$ for $k < n$. Then by induction $H_k(X, A) \in \mathcal{C}$ for $k < n$. We must show that $\rho : \pi_n(X, A) \to H_n(X, A)$ is a $\mathcal{C}$-isomorphism.

Let $f : PX \to X$ denote the path space fibration, and let $L = L(X, A) = f^{-1}(A)$. Thus we have a relative fibration

$$\Omega X \to (PX, L) \to (X, A).$$

Recall from page 154 in Chapter 6 that $\pi_{k-1}(L) \cong \pi_k(X, A)$ for all k. The Leray–Serre spectral sequence for this fibration has

$$E^2_{p,q} = H_p(X, A; H_q(\Omega X))$$

and converges to $H_{p+q}(PX, L)$. The coefficients are untwisted since X and A are simply connected.

We have

$$H_p(X, A; H_q(\Omega X)) = H_p(X, A) \otimes H_q(\Omega X) \oplus \operatorname{Tor}(H_{p-1}(X, A), H_q(\Omega X)) \in \mathcal{C}$$

for $p < n$. This follows from the fact that $H_p(X, A) \in \mathcal{C}$ for $p < n$ and Axiom 2B. (Don't let this point slip by you; this is where we needed Axiom 2B, which is stronger than 2A.)

Therefore (see the picture on page 272 again):

1. all differentials out of $E^r_{n,0}$ have range in $\mathcal{C}$, and so $H_n(X, A) \cong E^r_{n,0} \cong E^\infty_{n,0}$ mod $\mathcal{C}$ for all r, and
2. $E^\infty_{p,n-p} \in \mathcal{C}$ for $p > 0$, and so $H_n(PX, L) \cong E^\infty_{n,0}$ mod $\mathcal{C}$.

Arguing as above we have:

$$\begin{aligned} H_n(PX, L) &\cong E^\infty_{n,0} \bmod \mathcal{C} \\ &\cong H_n(X, A) \bmod \mathcal{C}. \end{aligned} \tag{10.4}$$

This $\mathcal{C}$-isomorphism is induced by the edge homomorphism and hence coincides with the homomorphism induced by $f : (PX, L) \to (X, A)$.

The diagram

$$\begin{array}{ccccc} \pi_n(X, A) & \xleftarrow{f_*} & \pi_n(PX, L) & \xrightarrow[\cong]{\partial} & \pi_{n-1}(L) \\ \downarrow{\scriptstyle \rho} & & \downarrow{\scriptstyle \rho} & & \downarrow{\scriptstyle \rho} \\ H_n(X, A) & \xleftarrow{f_*} & H_n(PX, L) & \xrightarrow[\cong]{\partial} & H_{n-1}(L) \end{array}$$

commutes, with the two right horizontal arrows isomorphisms by the long exact sequence of the pair in homology and homotopy groups and the fact that PX is contractible.

The top left horizontal arrow is an isomorphism since $f : PX \to X$ is a fibration (see Lemma 6.54). Since $\pi_{k-1}(L) = \pi_k(X, A) = 0$, $\pi_1(L) = 0$ and $\pi_k(L) \in \mathcal{C}$ for all $k < n - 1$. The absolute Hurewicz theorem implies that $\rho : \pi_{n-1}(L) \to H_{n-1}(L)$ is a $\mathcal{C}$-isomorphism.

Finally the bottom left horizontal map is a $\mathcal{C}$-isomorphism by (10.4). Moving around the diagram shows that the Hurewicz map $\rho : \pi_n(X, A) \to H_n(X, A)$ is a $\mathcal{C}$-isomorphism. This proves Part 2(a) of Theorem 10.5. As before, Part 2(b) follows from Part 2(a). □

10.2. Homotopy groups of spheres

In this section we will use the machinery of spectral sequences and Serre classes to obtain more non-trivial information about the elusive homotopy groups of spheres. An immediate consequence of Corollary 10.7 is the following.

Corollary 10.9. *The homotopy groups of spheres $\pi_k S^n$ are finitely generated abelian groups.* □

Here is a result which follows easily from Serre mod $\mathcal{C}$ theory.

Theorem 10.10. *If n is odd, $\pi_m S^n$ is finite for $m \neq n$.*

Proof. If $n = 1$, then $\pi_m S^1 = \pi_m \mathbf{R} = 0$ for $m \neq 1$. If $n > 1$ is odd, then a map $f : S^n \to K(\mathbf{Z}, n)$ inducing an isomorphism on π_n, induces an isomorphism on homology with rational coefficients by Proposition 9.25. Taking the Serre class $\mathcal{C}_\phi$ as in Exercise 161, we see f_* induces a $\mathcal{C}_\phi$-isomorphism on homology, and hence, by the mod $\mathcal{C}$ Whitehead theorem, also a $\mathcal{C}_\phi$-isomorphism on homotopy. Thus for all m, the kernel and cokernel of

$$f_* : \pi_m S^n \to \pi_m(K(\mathbf{Z}, n))$$

are torsion groups. However, the homotopy groups of spheres are finitely generated. The result follows.

□

Corollary 10.11. *The stable homotopy groups of spheres π_n^S are finite for $n > 0$.* □

Exercise 166. Prove that if n is even, then $\pi_k S^n$ is finite except for $k = n$ and $k = 2n-1$, and that $\pi_{2n-1} S^n$ is the direct sum of $\mathbf{Z}$ and a finite abelian group. (Hint: Let $S^{n-1} \to T \to S^n$ be the unit tangent bundle of S^n for n even. Show that $\pi_n T \to \pi_n S^n$ is not onto by, for example, showing that a lift of $\mathrm{Id} : S^n \to S^n$ leads to a nonzero vector field on S^n, and hence a homotopy from the identity to the antipodal map. Conclude that $\pi_n T \to \pi_n S^n$ is the zero map. By looking at the transgression, deduce that $H_n(T)$ is finite except for $H_0 T = H_{2n-1} T = \mathbf{Z}$. Find a map inducing an isomorphism $H_k(T) \cong H_k(S^{2n-1})$ mod $\mathcal{C}_\phi$. Then $\pi_k(T) \cong \pi_k(S^{2n-1})$ mod $\mathcal{C}_\phi$ via the Whitehead theorem. Then apply Theorem 10.10.)

For the next sequence of results, let $K(\mathbf{Z},2) \to X \to S^3$ be the fibration from (9.12). Thus X is the homotopy fiber of the map $S^3 \to K(\mathbf{Z},3)$ inducing an isomorphism on π_3. It follows that $\pi_n X = 0$ for $n \leq 3$ and $\pi_n X = \pi_n S^3$ for $n > 3$.

Lemma 10.12.

$$H_k X = \begin{cases} 0 & \textit{if } k \textit{ is odd,} \\ \mathbf{Z} & \textit{if } k = 0, \textit{ and} \\ \mathbf{Z}/n & \textit{if } k = 2n. \end{cases}$$

Proof. The integral cohomology ring of $K(\mathbf{Z},2) = \mathbf{C}P^\infty$ is a polynomial ring $H^*(K(\mathbf{Z},2);\mathbf{Z}) \cong \mathbf{Z}[c]$, where deg $c = 2$ (see Exercise 156).

Consider the cohomology spectral sequence for the fibration (9.12). This has

$$E_2 = H^*(S^3) \otimes \mathbf{Z}[c].$$

More precisely,

$$\begin{aligned} E_2^{p,q} &= H^p(S^3; H^q(K(\mathbf{Z},2))) \\ &= \begin{cases} 0 & \text{if } p = 1,2 \text{ or } p > 3, \text{ or if } q \text{ is odd,} \\ \mathbf{Z} \cdot c^k & \text{if } p = 0 \text{ and } q = 2k \text{ is even, and} \\ \mathbf{Z} \cdot \iota c^k & \text{if } p = 3 \text{ and } q = 2k \text{ is even,} \end{cases} \end{aligned}$$

where $\iota \in H^3(S^3)$ denotes the generator, using Theorem 2.33 (the universal coefficient theorem for cohomology). Notice also that since all the differentials d_2 are zero, $E_2 = E_3$.

Since $H^2 X = 0 = H^3 X$, the differential $d_3 : E_3^{0,2} \to E_3^{3,0}$ must be an isomorphism, and so $d_3 c = \iota$ (after perhaps replacing ι by $-\iota$). Thus

$$d(c^2) = (dc) \cdot c + c \cdot dc = \iota \cdot c + c \cdot \iota = 2\iota \cdot c.$$

More generally, one shows by an easy induction argument that

$$d(c^n) = n\iota \cdot c^{n-1}.$$

All other differentials in the spectral sequence are zero since either their domain or range is zero. Therefore $E_\infty^{3,2n-2} = \mathbf{Z}/n$ and hence $H^{2n+1}(X) = \mathbf{Z}/n$ if $n \geq 1$. The universal coefficient theorem now implies that $H_{2n}(X) = \mathbf{Z}/n$ for $n \geq 1$. $\square$

Corollary 10.13. *If p is a prime, the p-primary component of $\pi_i S^3$ is zero if $3 < i < 2p$, and is $\mathbf{Z}/p$ if $i = 2p$.*

Proof. We use the class $\mathcal{C}_p$. As before, let X be the space from the fibration (9.12). Lemma 10.12 implies that $H_i(X) \in \mathcal{C}_p$ for $0 < i < 2p$. Using the mod $\mathcal{C}$ Hurewicz theorem, we conclude that $\pi_i(X) \in \mathcal{C}_p$ for $0 < i < 2p$, and

$\mathbf{Z}/p = H_{2p}(X) \cong \pi_{2p}(X) \bmod \mathcal{C}_p$. This implies that the p-primary part of $\pi_{2p}(X)$ is $\mathbf{Z}/p$. The corollary now follows from the fact that $\pi_i(X) = \pi_i(S^3)$ for $i \neq 3$. $\square$

With a bit more work one can show that the p-primary component of π_n^S is trivial for $n < 2p-3$ and equals $\mathbf{Z}/p$ for $n = 2p-3$ (see [**36**]). Take a look at the table on page 208 to verify this in low dimensions. So for example, π_2^S has trivial p-primary part for $p > 2$ and $\pi_3^S = \mathbf{Z}/3 \oplus$ (2-primary subgroup).

We turn now to the computation of π_2^S.

In Theorem 9.28 we computed that $\pi_4 S^2 = \mathbf{Z}/2$. Consider the suspension map $s : S^2 \to \Omega S^3$, i.e. the adjoint of the identification $S(S^2) = S^3$. Let F be the homotopy fiber of s. The long exact sequence of homotopy groups of the fibration

$$F \to S^2 \xrightarrow{s} \Omega S^3 \tag{10.5}$$

shows that $\pi_1 F = \pi_2 F = 0$. Thus by the Hurewicz theorem $\pi_3 F = H_3 F$. The spectral sequence for the fibration (10.5) shows that the transgression $\tau : H_3 F \to H_4(\Omega S^3)$ is an isomorphism. In Chapter 9 we computed that $\mathbf{Z} \cong H_4(\Omega S^3)$ (see Equation (9.3)), and hence $\mathbf{Z} \cong H_3(F) = \pi_3(F)$.

The long exact sequence in homotopy groups for (10.5) is

$$\cdots \to \pi_4 S^2 \xrightarrow{s_*} \pi_4(\Omega S^3) \to \pi_3 F \to \pi_3 S^2 \to \pi_4 S^3 \to 0.$$

From the Hopf fibration $S^1 \to S^3 \to S^2$ we know that $\pi_3 S^2 = \mathbf{Z}$, and from Theorem 9.26 we know $\pi_4 S^3 = \mathbf{Z}/2$. Since $\pi_3 F = \mathbf{Z}$ it follows from this exact sequence that the suspension map $s_* : \pi_4 S^2 \to \pi_4(\Omega S^3) = \pi_5 S^3$ is onto. Therefore, $\pi_5 S^3$ is either 0 or $\mathbf{Z}/2$. We will show that $\pi_5 S^3 = \mathbf{Z}/2$.

Consider once again our friend the space X of the fibration (9.12). Since $\pi_4 X = \pi_4 S^3 = \mathbf{Z}/2$, let $f : X \to K(\mathbf{Z}/2, 4)$ be a map inducing an isomorphism on π_4 and let Y denote the homotopy fiber of f. Since Y is 4-connected, $H_5(Y; \mathbf{Z}) = \pi_5 Y = \pi_5 X = \pi_5 S^3$. Since $\pi_5 S^3$ is either 0 or $\mathbf{Z}/2$, the universal coefficient theorem implies that $\pi_5 S^3 = H^5(Y; \mathbf{Z}/2)$.

In the spectral sequence in $\mathbf{Z}/2$-cohomology for the fibration $Y \to X \to K(\mathbf{Z}/2, 4)$, the differential

$$H^5(F; \mathbf{Z}/2) = E_2^{0,5} = E_6^{0,5} \xrightarrow{d_6} E_6^{6,0} = E_2^{6,0} = H^6(K(\mathbf{Z}/2, 4); \mathbf{Z}/2)$$

is surjective. This follows from the fact that Y is 4-connected, $K(\mathbf{Z}/2, 4)$ is 3-connected, and from Lemma 10.12 which implies that $H^6(X; \mathbf{Z}/2) = 0$ (you should check this fact).

We will show in Section 10.5 below (Equation (10.12)) that

$$H^6(K(\mathbf{Z}/2, 4); \mathbf{Z}/2) = \mathbf{Z}/2.$$

Hence $H^5(F;\mathbf{Z}/2)$ surjects to $\mathbf{Z}/2$ and therefore equals $\mathbf{Z}/2$. Thus we have computed $\pi_5 S^3 = \mathbf{Z}/2$.

The homotopy exact sequence for the Hopf fibration $S^3 \to S^7 \to S^4$ shows that $\pi_2^S = \pi_6 S^4 \cong \pi_5 S^3$. In particular this shows that the sequence of suspension homomorphisms

$$\pi_2 S^0 \to \pi_3 S^1 \to \pi_4 S^2 \to \pi_5 S^3 \to \pi_6 S^4 = \pi_2^S$$

is

$$0 \to 0 \to \mathbf{Z}/2 \xrightarrow{\cong} \mathbf{Z}/2 \xrightarrow{\cong} \mathbf{Z}/2.$$

The long exact sequence of homotopy groups for the Hopf fibration $S^1 \to S^3 \to S^2$ shows that $\pi_5 S^2 = \pi_5 S^3$, and so $\pi_5 S^2 = \mathbf{Z}/2$ also.

10.3. Suspension, looping, and the transgression

For any space X, identify the reduced cone CX as a quotient of $[0,1] \times X$ with the inclusion $X \subset CX$ corresponding to $x \mapsto (1,x)$. Take the reduced suspension SX to be CX/X, and let $c : CX \to SX$ denote the quotient map. Unless otherwise specified, in this section H_k denotes (ordinary) homology with some fixed (untwisted) coefficients.

Now consider the two fundamental maps

$$s : X \to \Omega SX, \ x \mapsto (t \mapsto (t,x))$$

and

$$\ell : S\Omega X \to X, \ (t,\alpha) \mapsto \alpha(t).$$

Then $s : Y \to \Omega SY$ induces the suspension map

$$[X,Y]_0 \xrightarrow{s_*} [X, \Omega SX]_0 = [SX, SY]_0.$$

In particular, taking $X = S^k$ yields the suspension homomorphism

$$s_* : \pi_k(Y) \to \pi_{k+1}(SY).$$

The map $\ell : S\Omega X \to X$ induces the "looping" map

$$[X,Y]_0 \xrightarrow{\ell^*} [S\Omega X, Y]_0 = [\Omega X, \Omega Y]_0$$

which takes a function to the induced function on loop spaces. The purpose of this section is to prove the Freudenthal suspension theorem (Theorem 8.7) and to develop material for a dual result about stable cohomology operations given in the next section.

We will relate these maps to the transgression for the path space fibration. For simplicity we assume throughout that X and Y are simply connected.

We begin with the map s. Consider the path space fibration over SY:

$$\Omega SY \to PSY \xrightarrow{e} SY$$

where e evaluates a path at its end point. The transgression in homology for this fibration is the "composite" (with domain a submodule of $H_k(SY)$ and range a quotient module of $H_{k-1}(\Omega SY)$):

$$\tau : H_k(SY,*) \supset \operatorname{Im}(e_*) \xleftarrow{e_*} H_k(PSY, \Omega SY) \xrightarrow{\partial} H_{k-1}(\Omega SY)/\partial(\ker e_*).$$

Let $S_* : H_{k-1}(Y) \to H_k(SY)$ denote the suspension isomorphism, defined as the composite of isomorphisms

$$S_* : H_{k-1}(Y) \xleftarrow{\partial} H_k(CY, Y) \xrightarrow{c_*} H_k(SY,*) \cong H_k(SY).$$

Theorem 10.14. *The domain of τ is all of $H_k(Y)$. Moreover, the map $s_* : H_{k-1}(Y) \to H_{k-1}(\Omega SY)$ induced by s is a lift of $\tau \circ S_*$.*

In particular, if $\tau : H_k(SY) \to H_{k-1}(\Omega SY)$ is an isomorphism, then so is $s_* : H_{k-1}(Y) \to H_{k-1}(\Omega SY)$.

Proof. Consider the map $f : CY \to PSY$ defined by

$$f(t, y) = (r \mapsto (rt, y)).$$

Then $e \circ f : CY \to SY$ is just the map $(t, y) \mapsto (t, y)$, i.e. the natural collapse map c. Moreover, the restriction of f to $Y = \{1\} \times Y \subset CY$ is the map $y \mapsto (r \mapsto (r, y))$; this is just the map s. In other words, f induces a map of pairs $f : (CY, Y) \to (PSY, \Omega SY)$ whose restriction to the subspaces is s. Thus f induces a map of the long exact sequences of the pairs; since PSY and CY are contractible every third term vanishes and so we obtain commuting diagrams with the horizontal arrows isomorphisms:

$$\begin{array}{ccc} H_k(CY, Y) & \xrightarrow{\partial} & H_{k-1}(Y) \\ \downarrow{f_*} & & \downarrow{s_*} \\ H_k(PSY, \Omega SY) & \xrightarrow{\partial} & H_{k-1}(\Omega SY) \end{array}$$

Since $e \circ f = c$, the diagram

$$\begin{array}{ccc} H_k(CY, Y) & \xrightarrow{f_*} & H_k(PSY, \Omega SY) \\ & {\scriptstyle c_*}\searrow \quad \swarrow{\scriptstyle e_*} & \\ & H_k(SY,*) & \end{array}$$

commutes, with c_* an isomorphism. It follows that e_* is onto, so that the domain of τ is all of $H_k(SY)$. Moreover $\partial \circ f_* \circ (c_*)^{-1}$ is a lift of τ, and so using the definition of S_* and the commuting square above, we compute:

$$\tau \circ S_* = \partial \circ f_* \circ (c_*)^{-1} \circ S_* = \partial f_* \partial^{-1} = s_*.$$

$\square$

As an application of Theorem 10.14, we prove the Freudenthal suspension theorem (Theorem 8.7).

Suppose that Y is an $(n-1)$-connected space, with $n > 1$. Using the Hurewicz theorem and the suspension isomorphism $S_* : H_k(Y;\mathbf{Z}) \cong H_{k+1}(SY;\mathbf{Z})$, we see that SY is n-connected. Since $\pi_k(\Omega SY) = \pi_{k+1}(SY)$, ΩSY is $(n-1)$-connected, and hence its homology vanishes in dimensions less than n by the Hurewicz theorem.

The Serre exact sequence (Theorem 9.15) for the fibration

$$\Omega SY \to PSY \xrightarrow{e} SY$$

is

$$H_{2n}(\Omega SY;\mathbf{Z}) \to H_{2n}(PSY;\mathbf{Z}) \xrightarrow{e_*} H_{2n}(SY;\mathbf{Z}) \xrightarrow{\tau} H_{2n-1}(\Omega SY;\mathbf{Z}) \to \cdots,$$

and since the path space PSY is contractible it follows that the transgression $\tau : H_k(SY;\mathbf{Z}) \to H_{k-1}(\Omega SY;\mathbf{Z})$ is an isomorphism for all $k \leq 2n$. Theorem 10.14 implies that $s_* : H_{k-1}(Y;\mathbf{Z}) \to H_{k-1}(\Omega SY;\mathbf{Z})$ is an isomorphism for all $k \leq 2n$. Hence the relative homology groups $H_k(\Omega SY, Y;\mathbf{Z})$ vanish for $k \leq 2n-1$. From the relative Hurewicz theorem we conclude that $\pi_k(\Omega SY, Y) = 0$ for $k \leq 2n-1$ and so $s_* : \pi_k(Y) \to \pi_k(\Omega SY)$ is an isomorphism for $k < 2n-1$ and an epimorphism for $k = 2n-1$. The composite $\pi_k(Y) \xrightarrow{s_*} \pi_k(\Omega SY) = \pi_{k+1}(SY)$ is the suspension homomorphism, and so we have proven the Freudenthal suspension theorem.

We turn our attention to the map $\ell : S\Omega X \to X$. This induces a map

$$\ell_* : H_k(S\Omega X) \to H_k(X)$$

which we can precompose with the suspension isomorphism to obtain

$$\ell_* \circ S_* : H_{k-1}(\Omega X) \to H_k(X).$$

The transgression for the path space fibration over X,

$$\Omega X \to PX \xrightarrow{e} X,$$

is the homomorphism

$$\tau : H_k(X,*) \supset \mathrm{Im}(e_*) \xleftarrow{e_*} H_k(PX,\Omega X) \xrightarrow{\partial} H_{k-1}(\Omega X)/\partial(\ker e_*).$$

The following theorem is the analogue of Theorem 10.14 for the map ℓ.

Theorem 10.15. *The homomorphism $\ell_* \circ S_*$ is a left inverse for τ on its domain; i.e. $\ell_* \circ S_* \circ \tau(x) = x$ if x lies in the domain of τ. Hence τ is injective.*

In particular, if $\tau : H_k(X) \to H_{k-1}(\Omega X)$ is an isomorphism, then τ and $\ell_* \circ S_*$ are inverses.

Proof. This time we use the map $g : C\Omega X \to PX$ defined by

$$(t, \alpha) \mapsto (s \mapsto \alpha(st)).$$

Thus $g(0, \alpha)$ and $g(t, \text{const}_*)$ are both the constant path at $*$, so that this indeed gives a well defined map on the reduced cone. Moreover, $g(1, \alpha) = \alpha$, and so g defines a map of pairs $g : (C\Omega X, \Omega X) \to (PX, \Omega X)$ which restricts to the identity map on ΩX. Since both $C\Omega X$ and PX are contractible it follows that the maps between the long exact sequences of these pairs reduce to a commuting triangle of isomorphisms

$$\begin{array}{ccc} H_k(C\Omega X, \Omega X) & \xrightarrow{\quad g_* \quad} & H_k(PX, \Omega X) \\ & {\scriptstyle \partial} \searrow \quad \swarrow {\scriptstyle \partial} & \\ & H_{k-1}(\Omega X) & \end{array}$$

The composite $e \circ g : C\Omega X \to X$ is the map $(t, \alpha) \mapsto \alpha(t)$. Since ΩX is the fiber of the fibration e, $e \circ g$ factors through the suspension $S\Omega X$, and in fact the diagram

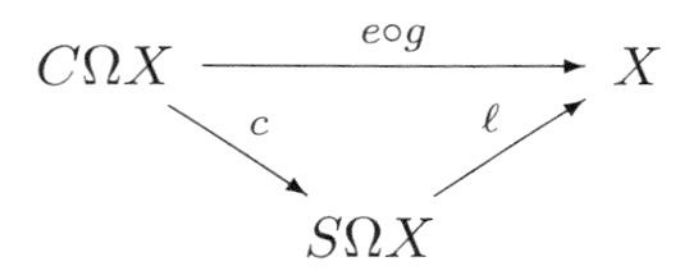

commutes.

Thus

$$\begin{aligned} \ell_* S_* \tau &= \ell_* S_* \partial (e_*)^{-1} = \ell_* S_* \partial (g_*)^{-1} (e_*)^{-1} \\ &= \ell_* S_* \partial (\ell_* c_*)^{-1} = \ell_* c_* (c_*)^{-1} (\ell_*)^{-1} = \text{Id}. \end{aligned}$$

One of course needs to be careful with domains and ranges in this calculation. □

Theorems 10.14 and 10.15 have their cohomology analogues, whose statements and proofs are given essentially by reversing all the arrows. We will use the cohomology analogue of Theorem 10.15 in our discussion of stable cohomology operations. Let $S^* : H^k(SY) \to H^{k-1}(Y)$ denote the suspension isomorphism in cohomology.

Corollary 10.16. *Suppose that the cohomology transgression*

$$\tau^* : H^{k-1}(\Omega X) \to H^k(X)$$

for the path space fibration $\Omega X \to PX \to X$ is an isomorphism.

Then the composite $S^* \circ \ell^*$ is the inverse of τ^*. □

The following exercise gives a "dual" description of the isomorphism S^*.

Exercise 167. From the perspective of the Eilenberg–MacLane spectrum, prove that $S^* : H^k(SY) \to H^{k-1}(Y)$ is the composite of the adjoint and the map induced by the homotopy equivalence $\Omega K(A,k) \sim K(A,k-1)$:

$$H^k(SY;A) \cong [SY, K(A,k)]_0 \cong [Y, \Omega K(A,k)]_0 \cong [Y, K(A,k-1)]_0 \cong H^{k-1}(Y;A).$$

10.4. Cohomology operations

We have seen that the cohomology of a space with coefficients in a ring has a natural ring structure. Cohomology operations are a further refinement of the structure of the cohomology of a space. We have already come across cohomology operations in Chapter 7 (see Definition 7.25 and Exercise 121).

10.4.1. Definition and simple examples.

We recall the definition.

Definition 10.17. If A, C are abelian groups, a *cohomology operation of type $(n, A; q, C)$* is a natural transformation of functors

$$\theta : H^n(-;A) \to H^q(-;C).$$

The set of all cohomology operations of type $(n, A; q, C)$ is denoted by $O(n, A; q, C)$.

The following are some standard examples.

Coefficient homomorphisms. If $h : A \to C$ is a homomorphism, then h induces homomorphisms

$$h_* : H^n(X;A) \to H^n(X;C)$$

for all n; these are natural, so h defines an operation h_* of type $(n, A; n, C)$ for any n.

Bockstein homomorphisms. If

$$0 \to A \to B \to C \to 0 \tag{10.6}$$

is a short exact sequence of abelian groups, then $0 \to \mathrm{Hom}(C_*X, A) \to \mathrm{Hom}(C_*X, B) \to \mathrm{Hom}(C_*X, C) \to 0$ is exact, where C_*X denotes the singular or cellular chain complex of X. Thus one obtains a long exact sequence in cohomology

$$\cdots \to H^k(X;A) \to H^k(X;B) \to H^k(X;C) \to H^{k+1}(X;A) \to \cdots.$$

The connecting homomorphisms

$$\beta_k : H^k(X;C) \to H^{k+1}(X;A)$$

are called the *Bockstein operators* associated to the short exact sequence (10.6). For each k this construction defines a cohomology operation β_k of type $(k, C; k+1, A)$.

Squaring. If R is a ring, let $\theta_n : H^n(X;R) \to H^{2n}(X;R)$ be the map

$$x \mapsto x \cup x.$$

Then θ_n is a natural transformation since $f^*(x \cup x) = f^*(x) \cup f^*(x)$, and hence a cohomology operation of type $(n, R; 2n, R)$.

Remark. At this point we would like to avoid using the symbol "$\cup$" for the product in the cohomology ring of a space and will use juxtaposition to indicate multiplication whenever it is convenient.

Notice that θ_n is *not a homomorphism*, since $(x + y)^2 \neq x^2 + y^2$ in general. In fact the definition of a cohomology operation does not require it to be a homomorphism.

Main example. Let A, C be abelian groups, and let

$$u \in H^q(K(A,n);C).$$

For CW-complexes Theorem 7.22 says that

$$[X, K(A,n)] \cong H^n(X;A), \text{ via } \big(f : X \to K(A,n)\big) \mapsto f^*(\iota),$$

where $\iota \in H^n(K(A,n);A)$ is the fundamental class of $K(A,n)$ (see Definition 7.21).

Thus, $u \in H^q(K(A,n);C)$ defines a map (up to homotopy)

$$f_u : K(A,n) \to K(C,q),$$

and hence u defines a cohomology operation θ_u as the composite

$$H^n(X;A) = [X;K(A,n)] \xrightarrow{(f_u)_*} [X, K(C,q)] = H^q(X;C).$$

So $u \in H^q(K(A,n);C)$ defines the operation θ_u of type $(n, A; q, C)$.

In Exercise 121 you showed that the correspondence $u \mapsto \theta_u$ gave a bijection between $H^q(K(A,n);C)$ and $O(n, A; q, C)$; the inverse map is $\theta \mapsto \theta(\iota)$.

10.4.2. Stable operations.

Definition 10.18. Given a cohomology operation $\theta \in O(n, A; q, C)$, the *suspension of* θ, $\sigma^*(\theta)$, is the operation of type $(n-1, A; q-1, C)$ defined by requiring the following diagram to commute:

$$\begin{array}{ccc} H^{n-1}(X;A) & \xrightarrow{\sigma^*(\theta)} & H^{q-1}(X;C) \\ \cong\uparrow & & \uparrow\cong \\ H^n(SX;A) & \xrightarrow{\theta} & H^q(SX;C) \end{array}$$

where the vertical maps are the usual suspension isomorphisms,

$$S^* : H^n(SX;A) \cong H^{n-1}(X;A) \text{ and } S^* : H^q(SX;C) \cong H^{q-1}(X;C).$$

(The name "suspension" is perhaps not the best choice for σ^*, since as we will see below it is induced by the "looping" map ℓ of Section 10.3 and so is more properly thought of as a desuspension. However, this is its traditional name, and so we will stick to it. The reader should not confuse it with the suspension isomorphism $S^* : H^k(SX) \to H^{k-1}(X)$, although they are of course related.)

Thus to any cohomology operation θ we can associate the sequence $\sigma^*(\theta), \sigma^*(\sigma^*(\theta)), \cdots$. This motivates the following definition.

Definition 10.19. A *stable cohomology operation of degree* r *and type* (A, C) is a sequence of operations $\theta = \{\theta_n\}$ where

$$\theta_n \in O(n, A; n+r, C) \quad \text{and} \quad \sigma^*(\theta_n) = \theta_{n-1}.$$

Thus a stable operation of degree r and type (A, C) is the same thing as an element in the inverse limit of the sequence

$$\cdots \xrightarrow{\sigma^*} H^{n+r}(K(A,n);C) \xrightarrow{\sigma^*} H^{n+r-1}(K(A,n-1);C) \xrightarrow{\sigma^*} \cdots \xrightarrow{\sigma^*} H^{r+1}(K(A,1);C).$$

Denote by $\mathcal{A}^r(A; C)$ the set of all stable cohomology operations $\theta = \{\theta_n\}$ of degree r and type (A, C).

To decide whether a cohomology operation $\theta \in O(n, A; q, C)$ forms a component of a stable cohomology operation, at the very least we need to know whether $\theta = \sigma^*(\theta')$ for some θ'. This is possible if θ is transgressive, as we now explain.

The map

$$\sigma^* : O(n, A; q, C) \to O(n-1, A; q-1, C)$$

is defined for each $(n, A; q, C)$. Using Exercise 121 we think of σ^* as a map

$$\sigma^* : H^q(K(A, n); C) \to H^{q-1}(K(A, n-1); C)$$

or, equivalently, a map

$$\begin{aligned} \sigma^* : [K(A, n), K(C, q)]_0 &\to [K(A, n-1), K(C, q-1)]_0 \\ &\xrightarrow{\cong} [\Omega K(A, n), \Omega K(C, q)]_0. \end{aligned} \tag{10.7}$$

Exercise 168. Show that in Equation (10.7), σ^* is given by "looping"; i.e.

$$\sigma^* = \Omega : [K(A, n), K(C, q)]_0 \to [\Omega K(A, n), \Omega K(C, q)]_0,$$

where $(\Omega f)(\alpha)(t) = f(\alpha(t))$. Conclude that the composite of σ^* with the isomorphism given by the adjoint

$$[\Omega K(A, n), \Omega K(C, q)]_0 = [S\Omega K(A, n), K(C, q)]_0$$

is ℓ^*, the map induced by $\ell : S\Omega K(A, n) \to K(A, n)$ of Section 10.3.

Since the suspension isomorphism $S^*: H^q(SX;C) \to H^{q-1}(X;C)$ is the composite of the adjoint and the identification $\Omega K(C,q) = K(C,q-1)$:

$$S^* : [SX, K(C,q)]_0 = [X, \Omega K(K,q)]_0 = [X, K(C,q-1)]_0,$$

it follows immediately from Exercise 168 that

$$\sigma^* = S^* \circ \ell^*. \tag{10.8}$$

Consider the cohomology transgression τ^* for the path space fibration over $K(A,n)$

$$K(A,n-1) \to P \xrightarrow{e} K(A,n). \tag{10.9}$$

Corollary 10.16 implies that the homomorphism σ^* is a left inverse of τ^*. This immediately implies the following.

Corollary 10.20. *If the class $\theta \in H^q(K(A,n);B)$ is transgressive (i.e. in the domain of the transgression τ), then $\sigma^*(\tau^*(\theta)) = \theta$.* □

What this corollary says is that $\theta \in O(n,A;q,C)$ is of the form $\sigma^*(\theta')$ if θ is transgressive for the fibration (10.9), when viewed as an element in $H^q(K(A,n);C)$.

Theorem 10.21. *If $n \geq 2$, the transgression for the fibration (10.9) induces isomorphisms*

$$\tau : H^{q-1}(K(A,n-1);C) \to H^q(K(A,n);C)$$

for $2n \geq q+2$.

Proof. Consider the Leray–Serre cohomology spectral sequence with C coefficients for the fibration (10.9). Since $E_2^{p,q} = H^p(K(A,n); H^q(K(A,n-1)),$ the $E_2^{p,q}$ terms vanish if $1 \leq p \leq n-1$ or if $1 \leq q \leq n-2$.

This implies that if $2n \geq q+2$,

$$H^{q-1}(K(A,n-1)) \cong E_2^{0,q-1} = E_q^{0,q-1}$$

and

$$H^q(K(A,n)) \cong E_2^{q,0} = E_q^{q,0}.$$

Since the total space is contractible, $E_\infty^{p,q} = 0$ if $p+q \neq 0$. Hence the differential $d_q : E_q^{0,q-1} \to E_q^{q,0}$ is an isomorphism. Theorem 9.13 states that this differential coincides with the transgression, and so we conclude that the transgression $\tau : H^{q-1}(K(A,n-1)) \to H^q(K(A,n))$ is an isomorphism for $2n \geq q+2$. □

Since σ^* and τ^* are inverses, if one starts with a cohomology operation $\theta \in H^q(K(A,n);C)$, then the sequence $\theta, \sigma^*(\theta), \sigma^*(\sigma^*(\theta)), \cdots$ can be extended to the left to give a stable operation provided θ, $\tau^*(\theta)$, $\tau^*(\tau^*(\theta))$, etc., are transgressive.

A convenient way to organize this information is the following. If we write

$$A^r(A;C) = \varprojlim_n H^{n+r}(K(A,n);C),$$

then Theorem 10.21 shows that the limit is attained at a finite stage:

$$A^r(A;C) = \varprojlim_n H^{n+r}(K(A,n);C) = H^{2r+1}(K(A,r+1);C) \tag{10.10}$$

and so a class $\theta \in H^{2r+1}(K(A,r+1);C)$ defines the stable operation

$$\cdots, \tau^*(\tau^*(\theta)), \tau^*(\theta), \theta, \sigma^*(\theta), \sigma^*(\sigma^*(\theta)), \cdots.$$

Exercise 169. Show that the composition of two stable cohomology operations is a stable cohomology operation.

The proof of the following proposition is easy and is left to the reader.

Proposition 10.22. *If G is an abelian group, then the sum and composition give $\mathcal{A}(G) = \oplus_r A^r(G,G)$ the structure of a graded, associative ring with unit.* □

Exercise 170. Prove Proposition 10.22.

Exercise 171. Use Exercise 168 to show that if θ is a stable cohomology operation, then $\theta(f+g) = \theta(f) + \theta(g)$. (Hint: if $f, g \in H^n(X;A) = [X, K(A,n)]_0 = [X, \Omega K(A,n-1)]_0$, then the group structure is given by taking composition of loops, which is preserved by σ^*.)

An interesting consequence of Proposition 10.22 and Exercise 171 is that for any space X, the cohomology $H^*(X;G)$ has the structure of a module over $\mathcal{A}(G)$. This additional structure is functorial.

Definition 10.23. Take $G = \mathbf{Z}/p$, p a prime. Then $\mathcal{A}_p = \mathcal{A}(\mathbf{Z}/p)$ is called the *mod p Steenrod algebra*. It is a graded algebra over $\mathbf{Z}/p$.

Thus the $\mathbf{Z}/p$ cohomology algebra of a space is a module over the mod p Steenrod algebra.

Exercise 172. Given two spectra $\mathbf{K}$ and $\mathbf{K}'$, define what a map of degree r from $\mathbf{K}$ to $\mathbf{K}'$ is, and what a homotopy of such maps is. Then show that taking $\mathbf{K}(A)$ (resp. $\mathbf{K}(B)$) to be the Eilenberg–MacLane spectrum for the abelian group A (resp. B),

$$\mathcal{A}^*(A,B) = [\mathbf{K}(A), \mathbf{K}(B)]_*.$$

Can you define stable cohomology operations for arbitrary generalized homology theories?

10.5. The mod 2 Steenrod algebra

The goal in this section is to explore the mod 2 Steenrod algebra. We will take an ad-hoc, hands-on approach. A systematic exposition of this important algebra can be found in many homotopy theory texts. The standard reference is [**31**].

The complete structure of the mod 2 Steenrod algebra is described in the following result, most of whose proof we will omit.

Theorem 10.24. *Let T be the (bigraded) tensor algebra over $\mathbf{Z}/2$ generated by the symbols Sq^i for $i \geq 0$. (Thus, if V is the graded $\mathbf{Z}/2$ vector space spanned by a basis $Sq^0, Sq^1, \cdots$, then $T = T(V) = \mathbf{Z}/2 \oplus V \oplus (V \otimes V) \oplus \cdots$). Let $I \subset T$ be the two-sided homogeneous ideal generated by:*

1. $1 + Sq^0$ *and*
2. (Adem relations)

$$Sq^a \otimes Sq^b + \sum_{c=0}^{[a/2]} \binom{b-c-1}{a-2c} Sq^{a+b-c} \otimes Sq^c$$

for all $0 < a < 2b$.

Then $\mathcal{A}_2$ is isomorphic to T/I. The identification takes the Sq^i to stable operations satisfying:

a. $Sq^i(x) = 0$ *if* $x \in H^{i-p}(X), p > 0$.
b. $Sq^i(x) = x^2$ *if* $x \in H^i(X)$.
c. Sq^1 *is the Bockstein associated to the short exact sequence*

$$0 \to \mathbf{Z}/2 \to \mathbf{Z}/4 \to \mathbf{Z}/2 \to 0.$$

d. (Cartan formula)

$$Sq^i(xy) = \sum_j Sq^j x \, Sq^{i-j} y.$$

□

In this theorem the Sq^i should be interpreted as a stable operations in the sense that $Sq^i = \{Sq^i_{(n)}\}$ where $Sq^i_{(n)} : H^n(X) \to H^{n+i}(X)$ and $\sigma^*(Sq^i_{(n)}) = Sq^i_{(n-1)}$.

Exercise 171 says that each component θ_n of a stable operation is a group homomorphism. Thus the Sq^i are additive; i.e. $Sq^i_{(n)} : H^n(X) \to$

$H^{n+i}(X)$ is a group homomorphism for all n. The operation Sq^i is not a ring homomorphism; this is clear from the Cartan formula. For example,

$$Sq^1(ab) = Sq^1 a \cdot b + a \cdot Sq^1 b,$$

so Sq^1 is a *derivation.* However, if we define the *total square* by the formula

$$Sq = \sum_{i=0}^{\infty} Sq^i$$

(on each element $x \in H^p(X)$ the sum is finite; $Sq(x) = x + Sq^1 x + \cdots + Sq^p x$ since $Sq^{p+k}x = 0$ for $k > 0$) then the Cartan formula simplifies to

$$Sq(xy) = Sq(x)Sq(y).$$

What this says is that the Sq^i are the *homogeneous components* of a ring endomorphism Sq of the cohomology algebra $H^*(X; \mathbf{Z}/2)$.

There are several ways of constructing the Sq^i and verifying their properties. We will not prove Theorem 10.24 in general, but instead will construct the operations Sq^i and focus on some special cases, taking the point of view that computing $\mathcal{A}_2^r$ is the same, using Equation (10.10), as computing the cohomology $H^{2r+1}(K(\mathbf{Z}/2, r+1); \mathbf{Z}/2)$.

As a first simple computation, notice that

$$\mathcal{A}_2^0 = H^1(K(\mathbf{Z}/2, 1); \mathbf{Z}/2) = \mathbf{Z}/2 = H^n(K(\mathbf{Z}/2, n); \mathbf{Z}/2).$$

The generator is

$$\iota_n \in H^n(K(\mathbf{Z}/2, n); \mathbf{Z}/2) = \operatorname{Hom}(H_n(K(\mathbf{Z}/2, n)), \mathbf{Z}/2);$$

i.e. the generator is the fundamental class, which corresponds to te identity map via the identification $H^n(K(\mathbf{Z}/2, n); \mathbf{Z}/2) = [K(\mathbf{Z}/2, n), K(\mathbf{Z}/2, n)]_0$. This implies that the corresponding cohomology operation is the identity operation, which we denote by Sq^0.

Next consider

$$\mathcal{A}_2^1 = H^3(K(\mathbf{Z}/2, 2), \mathbf{Z}/2) \cong H^{n+1}(K(\mathbf{Z}/2, n), \mathbf{Z}/2) \text{ for all } n \geq 2.$$

Recall that the isomorphism is given by the transgression in the spectral sequence for the (path space) fibration

$$K(\mathbf{Z}/2, n-1) \to * \to K(\mathbf{Z}/2, n).$$

Proposition 10.25. *The transgression*

$$\tau^* : H^2(K(\mathbf{Z}/2, 1); \mathbf{Z}/2) \to H^3(K(\mathbf{Z}/2, 2), \mathbf{Z}/2)$$

is an isomorphism.

Notice that the statement falls just out of the range of Theorem 10.21.

Before we prove this proposition, we draw some conclusions. Recall that infinite-dimensional real projective space $\mathbf{R}P^\infty$ has the homotopy type of a $K(\mathbf{Z}/2,1)$. Therefore the cohomology ring $H^*(K(\mathbf{Z}/2,1))$ is isomorphic to the polynomial ring $\mathbf{Z}/2[\iota_1]$. In particular,

$$H^2(K(\mathbf{Z}/2,1);\mathbf{Z}/2)=\mathbf{Z}/2,$$

generated by ι_1^2. Hence $\mathcal{A}_2^1 \cong \mathbf{Z}/2$. It is represented by an element which we denote by Sq^1.

Explicitly, define $Sq^1_{(n)} \in H^{n+1}(K(\mathbf{Z}/2,n);\mathbf{Z}/2)$ by

$$Sq^1_{(n)} = \begin{cases} \underbrace{\tau^* \circ \cdots \circ \tau^*}_{n-1}(i_1^2) & \text{if } n \geq 1, \\ 0 & \text{if } n = 0. \end{cases} \tag{10.11}$$

From earlier remarks it follows that $Sq^1_{(n)} = \sigma^*(Sq^1_{(n+1)})$ if $n \geq 1$. By construction it is immediate that

$$Sq^1_{(1)}: H^1(X;\mathbf{Z}/2) \to H^2(X;\mathbf{Z}/2)$$

is the map $x \mapsto x^2$. The following exercise shows that $\sigma^*(Sq^1_{(1)}) = 0 = Sq^1_{(0)}$, confirming that Equation (10.11) gives the correct definition for $Sq^1_{(0)}$.

Exercise 173. Show that for any space X, the map

$$H^n(SX) \to H^{2n}(SX)$$

given by $x \mapsto x^2$ is zero for $n > 0$. (Hint: consider the cup product $H^n(CX,X) \times H^n(CX) \to H^{2n}(CX,X)$ of Corollary 3.24.)

Proof of Proposition 10.25. Consider the cohomology spectral sequence for the path space fibration

$$K(\mathbf{Z}/2,1) \to * \to K(\mathbf{Z}/2,2).$$

The E_2 level in low degrees is tabulated in the following diagram.

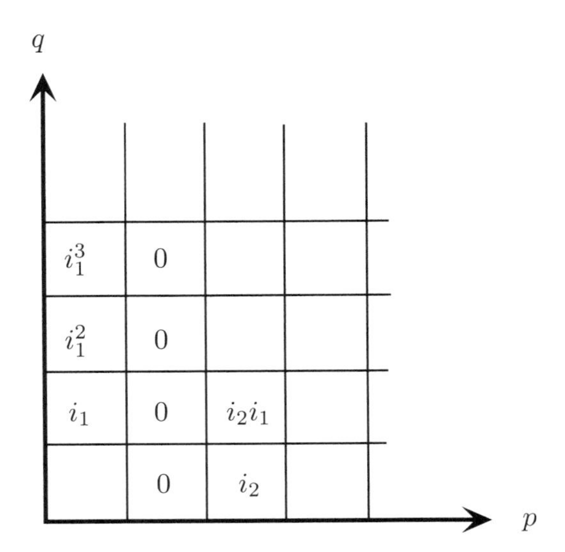

Since $d_2(i_1) = i_2$ it follows that $d_2(i_1^2) = d_2(i_1)i_1 + i_1 d_2 i_1 = i_2 i_1 + i_1 i_2 = 2i_1 i_2 = 0$. Thus i_1^2 survives to E_3, and hence is transgressive. Moreover $d_3 : E_3^{0,2} \to E_3^{3,0}$ must be an isomorphism, since $E_\infty^{p,q} = 0$ for all $(p,q) \neq (0,0)$. Thus i_1^2 transgresses to a generator $\tau(i_1^2)$ of $H^3(K(\mathbf{Z}/2,2);\mathbf{Z}/2)$. □

We next turn our attention to

$$\mathcal{A}_2^2 \cong H^5(K(\mathbf{Z}/2,3);\mathbf{Z}/2) \cong H^{n+2}(K(\mathbf{Z}/2,n),\mathbf{Z}/2)$$

for $n \geq 3$. The spectral sequence for the fibration

$$K(\mathbf{Z}/2,2) \to * \to K(\mathbf{Z}/2,3)$$

in low dimensions is described in the diagram:

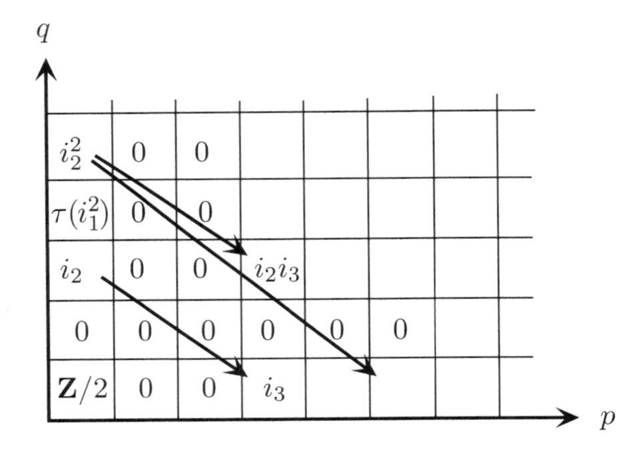

We have used the fact from the proof of Proposition 10.25 that $\tau(\iota_2^2)$ generates $H^3(K(\mathbf{Z}/2,2);\mathbf{Z}/2)$.

Three differentials are drawn. The differential $d_3 : E_3^{0,2} \to E_3^{3,0}$ is an isomorphism since $E_3^{0,2} = E_2^{0,2} = H^2(K(\mathbf{Z}/2,2)) = \mathbf{Z}/2\langle\iota_2\rangle$. Also $E_3^{3,0} = H^3(K(\mathbf{Z}/2,3)) = \mathbf{Z}/2\langle\iota_3\rangle$, and since $E_\infty^{p,q} = 0$ for all $(p,q) \neq (0,0)$ we conclude that $d_3\iota_2 = \iota_3$.

Now observe that since

$$E_2^{3,2} = H^3(K(\mathbf{Z}/2,3)) \otimes H^2(K(\mathbf{Z}/2,2)) = \mathbf{Z}/2\langle\iota_2\iota_3\rangle,$$

the differential $d_3 : E_3^{0,4} \to E_3^{3,2} = E_2^{3,2}$ takes ι_2^2 to $2\iota_2\iota_3 = 0$.

Lemma 10.26. *$E_3^{0,4}$ is isomorphic to $\mathbf{Z}/2$, spanned by ι_2^2.*

Assuming Lemma 10.26 for a moment, it follows that $d_5 : E_5^{0,4} \to E_5^{5,0}$ is an isomorphism, so that ι_2^2 is transgressive, and

$$E_5^{5,0} = E_2^{5,0} = H^5(K(\mathbf{Z}/2,3),\mathbf{Z}/2) = \mathbf{Z}/2\langle\tau(\iota_2^2)\rangle.$$

Thus

$$\mathbf{Z}/2 = \mathcal{A}_2^2 = H^{k+2}(K(\mathbf{Z}/2,k);\mathbf{Z}/2) \text{ for } k \geq 3, \tag{10.12}$$

spanned by $\tau(\iota_2^2)$. We *denote* $\tau(\iota_2^2)$ by Sq^2. More precisely, denote $\tau(\iota_2^2)$ by $Sq^2_{(3)}$. Then let $Sq^2 = \{Sq^2_{(n)}\}$ where $Sq^2_{(n)} = \tau^{(n-3)}(Sq^2_{(3)})$ for $n > 3$, $Sq^2_{(2)} = \iota_2^2$, and $Sq^2_{(1)} = Sq^2_{(0)} = 0$.

Since $Sq^2_{(2)} = \iota_2^2$, it follows that $Sq^2 x = x^2$ if $x \in H^2(X;\mathbf{Z}/2)$. Exercise 173 shows that $\sigma^*(Sq^2_{(2)}) = 0 = Sq^2_{(1)}$, so we have correctly defined the stable operation Sq^2.

Proof of Lemma 10.26. First,

$$E_3^{0,4} = E_2^{0,4} = H^0(K(\mathbf{Z}/2,3);H^4(K(\mathbf{Z}/2,2)) = H^4(K(\mathbf{Z}/2,2)).$$

Hence it must be shown that $H^4(K(\mathbf{Z}/2,2)) \cong \mathbf{Z}/2$, generated by ι_2^2.

Let $f : K(\mathbf{Z},2) \to K(\mathbf{Z}/2,2)$ be a map representing the generator c of $H^2(K(\mathbf{Z},2);\mathbf{Z}/2) = \mathbf{Z}/2$. Recall that $K(\mathbf{Z},2) \sim \mathbf{C}P^\infty$, so that its cohomology ring $H^*(K(\mathbf{Z},2);\mathbf{Z}/2)$ is isomorphic to $\mathbf{Z}/2[c]$, the polynomial ring with generator c. In particular, $c^2 \neq 0$. By definition of the map f, $f^*(\iota_2) = c$. So $f^*(\iota_2^2) = c^2 \neq 0$. Hence ι_2^2 is non-zero in $H^4(K(\mathbf{Z}/2,2);\mathbf{Z}/2)$. To complete the proof it therefore suffices to show that $H^4(K(\mathbf{Z}/2,2)) = \mathbf{Z}/2$.

Consider the spectral sequence for the path space fibration

$$K(\mathbf{Z}/2,1) \to * \to K(\mathbf{Z}/2,2).$$

The facts $d_2(\iota_1^2) = 0$ and $d_3(\iota_1^2) \neq 0$ established in the proof of Proposition 10.25 imply that $d_2 : E_2^{2,1} = \mathbf{Z}/2\langle\iota_2\iota_1\rangle \to E_2^{4,0}$ is an isomorphism. Thus $\mathbf{Z}/2 = E^{4,0} = H^4(K(\mathbf{Z}/2,2))$ as desired. □

The construction of Sq^1 and Sq^2 suggests we define Sq^n in the following way. Let $\iota_n \in H^n(K(\mathbf{Z}/2,n),\mathbf{Z}/2)$ denote the fundamental class. Let

$$y_0 = \iota_n^2 \in H^{2n}(K(\mathbf{Z}/2,n);\mathbf{Z}/2).$$

In the spectral sequence for the fibration

$$K(\mathbf{Z}/2,n) \to * \to K(\mathbf{Z}/2,n+1),$$

the differential d_{n+1} takes ι_n to ι_{n+1}. Thus $d_{n+1}(y_0) = 2\iota_n i_{n+1} = 0$, and hence y_0 is transgressive.

Let

$$y_1 = d_{2n+1}(y_0) = \tau^*(y_0) \in E_{2n+1}^{2n+1,0} = E_2^{2n+1,0} = H^{2n+1}(K(\mathbf{Z}/2,n+1);\mathbf{Z}/2).$$

Then $\sigma^*(y_1) = y_0$.

The transgressions are isomorphisms

$$\tau^* : H^{2n+k}(K(\mathbf{Z}/2,n+k);\mathbf{Z}/2) \to H^{2n+k+1}(K(\mathbf{Z}/2,n+k+1);\mathbf{Z}/2)$$

for $k \geq 1$. Thus we define

$$y_r = (\tau^*)^r(y_0) \in H^{2n+r}(K(\mathbf{Z}/2,n+r);\mathbf{Z}/2).$$

This defines a sequence $\{y_r\}$ with $\sigma^*(y_r) = y_{r-1}$ for $r \geq 1$.

If we extend this sequence by defining $x_r = 0$ for $r < 0$, then we still have $\sigma^*(y_r) = y_{r-1}$ by Exercise 173. Hence the sequence $\{y_r\}$ defines a stable cohomology operation $Sq^n_{(k)} = y_{n-k}$ in $\mathcal{A}_2^n$. By construction, $Sq^n x = x^2$ if $x \in H^n(X;\mathbf{Z}/2)$.

This completes the construction of the Steenrod operations Sq^n. This approach does not reveal much about their structure beyond showing that $Sq^n(x) = x^2$ if $x \in H^n(X;\mathbf{Z}/2)$. Showing that the Sq^n generate the Steenrod algebra $\mathcal{A}_2$, establishing the Adem relations, and proving the rest of Theorem 10.24 are more involved and require a more detailed analysis of the cohomology of the Eilenberg–MacLane spaces. We will content ourselves with proving part (c) of Theorem 10.24, identifying Sq^1 with the Bockstein operator.

Lemma 10.27. *The stable operation Sq^1 is the Bockstein associated to the exact sequence*

$$0 \to \mathbf{Z}/2 \to \mathbf{Z}/4 \to \mathbf{Z}/2 \to 0.$$

Proof. Consider the long exact sequence in cohomology of $K = K(\mathbf{Z}/2,k)$ associated to the short exact sequence $0 \to \mathbf{Z}/2 \to \mathbf{Z}/4 \to \mathbf{Z}/2 \to 0$. This is the sequence

$$\cdots \to H^{k-1}(K;\mathbf{Z}/2) \to H^k(K;\mathbf{Z}/2) \to$$
$$H^k(K;\mathbf{Z}/4) \to H^k(K;\mathbf{Z}/2) \xrightarrow{\beta} H^{k+1}(K;\mathbf{Z}/2) \to \cdots .$$

By definition the map labeled β is the Bockstein.

The Hurewicz and universal coefficient theorems show that

$$H^{k-1}(K;\mathbf{Z}/2)=0.$$

Also,

$$H^k(K;\mathbf{Z}/2)=\operatorname{Hom}(H_k(K),\mathbf{Z}/2)=\operatorname{Hom}(\mathbf{Z}/2,\mathbf{Z}/2)=\mathbf{Z}/2$$

and

$$H^k(K;\mathbf{Z}/4)=\operatorname{Hom}(H_k(K);\mathbf{Z}/4)=\operatorname{Hom}(\mathbf{Z}/2,\mathbf{Z}/4)=\mathbf{Z}/2.$$

It follows that β is an injection.

Induction using the spectral sequences for the fibrations

$$K(\mathbf{Z}/2,n-1)\to *\to K(\mathbf{Z}/2,n)$$

and the fact that the cellular chain complex for $K(\mathbf{Z}/2,1)=\mathbf{R}P^\infty$ is

$$\cdots\to\mathbf{Z}\xrightarrow{0}\mathbf{Z}\xrightarrow{\times 2}\mathbf{Z}\xrightarrow{0}\mathbf{Z}\to 0$$

show that

$$\begin{aligned}H_{k+1}(K;\mathbf{Z}) &\cong H_k(K(\mathbf{Z}/2,k-1);\mathbf{Z})\\ \cdots &\cong H_2(K(\mathbf{Z}/2,1);\mathbf{Z})\\ &= 0.\end{aligned}$$

Thus

$$\begin{aligned}H^{k+1}(K;\mathbf{Z}/2) &= \operatorname{Hom}(H_{k+1}(K,\mathbf{Z}),\mathbf{Z}/2))\oplus \operatorname{Ext}\,(H_kK,\mathbf{Z}/2)\\ &= 0\oplus\mathbf{Z}/2,\end{aligned}$$

so that β is an isomorphism.

By definition, the cohomology operation β corresponds to the element $\beta(i_k)$ in $H^{k+1}(K;\mathbf{Z}/2)=\mathbf{Z}/2$, but Sq^1 was constructed to be the generator of $H^{k+1}(K;\mathbf{Z}/2)$. Thus $\beta=Sq^1$. □

Here is an interesting application of the Steenrod squares to the homotopy groups of spheres. Consider the Hopf fibration $S^3\to S^7\xrightarrow{h}S^4$. Using h as an attaching map for an 8-cell to S^4, we obtain the quaternionic projective plane $X=\mathbf{H}P^2$. This has $\mathbf{Z}/2$-cohomology $\mathbf{Z}/2$ in dimensions $0,4$ and 8. Poincaré duality implies that the intersection form on fourth cohomology is non-degenerate. Therefore, (using $\mathbf{Z}/2$-coefficients) if $x\in H^4(X)$, $x^2=Sq^4(x)\in H^8(X)=\mathbf{Z}/2$ is non-zero.

If we use the suspension $Sh:S^8\to S^5$ to attach a 9-cell to S^5, we obtain the suspension SX (prove this). We will show that Sh is not nullhomotopic, and hence the suspension homomorphism $\pi_7S^4\to\pi_8S^5=\pi_3^S$ (which is onto by the Freudenthal suspension theorem) is non-zero.

Let $y \in H^5(SX)$ denote the non-zero element. Suppose to the contrary that Sh is nullhomotopic. Then SX is homotopy equivalent to the wedge $S^5 \vee S^8$. In particular the map $Sq^4 : H^5(SX) \to H^8(SX)$ is trivial, since if y is the non-zero element of $H^5(S^5 \vee S^8)$, then y is pulled back from $H^5(S^5)$ via the projection $S^5 \vee S^8 \to S^5$, but $H^8(S^5) = 0$ and so by naturality $Sq^4(y) = 0$.

But, since Sq^4 is a stable operation, the diagram

$$\begin{array}{ccc} H^4(X) & \xrightarrow{Sq^4} & H^8(X) \\ \cong \uparrow & & \cong \uparrow \\ H^5(SX) & \xrightarrow{Sq^4} & H^9(SX) \end{array}$$

commutes, and so $Sq^4(y) \neq 0$.

Thus $Sh : S^8 \to S^5$ is non-nullhomotopic, and so π_3^S is non-zero.

A similar argument, using the Hopf fibration $S^7 \to S^{15} \to S^8$ solves the following exercise.

Exercise 174. Show that the suspension

$$\pi_{15}S^8 \to \pi_{16}S^9 = \pi_7^S$$

is non-trivial on the homotopy class of the Hopf map $S^{15} \to S^8$.

10.6. The Thom isomorphism theorem

The Thom isomorphism theorem is a generalization of the fact that suspension induces an isomorphism $H^n(B) \cong H^{k+n}(S^k B)$. Roughly speaking, the Thom isomorphism theorem says that the suspension isomorphism continues to hold when one "twists" the suspension construction. More precisely, the k-fold suspension $S^k B$ can be considered as the quotient $(B \times D^k)/(B \times S^{k-1})$. This is generalized by replacing the space $B \times D^k$ by the disk bundle of some vector bundle over B and replacing $B \times S^{k-1}$ by the corresponding sphere bundle.

Recall that the Thom space $M(E)$ of a vector bundle $E \to B$ with metric is identified with $D(E)/S(E)$, where $S(E)$ denotes the unit sphere bundle and $D(E)$ denotes the unit disk bundle of $E \to B$. The collapsing map defines, via excision, an isomorphism (with any coefficients)

$$H^m(E, E_0) \to H^m(M(E), p)$$

where E_0 denotes the complement in E of the zero section. The inclusion of a fiber $\mathbf{R}^k \cong E_b \subset E$ can be viewed as a vector bundle map from a bundle over the point $b \in B$ to E, and so induces a restriction homomorphism (for any choice of coefficients C)

(10.13)
$$H^m(E, E_0; C) \to H^m(E_b, (E_0)_b; C) \cong H^m(\mathbf{R}^k, \mathbf{R}^k - \{0\}; C) \cong \begin{cases} C & \text{if } m = k, \\ 0 & \text{if } m \neq k. \end{cases}$$

We will be concerned only with the two fundamental cases of $C = \mathbf{Z}/2$ and $C = \mathbf{Z}$.

Exercise 175. Prove that a vector bundle $\mathbf{R}^k \hookrightarrow E \to B$ is orientable (i.e. its structure group reduces from $O(k)$ to $SO(k)$; see Definition 4.10) if and only if the the local coefficient system

$$\pi_1(B) \to \operatorname{Aut}(H^k(\mathbf{R}^k, \mathbf{R}^k - \{0\}; \mathbf{Z})) = \operatorname{Aut}(\mathbf{Z}) \cong \mathbf{Z}/2$$

(determined by Corollary 6.13) is trivial. Notice that by pulling back bundles, it is enough to prove this for $B = S^1$. Prove that a vector bundle over S^1 is orientable if and only if it is trivial. (Hint: use the clutching construction and the fact that $O(k)$ has exactly two path components.)

In preparation for the statement of the Thom isomorphism theorem, notice that given a vector bundle $\mathbf{R}^k \hookrightarrow E \xrightarrow{p} B$, there is a cup product (with any ring coefficients)

$$H^p(B) \times H^q(E, E_0) \xrightarrow{\cup} H^{p+q}(E, E_0) \tag{10.14}$$

obtained by pre-composing the cup product

$$H^p(E) \times H^q(E, E_0) \to H^{p+q}(E, E_0)$$

(see Corollary 3.24) with the isomorphism $p^* : H^p(B) \to H^p(E)$ induced by the bundle projection.

Theorem 10.28 (Thom isomorphism theorem). *Given a vector bundle*

$$\mathbf{R}^k \hookrightarrow E \to B$$

with $k \geq 1$,

1. *There exists a unique class $u \in H^k(E, E_0; \mathbf{Z}/2)$ so that for each $b \in B$, the restriction to the fiber over b,*

$$H^k(E, E_0; \mathbf{Z}/2) \to H^k(E_b, (E_0)_b; \mathbf{Z}/2) \cong \mathbf{Z}/2$$

(see Equation (10.13)) takes u to the unique non-zero element. This class u has the property that the homomorphism

$$\Phi : H^n(B; \mathbf{Z}/2) \to H^{k+n}(E, E_0; \mathbf{Z}/2), \quad x \mapsto x \cup u$$

(using the cup product (10.14)) is an isomorphism for all n.

2. *If the vector bundle $E \to B$ is orientable, then there exists a class $\tilde{u} \in H^k(E, E_0; \mathbf{Z})$ so that for each $b \in B$ the restriction to the fiber over b,*
$$H^k(E, E_0; \mathbf{Z}) \to H^k(E_b, (E_0)_b; \mathbf{Z}) \cong \mathbf{Z}$$
takes $\tilde{u}$ to a generator. The class $\tilde{u}$ is unique up to sign. It has the property that the homomorphism
$$\tilde{\Phi} : H^n(B; \mathbf{Z}) \to H^{k+n}(E, E_0; \mathbf{Z}), \quad x \mapsto x \cup \tilde{u}$$
is an isomorphism for all n. Moreover the coefficient homomorphism $H^k(E, E_0; \mathbf{Z}) \to H^k(E, E_0; \mathbf{Z}/2)$ takes u to $\tilde{u}$.

The classes u and $\tilde{u}$ are natural with respect to pulling back vector bundles: if $f : B' \to B$ is a continuous map, and $E' = f^*(E)$ the pulled back bundle, then (with the obvious notation) $u' = f^*(u)$ and $\tilde{u}' = f^*(\tilde{u})$.

Definition 10.29. The cohomology class $u \in H^k(E, E_0, \mathbf{Z}/2)$ (resp. $\tilde{u} \in H^k(E, E_0, \mathbf{Z})$) whose existence is assured by Theorem 10.28 is called the *Thom class for* $\mathbf{R}^k \hookrightarrow E \to B$. If we wish to emphasize the bundle $E \to B$, we will denote its Thom class by u_E (resp. $\tilde{u}_E$).

Proof of Theorem 10.28. We prove the two cases ($\mathbf{Z}$ and $\mathbf{Z}/2$) simultaneously. To simplify the notation we replace the pair (E, E_0) by the pair $(D(E), S(E))$ where $D(E)$ denotes the unit disk bundle and $S(E)$ the unit sphere bundle of E for some riemannian metric. We lose no information since the inclusions $(D(E), S(E)) \hookrightarrow (E, E_0)$ and $(D^k, S^{k-1}) \hookrightarrow (E_b, (E_0)_b)$ are deformation retracts.

Consider the spectral sequence for the relative fibration
$$(D^k, S^{k-1}) \to (D(E), S(E)) \to B$$
with $\mathbf{Z}/2$ or $\mathbf{Z}$ coefficients.

Using the universal coefficient theorem (and, for the orientable case, the fact that the coefficients are untwisted),
$$\begin{aligned} E_2^{p,q} &= H^p(B; H^q(D^k, S^{k-1})) \\ &\cong H^p(B) \otimes H^q(D^k, S^{k-1}) \\ &\cong \begin{cases} 0 & \text{if } q \neq k \\ H^p(B) & \text{if } q = k. \end{cases} \end{aligned}$$

The isomorphism $H^p(B) \to E_2^{p,k}$ is given by $\gamma \mapsto \gamma \otimes \tau$ where τ generates $H^k(D^k, S^{k-1})$.

Clearly
$$E_2^{p,q} = E_\infty^{p,q} = \begin{cases} H^{p+k}(D(E), S(E)) & \text{if } q = k, \\ 0 & \text{otherwise.} \end{cases}$$

Thus $H^p(B) \cong H^{p+k}(D(E), S(E))$.

Let u (resp. $\tilde{u}$) generate $H^k(D(E), S(E)) \cong H^0(X)$, which is isomorphic to $\mathbf{Z}/2$ (resp. $\mathbf{Z}$). Clearly u is unique and $\tilde{u}$ is unique up to sign. Then the multiplicative properties of the spectral sequence imply that the isomorphism $H^p(B) \cong H^{p+k}(D(E), S(E))$ is given by $\gamma \mapsto \gamma \cup u$ (resp. $\gamma \cup \tilde{u}$).

Naturality of the classes u and $\tilde{u}$ follows from the naturality properties of the spectral sequences with respect to cellular maps and the cellular approximation theorem.

The fact that u (resp. $\tilde{u}$) restricts to the generator in each fiber follows from naturality by considering the inclusion of the fibrations induced by the inclusion of a point b in B. The fact that $\tilde{u}$ is mapped to u via $H^k(D(E), S(E); \mathbf{Z}) \to H^k(D(E), S(E); \mathbf{Z}/2)$ follows from uniqueness of the Thom class and commutativity of the diagram

$$\begin{array}{ccc} H^k(D(E), S(E); \mathbf{Z}) & \longrightarrow & H^k(D(E), S(E); \mathbf{Z}/2) \\ \downarrow & & \downarrow \\ H^k(D^k, S^{k-1}; \mathbf{Z}) & \longrightarrow & H^k(D^k, S^{k-1}; \mathbf{Z}/2) \end{array}$$

where the vertical arrows are induced by the inclusions and the horizontal arrows by the coefficient homomorphism $\mathbf{Z} \to \mathbf{Z}/2$. □

The Thom isomorphism theorem has a homology counterpart. We will use the following corollary.

Corollary 10.30. *Taking the cap product with $u \in H^k(E, E_0; \mathbf{Z}/2)$ induces isomorphisms*

$$u \cap \; : H_{n+k}(E, E_0; \mathbf{Z}/2) \to H_n(B; \mathbf{Z}/2).$$

Proof. (Use $\mathbf{Z}/2$-coefficients throughout the proof.) For any space X, $H_n(X)$ is isomorphic to $\mathrm{Hom}(H^n(X), \mathbf{Z}/2)$ via $x \mapsto \langle -, x \rangle$, where $\langle \;,\; \rangle$ denotes the Kronecker pairing.

The composite isomorphism

$$H_{n+k}(E, E_0) \cong \mathrm{Hom}(H^{n+k}(E, E_0), \mathbf{Z}/2) \xrightarrow{(\cup u)^*} \mathrm{Hom}(H^n(B), \mathbf{Z}/2) \cong H_n(B)$$

is given by $x \mapsto z$, where x and z are related by

$$\langle \beta, z \rangle = \langle \beta \cup u, x \rangle \quad \text{for all} \quad \beta \in H^n(B).$$

Since $\langle \beta \cup u, x \rangle = \beta \cap (u \cap x)$, we conclude that $z = u \cap x$. □

10.7. Intersection theory

One very useful consequence of the Thom isomorphism theorem is the identification of intersection numbers with cup products in manifolds. For simplicity we will discuss only the case of smooth compact manifolds, but everything we say holds in greater generality (with trickier proofs). In this section all homology and cohomology is taken with $\mathbf{Z}$ coefficients.

Resolving the various notions of orientability is necessary, and so our exposition will involve a sequence of exercises to relate the various notions. These exercises are all straightforward, but they can be a bit confusing. A mastery of orientation issues is quite useful for a working mathematician, and you should keep in mind that such a mastery comes only from a thorough understanding of the equivalence between different points of view. (In other words: solve these exercises!)

Recall that an orientation of a real finite-dimensional vector space V is an equivalence class of bases of V where two bases are considered equivalent if the determinant of the change of basis matrix is positive. Notice that a choice of basis identifies V with $\mathbf{R}^m$ for some m. This in turn induces an isomorphism

$$H_m(V, V - \{0\}) \xrightarrow{\cong} H_m(\mathbf{R}^m, \mathbf{R}^m - \{0\}) \cong \mathbf{Z}. \tag{10.15}$$

Exercise 176. Show that changing the orientation of V changes the identification of Equation (10.15) by a sign.

It follows from this exercise that an orientation of V can be defined as a choice of generator of $H_m(V, V - \{0\})$. By choosing the dual generator an orientation of V can also be defined as a choice of orientation of $H^m(V, V - \{0\})$.

We have come across several notions of orientability for smooth manifolds. One notion is that a smooth manifold M is orientable if its tangent bundle is orientable; i.e. the structure group of TM can be reduced from $O(m)$ to $SO(m)$. An orientation is a choice of such a reduction.

Exercise 177. Show that an orientation in this sense determines an equivalence class of bases at each tangent space T_pM. (Hint: use Exercise 175.) More generally show that a reduction of the structure group of a vector bundle E from $O(n)$ to $SO(n)$ determines an equivalence class of bases in each fiber E_x so that these equivalence classes are compatible with the local trivializations of E; i.e. the homeomorphism $E|_U \cong U \times \mathbf{R}^n$ takes the orientation of E_x to the same orientation of $\mathbf{R}^n$ for all $x \in U$.

Another notion of orientability says that a compact, connected manifold M is orientable if $H_m(M, \partial M) \cong \mathbf{Z}$ and that an orientation is a choice

$[M, \partial M] \in H_m(M, \partial M)$ of generator, called the *fundamental class* of the oriented manifold M. In the course of the proof of the Poincaré duality theorem one shows that if $[M, \partial M]$ is the fundamental class of M, then for each $p \in M$ the inclusion $H_m(M, \partial M) \to H_m(M, M - p)$ is an isomorphism.

Given $p \in M$ (and a choice of riemannian metric on M) the exponential map $\exp : T_pM \to M$ restricts to a diffeomorphism in a small ball $W \subset T_pM$, exp: $W \to U \subset M$ and hence gives isomorphisms (the first and third are excision isomorphisms)

$$H_m(T_pM, T_pM - 0) \cong H_m(W, W - 0) \xrightarrow{\exp} H_m(U, U - p) \cong H_m(M, M - p).$$

This shows that the choice of fundamental class $[M, \partial M] \in H_m(M, \partial M)$ orients the tangent space T_pM.

Exercise 178. Prove that this sets up an identification between the two notions of an orientation of a smooth manifold (the choice of $[M, \partial M]$ and an orientation of the vector bundle TM).

The fundamental class $[M, \partial M] \in H_m(M, \partial M)$ of an oriented manifold determines the dual *cohomology fundamental class* $[M, \partial M]^* \in H^m(M, \partial M)$ (and conversely) by the equation

$$\langle [M, \partial M]^*, [M, \partial M] \rangle = 1$$

where $\langle\ ,\ \rangle$ denotes the Kronecker pairing.

Suppose that V and W are oriented subspaces of an oriented vector space Z and that $\dim(V) + \dim(W) = \dim(Z)$. Suppose that V and W are transverse, i.e. $V \cap W = 0$. Then the *intersection number* of V and W is the number in $\{\pm 1\}$ defined to be the sign of the determinant of the change of basis matrix from the (ordered) basis $\{\mathbf{b}_V, \mathbf{b}_W\}$ to $\mathbf{b}_Z$, where $\mathbf{b}_V, \mathbf{b}_W$, and $\mathbf{b}_Z$ denote bases in the given equivalence classes. Notice that reversing the order of V and W changes the intersection number by $(-1)^{\dim(V)\dim(W)}$.

Now suppose that A and B are smooth, compact, connected, oriented submanifolds of dimensions a and b of a compact oriented manifold M of dimension m. (A smooth manifold is oriented if its tangent bundle is oriented.) Assume that A is properly embedded; i.e. the boundary of A is embedded in the boundary of M. Assume also that the boundary of B is empty and that B is contained in the interior of M. Finally assume that A and B are transverse. This means that at each point $p \in A \cap B$, the tangent subspaces T_pA and T_pB span T_pM.

Definition 10.31. Suppose that $a + b = m$. Then since A and B are transverse and compact, their intersection consists of a finite number of points. Because A, B, and M are oriented, we can assign an intersection number ϵ_p to each intersection point $p \in A \cap B$ by taking the intersection number (as above) of the oriented subspaces T_pA and T_pB in T_pM. Then define the *intersection number of* A *and* B to be the integer

$$A \cdot B = \sum_{p \in A \cap B} \epsilon_p.$$

Notice that $A \cdot B = (-1)^{ab} B \cdot A$.

Since A and B are oriented, they have fundamental classes $[A, \partial A] \in H_a(A, \partial A)$, $[B] \in H_b(B)$. Let $e_A : A \subset M$ and $e_B : B \subset M$ denote the inclusions. Then $e_A([A, \partial A]) \in H_a(M, \partial M)$ and $e_B([B]) \in H_b(M)$.

Theorem 10.32. *Let* $\alpha \in H^b(M)$ *be the Poincaré dual to* $e_A([A, \partial A])$, *and* $\beta \in H^a(M, \partial M)$ *the Poincaré dual to* $e_B([B])$, *i.e.*

$$\alpha \cap [M, \partial M] = e_A([A, \partial A]) \quad \text{and} \quad \beta \cap [M, \partial M] = e_B([B]).$$

Then

$$A \cdot B = \langle \alpha \cup \beta, [M, \partial M] \rangle \tag{10.16}$$

where $\langle \ , \ \rangle$ *denotes the Kronecker pairing.*

Theorem 10.32 justifies the terminology "intersection pairing" for the cup product

$$H^b(M) \times H^a(M, \partial M) \xrightarrow{\cup} H^m(M, \partial M) \xrightarrow{\cong} \mathbf{Z}$$

(see Section 3.6.2). Moreover, it implies that the intersection number $A \cdot B$ depends only on the homology classes $e_A([A, \partial A])$ and $e_B([B])$. In particular, given any not necessarily transverse submanifolds A and B as above, the transversality theorems imply that B can be isotoped to be transverse to A. This preserves the class $e_B([B])$, and so the resulting intersection number $A \cdot B$ is independent of the choice of the isotopy.

With more work one can define $x \cdot y$ for any classes $x \in H_a(M, \partial M)$ and $y \in H_b(M)$, or even on the chain level for a simplicial complex and its dual cell complex (by thinking of a simplex as a submanifold). Theorem 10.32 is true in this greater generality. Alternatively, this approach can be reversed to give a proof of Poincaré duality and a definition of cup products in terms of intersections.

There is also a mod 2 version of Theorem 10.32 in which orientation issues do not play a role; one defines $A \cdot_2 B$ to be the number of intersection points of A and B when A and B are transverse. It holds in greater generality since none of the manifolds need be orientable, but the conclusion is correspondingly weaker; it only holds mod 2. To help you digest the following argument, you might first consider the mod 2 case, thereby avoiding sign and orientation issues which complicate the proof.

Proof of Theorem 10.32. Let E denote the normal bundle of $B \subset M$. The tubular neighborhood theorem implies that E can be embedded as a neighborhood of B in M, with B itself corresponding to the zero section. Give E a metric.

It is a straightforward consequence of the fact that A and B are transverse that for $\varepsilon > 0$ small enough, the disk bundle $D(E)$ of vectors of E of length less than or equal to ε intersects A in a finite number of disks D_p, one for each p in $A \cap B$, with each D_p isotopic rel boundary in $(D(E), \partial D(E))$ to a fiber of the disk bundle $D(E) \to B$. Using the isotopy extension theorem (and maybe making ε smaller if necessary), we may assume A intersects $D(E)$ exactly in a union of fibers, one for each point $p \in A \cap B$. In other words, after an isotopy supported in E which fixes each $p \in A \cap B$,

$$A \cap D(E) = \bigcup_{p \in A \cap B} D(E)_p.$$

For convenience we simplify notation by setting $D = D(E)$. Thus the boundary ∂D is the ε-sphere bundle of E. The submanifold A intersects D in a union of disks D_p, one for each $p \in A \cap B$. The situation is illustrated in the following figure.

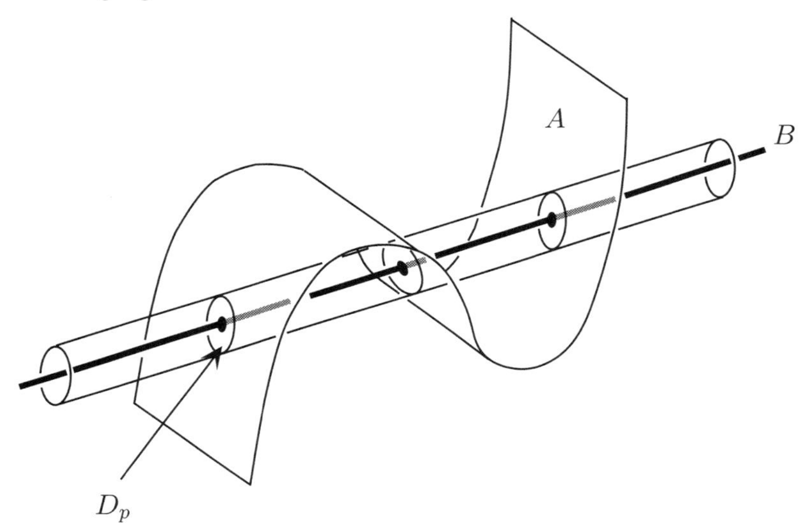

The manifold D is oriented as a submanifold of M. To see this, notice that there are excision isomorphisms

$$H_m(D, \partial D) \cong H_m(D, D - x) \cong H_m(M, M - x) \cong H_m(M, \partial M) \tag{10.17}$$

for any $x \in \text{Int}(D)$. Moving from left to right orients D compatibly with M. Alternatively, $TD = TM|_D$, so the orientation of the tangent bundle of M orients the tangent bundle of D.

The normal bundle $E \to B$ is orientable. One way to see this is to use the Whitney sum decomposition

$$E \oplus TB \cong TM|_B$$

(since $E = \nu(B \subset M)$). The fact that TB and TM are orientable implies that E is orientable. Orient E so that the intersection number of a fiber E_x of E with the zero section B equals 1:

$$E_x \cdot B = 1.$$

Exercise 179. Show that this condition uniquely specifies an orientation of the normal bundle E.

This orients the fibers D_x; i.e. for each $x \in B$ one gets a preferred generator $[D_x, \partial D_x]$ of $H_a(D_x, \partial D_x)$.

The Thom isomorphism theorem says that there is a unique Thom class $\tilde{u} \in H^a(D, \partial D)$ so that $\cup \tilde{u} : H^k(B) \to H^{k+a}(D, \partial D)$ is an isomorphism for all k so that the restriction of $\tilde{u}$ to the fiber D_x satisfies $\tilde{u}|_{D_x} = [D_x, \partial D_x]^*$, i.e. $\langle \tilde{u}, [D_x, \partial D_x] \rangle = 1$.

Now $[B]^* \cup \tilde{u}$ generates $H^m(D, \partial D)$ and so equals $[D, \partial D]^*$ up to sign. The sign is equal to $(-1)^{ab}$. To see this one can use naturality of the Thom class and work over a small open set in B diffeomorphic to a ball of dimension a.

Exercise 180. Prove that $\tilde{u} \cup [B]^* = [D, \partial D]^*$ (and hence, by Theorem 3.13 $[B]^* \cup \tilde{u} = (-1)^{ab}[D, \partial D]^*$) by pulling $E \to B$ back over a small neighborhood $U \subset B$.

Thus, using Exercise 37 and Proposition 3.21,

$$\begin{aligned}
(-1)^{ab} &= \langle [B]^* \cup \tilde{u}, [D, \partial D] \rangle \\
&= ([B]^* \cup \tilde{u}) \cap [D, \partial D] \\
&= [B]^* \cap (\tilde{u} \cap [D, \partial D]) \\
&= \langle [B]^*, \tilde{u} \cap [D, \partial D] \rangle
\end{aligned}$$

and so, since $H^b(D) \cong \mathbf{Z}$ and $H_b(D) \cong \mathbf{Z}$,

$$\tilde{u} \cap [D, \partial D] = (-1)^{ab}[B].$$

In other words, $(-1)^{ab}\tilde{u}$ is the Poincaré dual to $[B]$ in D.

The inclusion $i_1 : (D, \partial D) \subset (M, M - \mathrm{Int}(D))$ induces excision isomorphisms in homology and cohomology. Hence $H_n(M, M - \mathrm{Int}(D))$ is isomorphic to $\mathbf{Z}$. The inclusion $i_2 : (M, \partial M) \subset (M, M - \mathrm{Int}(D))$ is not an excision, but induces an isomorphism

$$H_n(M, \partial M) \to H_n(M, M - \mathrm{Int}(D))$$

since both groups are isomorphic to $\mathbf{Z}$ and the inclusions of both pairs to $(M, M-x)$ are excisions. Because the orientations were chosen compatibly,

$$i_1([D, \partial D]) = [M, M - \mathrm{Int}(D)] = i_2([M, \partial M]).$$

It follows from naturality of the cap product

$$H^p(X, Y) \times H_q(X, Y) \xrightarrow{\cap} H_{q-p}(X)$$

for any pair (X, Y) that the diagram

$$\begin{array}{ccc}
H^p(D, \partial D) & \xrightarrow{\cap[D,\partial D]} & H_{n-p}(D) \\
\uparrow{\scriptstyle i_1} & & \downarrow{\scriptstyle i_1} \\
H^p(M, M - \mathrm{Int}(D)) & \xrightarrow{\cap i_1([D,\partial D])} & H_{n-p}(M) \\
\downarrow{\scriptstyle i_2} & & \uparrow{\scriptstyle \mathrm{Id}} \\
H^p(M, \partial M) & \xrightarrow{\cap[M,\partial M]} & H_{n-p}(M)
\end{array}$$

commutes.

Denote by j^* the composite $i_2 \circ (i_1)^{-1} : H^p(D, \partial D) \to H^p(M, \partial M)$. The diagram above shows that if $x \in H^p(D, \partial D)$, $i_1(x \cap [D, \partial D]) = j^*(x) \cap [M, \partial M]$. Taking $x = \tilde{u}$ and using the notation $[B]$ for the image of the fundamental class of B in either D or M, we conclude that

$$j^*(\tilde{u}) \cap [M, \partial M] = i_1(\tilde{u} \cap [D, \partial D]) = (-1)^{ab}[B].$$

In particular

$$\beta = (-1)^{ab} j^*(\tilde{u}).$$

We can think of the homomorphism j^* as being induced by the quotient map $j : M/\partial M \to D/\partial D$. Thus we have a corresponding homomorphism $j_* : H_p(M, \partial M) \to H_*(D, \partial D)$. Using the notation $[A, \partial A]$ for the fundamental class of A in $H_a(M, \partial M)$, we see the class $j_*([A, \partial A])$ is represented by the union of fibers D_p, one for each $p \in A \cap B$, but oriented according to the local intersection number of A and B at p. Precisely:

$$j_*([A, \partial A]) = \sum_{p \in A \cap B} \varepsilon_p [D_p, \partial D_p],$$

where $\varepsilon_p = 1$ or -1 according to whether or not the two orientations of $D_p \subset (A \cap D)$ given by

1. restricting the orientation of A to D_p, and
2. the orientation of D_p as a fiber of the normal bundle D

agree. This is because the map $j : M/\partial M \to D/\partial D$ takes $A/\partial A$ to

$$\cup_p(D_p/\partial D_p) = \vee_p(D_p/\partial D_p).$$

By definition,

$$\sum_p \varepsilon_p = A \cdot B.$$

We now compute:

$$\begin{aligned}
\langle \alpha \cup \beta, [M, \partial M]\rangle &= \langle \alpha \cup (-1)^{ab} j^*(\tilde{u}), [M, \partial M]\rangle \\
&= \langle j^*(\tilde{u}) \cup \alpha, [M, \partial M]\rangle \\
&= \big(j^*(\tilde{u}) \cup \alpha\big) \cap [M, \partial M] \\
&= j^*(\tilde{u}) \cap \big(\alpha \cap [M, \partial M]\big) \\
&= j^*(\tilde{u}) \cap [A, \partial A] \\
&= \langle j^*(\tilde{u}), [A, \partial A]\rangle \\
&= \langle \tilde{u}, j_*([A, \partial A])\rangle \\
&= \sum_{p \in A \cap B} \varepsilon_p \langle \tilde{u}, [D_p, \partial D_p]\rangle \\
&= A \cdot B.
\end{aligned}$$

□

Exercise 181. State (and prove) the mod 2 version of Theorem 10.32.

During the proof of Theorem 10.32 we also proved the following.

Corollary 10.33. *Let $e : B \subset M$ be an embedding of a smooth, closed, oriented manifold in a compact, oriented manifold. Let D denote a closed tubular neighborhood of B in M, with Thom class $\tilde{u} \in H^{n-b}(D, \partial D)$, and let $j : M/\partial M \to D/\partial D$ denote the collapse map. Then $j^*(\tilde{u})$ is the Poincaré dual to $e_*([B])$ (up to sign).* □

The sign ambiguity in Corollary 10.33 comes from the fact that there are two possible choices of Thom classes $\tilde{u}$; during the proof of Corollary 10.33 we made a specific choice by requiring that $\langle \tilde{u}, [D_x, \partial D_x]\rangle = D_x \cdot B$.

We describe the usual way that a geometric topologist thinks of the Poincaré dual $\beta \in H^a(M, \partial M)$ to a cycle represented by a submanifold $B \subset M$. Given a cycle $x \in H_a(M, \partial M)$ represented by an oriented submanifold $(A, \partial A) \subset (M, \partial M)$, the class β is represented by the cochain whose value on x is given by the formula

$$\langle \beta, x\rangle = B \cdot A.$$

In brief, "the Poincaré dual β to B is given by intersecting with B."

To see why this is true, we compute:

$$\begin{aligned} B \cdot A &= (-1)^{ab} A \cdot B \\ &= (-1)^{ab} \langle \alpha \cup \beta, [M, \partial M] \rangle \\ &= (-1)^{ab} (\alpha \cup \beta) \cap [M, \partial M] \\ &= (\beta \cup \alpha) \cap [M, \partial M] \\ &= \beta \cap (\alpha \cap [M, \partial M]) \\ &= \beta \cap [A, \partial A] \\ &= \langle \beta, [A, \partial A] \rangle. \end{aligned}$$

Exercise 182. Show that if A and B are closed submanifolds of S^n intersecting transversally in finitely many points, then they intersect in an even number of points.

Exercise 183. Let M be a closed manifold and $f : M \to M$ a smooth map. Let $\Delta \subset M \times M$ be the diagonal and

$$G(f) = \{(m, f(m))\} \subset M \times M$$

the graph of f. Show that if $\Delta \cdot G(f)$ is no-zero, then any map homotopic to f has a fixed point. Can you show that $\Delta \cdot G(\mathrm{Id})$ equals the Euler characteristic of M or, more generally, that $\Delta \cdot G(f)$ equals the Lefschetz number of f?

Exercise 184. Think about how to modify the proof of Theorem 10.32 to handle the situation when A and B are only immersed instead of embedded.

A more ambitious exercise is the following, which says that the intersection of submanifolds is identified with the cup product even when the dimensions are not complementary.

Exercise 185. Show that if A and B are embedded, transverse, but $a + b > m$, then the intersection $A \cap B$ is an oriented, closed submanifold of dimension $m - a - b$. Prove that the Poincaré dual of $[A \cap B]$ is the class $\alpha \cup \beta$. Use the fact that $A \cap D$ is the pull-back of the disk bundle $D \to B$ over $A \cap B$, use naturality of the Thom class, and apply Corollary 10.33. (You might try the $\mathbf{Z}/2$ version first, to avoid orientation issues.)

10.8. Stiefel–Whitney classes

Denote by u_k the "universal" Thom class for the universal vector bundle over $BO(k)$, $V_k \to BO(k)$. Recall that the inclusion of matrices

$$A \mapsto \begin{pmatrix} A & 0 \\ 0 & 1 \end{pmatrix}$$

induces a map of vector bundles $V_{k-1} \oplus \varepsilon \to V_k$ which is an isomorphism on fibers and covers the natural map $BO(k-1) \to BO(k)$. The Thom isomorphism theorem implies that the cup product

$$\Phi_k = \cup u_k : H^n(BO(k); \mathbf{Z}/2) \to \widetilde{H}^{n+k}(MO(k); \mathbf{Z}/2)$$

is an isomorphism for all n.

Proposition 10.34. *The diagram*

$$\begin{array}{ccc} H^n(BO(k)) & \xrightarrow{\Phi_k} & H^{n+k}(MO(k)) \\ \downarrow & & \downarrow \\ H^n(BO(k-1)) & \xrightarrow{\Phi_{k-1}} & H^{n+k-1}(MO(k-1)) \end{array}$$

commutes, where the left vertical map is induced by the inclusion

$$BO(k-1) \to BO(k)$$

and the right vertical map is induced by the composite of the inclusion

$$H^{n+k}(MO(k)) \to H^{n+k}(M(V_{k-1} \oplus \varepsilon)) = H^{n+k}(SMO(k-1))$$

and the suspension isomorphism

$$S^* : H^{n+k}(SMO(k-1)) \cong H^{n+k-1}(MO(k-1)).$$

Proof. Notice that $M(E \oplus \varepsilon) = SM(E)$. Restricting to a fiber corresponds to the suspension $S(D^k, S^{k-1}) = (D^{k+1}, S^k)$. It follows by naturality of the suspension isomorphism that the Thom class for $E \oplus \varepsilon$ is the suspension of the Thom class for E.

If

$$\begin{array}{ccc} E & \xrightarrow{\tilde{f}} & E' \\ \downarrow & & \downarrow \\ X & \xrightarrow{f} & X' \end{array}$$

is a map of $\mathbf{R}^k$-vector bundles which is an isomorphism on each fiber (equivalently, $E \cong f^*(E')$), then the Thom class pulls back, $u_E = \tilde{f}^*(u_{E'})$; this follows again by uniqueness of the Thom class and by restricting to fibers.

The corollary now follows from these observations and the fact that if $i : BO(k-1) \to BO(k)$ is the map induced by inclusion, then $i^*(V_k) \cong V_{k-1} \oplus \varepsilon$. □

Proposition 10.35. *The homomorphism*

$$H^n(BO(k);\mathbf{Z}/2) \to H^n(BO(k-1);\mathbf{Z}/2)$$

induced by the natural map $BO(k-1) \to BO(k)$ *is an isomorphism for* $n < k-1$.

Proof. Consider the fibration

$$O(k-1) \hookrightarrow O(k) \to S^{k-1}$$

taking a matrix in $O(k)$ to its last column. This deloops twice (see Theorem 6.40) to give a fibration

$$S^{k-1} \to BO(k-1) \to BO(k).$$

The spectral sequence for this fibration (or the Gysin sequence) shows that the sequence

$$\cdots \to H^{n-k}(BO(k)) \to H^n(BO(k)) \to H^n(BO(k-1)) \to H^{n-k+1}(BO(k)) \to \cdots$$

is exact. Thus if $n-k+1 < 0$, $H^n(BO(k)) \to H^n(BO(k-1))$ is an isomorphism. □

The Stiefel–Whitney classes can now be defined using the Steenrod operations and the Thom isomorphism theorem.

Definition 10.36. Define the *nth (universal) Stiefel–Whitney class* to be

$$w_n = \Phi_k^{-1}(Sq^n(u_k)) \in H^n(BO(k);\mathbf{Z}/2),$$

where $u_k \in H^k(MO(k);\mathbf{Z}/2)$ denotes the Thom class and

$$\Phi_k = \cup u_k : H^n(BO(k);\mathbf{Z}/2) \to H^{k+n}(MO(k);\mathbf{Z}/2)$$

denotes the Thom isomorphism.

Proposition 10.34, Proposition 10.35, and naturality of the Steenrod operations imply

Proposition 10.37. *The restriction*

$$H^n(BO(k);\mathbf{Z}/2) \to H^n(BO(k-r);\mathbf{Z}/2)$$

takes w_n *for* $BO(k)$ *to* w_n *for* $BO(k-r)$. □

Hence the notation w_k is unambiguous. Notice that if $n > k$, then $Sq^n(u_k) = 0$, and so $w_n = 0$ in $H^n(BO(k);\mathbf{Z}/2)$ for $n > k$.

Definition 10.38. The *nth Stiefel–Whitney class* of an $\mathbf{R}^k$-vector bundle $E \to B$ is the class

$$w_n(E) = f_E^*(w_n) \in H^n(B; \mathbf{Z}/2)$$

where $f_E : B \to BO(k)$ denotes the classifying map for E (see Theorem 8.22).

It follows immediately from their definition that the Stiefel–Whitney classes are natural with respect to pulling back bundles. In particular, if E and E' are isomorphic bundles over B, then $w_n(E) = w_n(E')$ for all n. Moreover, since the w_n are compatible with respect to the maps $BO(k) \to BO(k+1)$,

$$w_n(E \oplus \epsilon) = w_n(E).$$

In other words, the Stiefel–Whitney classes are invariants of the stable equivalence class of a vector bundle.

Exercise 186. Show that a vector bundle E is orientable if and only if $w_1(E) = 0$. (Hint: first relate Sq^1 to the Bockstein associated to the exact sequence $0 \to \mathbf{Z} \xrightarrow{\times 2} \mathbf{Z} \to \mathbf{Z}/2 \to 0$.)

The Cartan formula (see Theorem 10.24) easily implies the following theorem.

Theorem 10.39. *The Stiefel–Whitney numbers of a Whitney sum of vector bundles satisfy*

$$w_k(E \oplus F) = \sum_n w_n(E) \cup w_{k-n}(F).$$

□

Exercise 187. Suppose that E and F are vector bundles over a finite-dimensional CW-complex so that $E \oplus F$ is trivial (i.e. E and F are stable inverses. For example, take E to be the tangent bundle of a smooth compact manifold and F its normal bundle for some embedding in S^n.) Use Theorem 10.39 to prove that

$$w_1(F) = w_1(E), w_2(F) = w_1(E)^2 + w_2(E), w_3(F) = w_1(E)^3 + w_3(E),$$

and, in general, that

$$w_n(F) = \sum_{i_1 + 2i_2 + \cdots + ki_k = n} \frac{(i_1 + \cdots + i_k)!}{i_1! \cdots i_k!} w_1(E)^{i_1} \cdots w_k(E)^{i_k}.$$

The Stiefel–Whitney classes generate the cohomology ring of $BO(k)$, as the following theorem shows.

Theorem 10.40. *The $\mathbf{Z}/2$-cohomology ring of $BO(k)$ is a polynomial ring on the Stiefel–Whitney classes of degree less than or equal to k:*

$$H^*(BO(k);\mathbf{Z}/2) = \mathbf{Z}/2[w_1, w_2, \cdots, w_k]$$

where $w_i \in H^i(BO(k);\mathbf{Z}/2)$ denotes the ith Stiefel–Whitney class.

Proof. (Use $\mathbf{Z}/2$ coefficients.) First we show that $w_k \in H^k(BO(n))$ is non-zero if $k \leq n$. To see this it suffices by naturality to find one $\mathbf{R}^n$-bundle with w_k non-zero. Let $\mathbf{R}^1 \hookrightarrow E \to S^1$ denote the "Möbius band" bundle over S^1, i.e. the bundle with clutching function $S^0 \to O(1)$ the non-constant map. This bundle has $w_1 \neq 0$ (for example, it is not orientable). Thus w_1 is non-zero in $H^1(BO(1))$, and since the restrictions

$$H^k(BO(n)) \to H^k(BO(n-1))$$

preserve the w_i by Proposition 10.37, w_1 is non-zero in $H^1(BO(n))$ for all $n \geq 1$.

Since $BO(1) = K(\mathbf{Z}/2, 1) = \mathbf{R}P^\infty$ and $w_1 \neq 0$, $H^*(BO(1)) = \mathbf{Z}/2[w_1]$. Let $E \to BO(1)$ be a bundle with $w_1(F) = w_1$. Then Theorem 10.39 (and induction) shows that

$$w_k(\underbrace{F \oplus \cdots \oplus F}_{k \text{ times}}) = w_1(F)^k,$$

which is non-zero in $H^k(BO(1))$. Therefore $w_k \in H^k(BO(n))$ is non-zero for all $n \geq k$.

We prove the theorem by induction. The case $n = 1$ is contained in the previous paragraph. Let $i : BO(n-1) \to BO(n)$ denote the inclusion. The induced map $i^* : H^*(BO(n)) \to H^*(BO(n-1))$ is surjective since by induction $H^*(BO(n-1)$ is generated by the w_i for $i \leq n-1$, and these are in the image of i^*.

The fiber of $i : BO(n-1) \to BO(n)$ is S^{n-1}; in fact the fibration obtained by taking an orthogonal matrix to its last column

$$O(n-1) \hookrightarrow O(n) \to S^{n-1}$$

deloops twice to give the fibration

$$S^{n-1} \to BO(n-1) \to BO(n).$$

Consider the cohomology spectral sequence for this fibration. It has $E_2^{p,q} = H^p(BO(n)) \otimes H^q(S^{n-1})$ which is zero if $q \neq 0$ or $n-1$. Hence

$$E_k^{p,q} = \begin{cases} E_2^{p,q} = H^p(BO(n)) \otimes H^q(S^{n-1}) & \text{if } k \leq n, \\ E_\infty^{p,q} & \text{if } k > n. \end{cases}$$

This leads to the exact sequence (this is just the Gysin sequence in cohomology)

$$\cdots \to H^{k-1}(BO(n)) \to H^{k-1}(BO(n-1)) \to$$
$$H^{n-1}(S^{n-1}) \otimes H^{k-1}(BO(n)) \xrightarrow{d_n} H^k(BO(n)) \to H^k(BO(n-1)) \to \cdots$$

which reduces to short exact sequences

(10.18)
$$0 \to H^{n-1}(S^{n-1}) \otimes H^{k-1}(BO(n)) \xrightarrow{d_n} H^k(BO(n)) \to H^k(BO(n-1)) \to 0$$

since $H^*(BO(n)) \to H^*(BO(n-1))$ is onto. The map labelled d_n in (10.18) is the differential $d_n : E_n^{q,n-1} \to E_n^{q+n,0}$.

Taking $k = n$ in the sequence (10.18), we obtain

$$0 \to H^{n-1}(S^{n-1}) \xrightarrow{d_n} H^n(BO(n)) \to H^n(BO(n-1)) \to 0.$$

Since $H^{n-1}(S^{n-1}) = \mathbf{Z}/2$, generated by the fundamental class $[S^{n-1}]^*$, and since $w_n \in H^n(BO(n))$ is non-zero and in the kernel of the restriction $H^n(BO(n)) \to H^n(BO(n-1))$, it follows that $d_n([S^{n-1}]^*) = w_n$.

Applying the sequence (10.18), the fact that $d_n([S^{n-1}]^* \cup \alpha) = w_n \cup \alpha$ for $\alpha \in H^{k-1}(BO(n))$, and induction completes the proof, since this sequence shows that any element in $H^k(BO(n))$ can be written uniquely as a sum of classes of the form

$$w_1^{i_1} \cdots w_{n-1}^{i_{n-1}} \text{ with } i_1 + 2i_2 + \cdots + (n-1)i_{n-1} = k$$

and classes of the form

$$d_n([S^{n-1}]^*)\alpha = w_n \alpha$$

for some $\alpha \in H^{k-n}(BO(n))$. □

Exercise 188. Show that if $E \to B$ is an $\mathbf{R}^k$-vector bundle, then $w_k(E)$ is the image of the Thom class under the composite

$$H^k(E, E_0; \mathbf{Z}/2) \xrightarrow{i^*} H^k(E; \mathbf{Z}/2) \xrightarrow{z^*} H^k(B; \mathbf{Z}/2)$$

where $z : B \to E$ denotes the zero section.

If $E \to B$ is an oriented $\mathbf{R}^k$-vector bundle, the class

$$e(E) = z^*(i^*(\tilde{u})) \in H^k(B; \mathbf{Z})$$

is called the Euler class of E and reduces to $w_k(E)$ mod 2. Compare this with the definition we gave of Euler class in Section 7.11.

10.9. Localization

Given a subset P of the set of prime numbers, let $\mathbf{Z}_{(P)}$ denote the integers localized at P. This is the subring of the rationals consisting of all fractions whose denominator is relatively prime to each prime in P:

$$\mathbf{Z}_{(P)} = \{\tfrac{r}{s} \mid (r,s) = 1 \text{ and } (s,p) = 1 \text{ for each prime } p \in P\}.$$

Thus

$$\mathbf{Z} \subset \mathbf{Z}_{(P)} \subset \mathbf{Q}.$$

If P consists of a single prime p, we write $\mathbf{Z}_{(P)} = \mathbf{Z}_{(p)}$.

Definition 10.41. Given a set P of prime numbers, an abelian group A is called *P-local* if the homomorphism

$$A \to A \otimes_{\mathbf{Z}} \mathbf{Z}_{(P)}, \quad a \mapsto a \otimes 1$$

is an isomorphism.

If p is a prime and $r > 0$,

$$\mathbf{Z}/p^r \otimes \mathbf{Z}_{(P)} = \begin{cases} \mathbf{Z}/p^r & \text{if } p \in P, \\ 0 & \text{if } p \notin P. \end{cases} \tag{10.19}$$

More generally, if $A \in \mathcal{C}_P$, then $A \otimes \mathbf{Z}_{(P)} = 0$. This is because if $a \in A$, choose $r > 0$ relatively prime to each $p \in P$ so that $ra = 0$. Then r is invertible in $\mathbf{Z}_{(P)}$, and so for each $z \in \mathbf{Z}_{(P)}$,

$$a \otimes z = a \otimes \tfrac{zr}{r} = ra \otimes \tfrac{z}{r} = 0.$$

Since $\mathbf{Z}_{(P)}$ is torsion free, it is flat as an abelian group (see Exercise 27); i.e. the functor $- \otimes_{\mathbf{Z}} \mathbf{Z}_{(P)}$ is exact. In particular, if $f : A \to B$ is a $\mathcal{C}_P$ isomorphism, then tensoring the exact sequence

$$0 \to \ker f \to A \xrightarrow{f} B \to \operatorname{coker} f \to 0$$

with $\mathbf{Z}_{(P)}$ and using the fact that

$$\ker f \otimes \mathbf{Z}_{(P)} = 0 = \operatorname{coker} f \otimes \mathbf{Z}_{(P)},$$

we conclude that

$$f \otimes 1 : A \otimes \mathbf{Z}_{(P)} \to B \otimes \mathbf{Z}_{(P)}$$

is an isomorphism. This implies that if A and B are $\mathcal{C}_P$-isomorphic, then $A \otimes \mathbf{Z}_{(P)}$ is isomorphic to $B \otimes \mathbf{Z}_{(P)}$. Conversely, suppose that A and B are finitely generated abelian groups so that $A \otimes \mathbf{Z}_{(P)}$ is isomorphic to $B \otimes \mathbf{Z}_{(P)}$. Then A and B have the same rank and their p-primary subgroups are isomorphic for $p \in P$. Thus there is a $\mathcal{C}_P$-isomorphism from A to B.

The (relative) Hurewicz theorem mod $\mathcal{C}$ implies the following result, when applied to $\mathcal{C}_P$.

Theorem 10.42. *Let A, X be spaces such that $H_i(A)$ and $H_i(X)$ are finitely generated for each i, such that $\pi_1(A) = \pi_1(X) = 0$.*

Let $f : A \to X$ be a map with $\pi_2(X, A) = 0$. Then the statements:

1. $f_* : H_i(A; \mathbf{Z}_{(P)}) \to H_i(X; \mathbf{Z}_{(P)})$ *is an isomorphism for $i < n$ and an epimorphism for $i = n$,*
2. $H_i(X, A; \mathbf{Z}_{(P)}) = 0$ *for $i \leq n$,*
3. $H_i(X, A; \mathbf{Z}) \in \mathcal{C}_P$ *for $i \leq n$,*
4. $\pi_i(X, A) \in \mathcal{C}_P$ *for $i \leq n$,*
5. $\pi_i(A) \to \pi_i(X)$ *is a $\mathcal{C}_P$-isomorphism for $i < n$ and a $\mathcal{C}_P$-epimorphism for $i = n$,*
6. $\pi_i(A) \otimes \mathbf{Z}_{(P)} \to \pi_i(X) \otimes \mathbf{Z}_{(P)}$ *is an isomorphism for $i < n$ and an epimorphism for $i = n$*

are equivalent and imply that if $i < n$, then $\pi_i(A)$ and $\pi_i(X)$ have equal rank and isomorphic p-primary components for each $p \in P$.

Proof. Since $\mathbf{Z}_{(P)}$ is flat, the universal coefficient theorem (Corollary 2.35) implies that $H_k(Y; \mathbf{Z}_{(P)}) = H_k(Y; \mathbf{Z}) \otimes \mathbf{Z}_{(P)}$ for any space Y and any k. Since X and A have finitely generated $\mathbf{Z}$-homology it follows from the discussion preceding this theorem that the second and third assertions are equivalent. The long exact sequence in homology and homotopy for a pair and the relative Hurewicz theorem mod $\mathcal{C}_P$ imply that (1) through (6) are equivalent.

The Hurewicz theorem mod $\mathcal{C}_{FG}$ implies that $\pi_i(A)$ and $\pi_i(X)$ are finitely generated for each i. Thus (6) and Equation (10.19) imply that $\pi_i(A)$ and $\pi_i(X)$ have isomorphic p-primary components and equal rank for $i < n$. □

An application of the universal coefficient theorem shows that a map $f : A \to X$ induces a $\mathbf{Z}_{(P)}$-homology isomorphism in all degrees if and only if it induces a $\mathbf{Z}_{(P)}$-cohomology isomorphism in all degrees.

Theorem 10.42 can be used to construct a functor (called the *localization of a space at P*)

$$L_{(P)} : \left\{ \begin{array}{l} \text{simply connected spaces with} \\ \text{finitely generated homology} \end{array} \right\} \to \{ \text{ simply connected spaces } \}$$

so that:

1. there exists a natural transformation from the identity functor to $L_{(P)}, \Phi : \mathrm{Id} \to L_{(P)}$,
2. for each X, $\Phi : X \to L_{(P)}(X)$ induces an isomorphism in $\mathbf{Z}_{(P)}$-homology, and
3. $H_*(L_{(P)}(X); \mathbf{Z}_{(P)}) = H_*(L_{(P)}(X); \mathbf{Z})$.

We write

$$L_{(P)}(X) = X_{(P)}.$$

The space $X_{(P)}$ is a good enough approximation to X to compute the p-primary part of its homotopy groups for $p \in P$; i.e. the p-primary part of $\pi_n X$ is isomorphic to the p-primary part of $\pi_n(X_{(P)})$ for $p \in P$ and the q-primary part of $\pi_n(X_{(P)}) = 0$ for $q \notin P$. In this manner one can study the algebraic topology of spaces one prime at a time, by taking $P = \{p\}$, and also the *rational homotopy* of a space, by taking P empty.

Such a functor $L_{(P)} : X \mapsto X_{(P)}$ exists and can be constructed by first constructing it for an Eilenberg–MacLane space $K(\pi, n)$ and then using a Postnikov decomposition of an arbitrary space into $K(\pi, n)s$.

We outline how to construct the localization functor $L_{(P)}$. For $K(\pi, n)$ with π a finitely generated abelian group, one just replaces π by $\pi \otimes \mathbf{Z}_{(P)}$. The natural map $\pi \to \pi \otimes \mathbf{Z}_{(P)}$ defines a (homotopy class of) map $K(\pi, n) \to K(\pi \otimes \mathbf{Z}_{(P)}, n)$. Thus we define

$$K(\pi, n)_{(P)} = K(\pi \otimes \mathbf{Z}_{(P)}, n).$$

For a general space one constructs $X_{(P)}$ inductively by assembling the pieces of its Postnikov tower, pulling back its k-invariants using the $\mathbf{Z}_{(P)}$ cohomology isomorphisms. Thus, if X has Postnikov system

$$\big(\pi_n = \pi_n(X), X_n, p_n : X_n \to X_{n-1}, k^n \in H^n(X_{n-1}; \pi_{n-1})\big),$$

then first define $(X_2)_{(P)} = K(\pi_2 \otimes \mathbf{Z}_{(P)}, 2)$. Since $X_2 = K(\pi_2, 2)$, the homomorphism $\pi_2 \to \pi_2 \otimes \mathbf{Z}_{(P)}$ induces a map $X_2 \to (X_2)_{(P)}$. The fibration $p_3 : X_3 \to X_2$ is obtained by pulling back the path space fibration $K(\pi_3, 3) \to * \to K(\pi_3, 4)$ via $k^4 \in H^4(X_2; \pi_3) = [X_2, K(\pi_3, 4)]$. Since the map $X_2 \to (X_2)_{(P)}$ induces an isomorphism

$$H^4((X_2)_{(P)}; \pi_3 \otimes \mathbf{Z}_{(P)}) \to H^4(X_2; \pi_3 \otimes \mathbf{Z}_{(P)}) \tag{10.20}$$

(using the universal coefficient theorem), it follows that there is a unique $k^4_{(P)} \in H^4((X_2)_{(P)}; \pi_3 \otimes \mathbf{Z}_{(P)})$ so that the image of $k^4_{(P)}$ via the homomorphism of Equation (10.20) coincides with the image of k^4 under the coefficient homomorphism

$$H^4(X_2; \pi_3) \to H^4(X_2; \pi_3 \otimes \mathbf{Z}_{(P)}).$$

Inductively, if $(X_k)_{(P)}$ and fibrations $(X_k)_{(P)} \to (X_{k-1})_{(P)}$ with fiber $K(\pi_k \otimes \mathbf{Z}_{(P)}, k)$ classified by $k^{k+1}_{(P)} \in H^{k+1}((X_{k-1})_{(P)}; \pi_k \otimes \mathbf{Z}_{(P)})$ have been defined for $k \leq n$, define $k^{n+2}_{(P)} \in H^{n+2}((X_n)_{(P)}; \pi_{n+1} \otimes \mathbf{Z}_{(P)})$ to be the image of the $(n+2)$-nd Postnikov invariant of X, k^{n+2}, under the composite

$$H^{n+2}(X_n; \pi_{n+1}) \to H^{n+2}(X_n; \pi_{n+1} \otimes \mathbf{Z}_{(P)}) \cong H^{n+2}((X_n)_{(P)}; \pi_{n+1} \otimes \mathbf{Z}_{(P)}).$$

Then take $(X_{n+1})_{(P)}$ to be total space in the fibration pulled back from the path space fibration $K(\pi_{n+1} \otimes \mathbf{Z}_{(P)}, n+1) \to * \to K(\pi_{n+1} \otimes \mathbf{Z}_{(P)}, n+2)$ using $k_{(P)}^{n+2} \in H^{n+2}((X_n)_{(P)}; \pi_{n+1} \otimes \mathbf{Z}_{(P)}) = [(X_n)_{(P)}, K(\pi_{n+1} \otimes \mathbf{Z}_{(P)}, n+2)]$.

Notice that the construction also gives a map $X_n \to (X_n)_{(P)}$ inducing the homomorphisms $\pi_k(X) = \pi_k(X_n) \to \pi_k(X_n) \otimes \mathbf{Z}_{(P)} = \pi_k((X_n)_{(P)})$ for all $k \leq n$. Thus if $X_{(P)}$ denotes the space determined by the Postnikov system $(X_n)_{(P)}$ with k-invariants $k_{(P)}^{n+1}$, there is a map $X \to X_{(P)}$ (this gives the natural transformation Φ) so that the induced map $\pi_n(X) \to \pi_n(X_{(P)})$ coincides with

$$\pi_n(X) \to \pi_n(X) \otimes \mathbf{Z}_{(P)}, \quad a \mapsto a \otimes 1.$$

From Theorem 10.42 we conclude that $X \to X_{(P)}$ induces an isomorphism on homology with $\mathbf{Z}_{(P)}$ coefficients (and so also on cohomology with $\mathbf{Z}_{(P)}$ coefficients). The facts that localization is functorial and that $X \to X_{(P)}$ defines a natural transformation $\Phi : \mathrm{Id} \to L_{(P)}$ can be proven by carrying out the construction we gave in a systematic fashion.

Here are some examples with $P = \phi$ to show you why localization is useful. The space $X_{(\phi)}$ is usually denoted by $X_{(0)}$ and is called the *rationalization of* X.

From Proposition 9.25 it follows that if n is odd, the map $S^n \to K(\mathbf{Z}, n)$ generating $H^n(S^n)$ induces an isomorphism on rational cohomology, and hence a homotopy equivalence $S^n_{(0)} \to K(\mathbf{Q}, n) = K(\mathbf{Z}, n)_{(0)}$. Therefore

$$\pi_k(S^n) \otimes \mathbf{Q} = \pi_k(K(\mathbf{Q}, n)) = 0 \text{ for } q \neq n.$$

This implies that $\pi_k(S^n)$ is finite for $k \neq n$.

For n even, $S^n \to K(\mathbf{Q}, n)$ induces an isomorphism in rational homology through dimensions $2n-1$. Hence $\pi_k(S^n)$ is finite for $k \leq 2n-1, k \neq n$. We can do better by taking E to be the homotopy fiber of the map $K(\mathbf{Q}, n) \to K(\mathbf{Q}, 2n)$ representing $\iota_n^2 \in H^{2n}(K(\mathbf{Q}, n))$. The map $S^n \to K(\mathbf{Q}, n)$ lifts to E since $H^{2n}(S^n; \mathbf{Q}) = 0$. The long exact sequence in homotopy shows that

$$\pi_k(E) = \begin{cases} \mathbf{Q} & \text{if } k = n, 2n-1 \\ 0 & \text{otherwise.} \end{cases}$$

Again, a simple application of the Leray–Serre spectral sequence for the fibration $K(\mathbf{Q}, 2n-1) \to E \to K(\mathbf{Q}, n)$ and Proposition 9.25 shows that $H^*(E; \mathbf{Q}) = H^*(S^n; \mathbf{Q})$; the isomorphism is induced by the map $S^n \to E$. Thus $S^n_{(0)} = E$ and so $\pi_k(S^n) \otimes \mathbf{Q} = \pi_k(E)$. This shows that $\pi_k(S^n)$ is finite for $k \neq n, 2n-1$ and that the rank of $\pi_k(S^n)$ is 1 for $k = n$ or $2n-1$.

These two calculations were obtained in Theorem 10.10 and Exercise 166 by similar arguments; the point is that the argument using localization is conceptually much simpler since calculating with the Leray-Serre spectral sequence using rational coefficients is easier than using integer coefficients; for example $E_2^{p,q} = H^p(B) \otimes H^q(F)$. Moreover, the rational cohomology of $K(\mathbf{Q}, n)$ is simple, and so constructing rational Postnikov systems which do what we want is a more manageable problem than constructing an arbitrary Postnikov system.

As a new example, consider the space $\mathbf{C}P^n$. The (rational) cohomology ring of $\mathbf{C}P^n$ is a truncated polynomial ring, and the cohomology of $\mathbf{C}P^\infty$ is a polynomial ring. The inclusion $\mathbf{C}P^n \to \mathbf{C}P^\infty = K(\mathbf{Z}, 2)$ induces isomorphisms on (rational) cohomology through dimension $2n$. Let $c \in H^2(K(\mathbf{Q}, 2))$ denote the generator. Think of $c^{n+1} \in H^{2n+2}(K(\mathbf{Q}, 2))$ as a map $c^{n+1} : K(\mathbf{Q}, 2) \to K(\mathbf{Q}, 2n+2)$ and let E be its homotopy fiber. The map $\mathbf{C}P^n \to K(\mathbf{Q}, 2)$ lifts to E since $H^{2n+2}(\mathbf{C}P^n) = 0$. The spectral sequence for the fibration $K(\mathbf{Q}, 2n+1) \to E \to K(\mathbf{Q}, 2)$ and the calculation of Proposition 9.25 shows that $\mathbf{C}P^n \to E$ induces an isomorphism on rational cohomology.

Exercise 189. Prove this to see how easy it is.

Using the long exact sequence in homotopy, we conclude that

$$\pi_k(\mathbf{C}P^n) \otimes \mathbf{Q} = \pi_k(E) = \begin{cases} \mathbf{Q} & \text{if } k = 2, 2n+1, \\ 0 & \text{otherwise.} \end{cases}$$

Since $\mathbf{C}P^n$ is a finite complex this shows that $\pi_k(\mathbf{C}P^n)$ is finite for $k \neq 2, 2n+1$ and has rank 1 for $k = 2$ and $k = 2n+1$.

Another application is to Chern classes and Bott periodicity for the unitary group. First, we have the following complex analogue of Theorem 10.40.

Theorem 10.43. *Let $BU(n)$ denote the classifying space for $U(n)$. Then the cohomology ring of $BU(n)$ is a polynomial ring:*

$$H^*(BU(n); \mathbf{Z}) = \mathbf{Z}[c_1, c_2, \cdots, c_n]$$

where the generators c_k have degree $2k$. The inclusion $U(n-1) \to U(n)$ induces a map $H^*(BU(n)) \to H^*(BU(n-1))$ which preserves the c_k. □

Exercise 190. Prove Theorem 10.43 using induction and the multiplicative properties of the Leray–Serre spectral sequence for the fibration

$$S^{2n-1} \to BU(n-1) \to BU(n)$$

obtained by delooping the fibration $U(n-1) \to U(n) \to S^{2n-1}$ twice. You may use the proof of Theorem 10.40 as a guide, but the argument in this case is much simpler.

The class $c_k \in H^{2k}(BU(n); \mathbf{Z})$ is called the kth *Chern class*. Since isomorphism classes of $\mathbf{C}^n$-vector bundles are classified by homotopy classes of maps to $BU(n)$, the Chern classes determine characteristic classes of complex vector bundles. By construction, c_k is a stable class; i.e. if E is a complex vector bundle and $\epsilon_{\mathbf{C}}$ denotes the trivial 1-dimensional complex vector bundle, then $c_k(E \oplus \epsilon_{\mathbf{C}})$ is sent to $c_k(E)$ by the map $H^k(BU(n+1)) \to H^k(BU(n))$.

Now consider the map $BU(n) \to \prod_{k=1}^n K(\mathbf{Q}, 2k)$ given by the product of the Chern classes, thinking of $c_k \in H^{2k}(BU(n); \mathbf{Q}) = [BU(n); K(\mathbf{Q}, 2k)]$. By the Künneth theorem and Proposition 9.25 this map induces an isomorphism on rational cohomology. Therefore the rationalization of $BU(n)$ is $\prod_{k=1}^n K(\mathbf{Q}, 2k)$. Since $\Omega(X \times Y) = \Omega X \times \Omega Y$, $\Omega Z = \text{pt}$ if Z is discrete, and $\Omega K(G, n) \sim K(G, n-1)$, we see that

$$\Omega^2 BU(n)_{(0)} = \Omega^2 (\prod_{k=1}^{n} K(\mathbf{Q}, 2k)) \sim \prod_{k=0}^{n-1} K(\mathbf{Q}, 2k) \sim \mathbf{Q} \times BU(n-1)_{(0)}.$$

In particular, letting n go to infinity we obtain a proof of the rational form of Bott periodicity:

$$\Omega^2(\mathbf{Q} \times BU_{(0)}) \sim \mathbf{Q} \times BU_{(0)}.$$

10.10. Construction of bordism invariants

We finish this chapter with some comments on Thom's computation of the unoriented bordism groups. An *invariant* of unoriented bordism is a homomorphism $w : \Omega_n^O \to G$ for some abelian group G. Since $2M = \partial(M \times I)$, the group Ω_n^O is a 2-torsion abelian group. Thus to construct bordism invariants one might as well restrict to constructing homomorphisms $w : \Omega_n^O \to \mathbf{Z}/2$. Thom computed Ω_n^O in this fashion for all n in his famous 1954 paper [**40**]. He did this by finding enough bordism invariants (the *Stiefel-Whitney* numbers) $w_\alpha : \Omega_n^O \to \mathbf{Z}/2$ so that the sum

$$\bigoplus_{\alpha} w_\alpha : \Omega_n^O \to \bigoplus_{\alpha} \mathbf{Z}/2$$

is an isomorphism.

We will outline some of the ingredients in Thom's arguments.

Proposition 10.44. *Let*

$$w = w_1^{i_1} \cdots w_n^{i_n} \in H^n(BO(n)),$$

so $i_1 + 2i_2 + \cdots + n i_n = n$. *If* M *is a smooth* n*-manifold, then the number*

$$\langle w(TM), [M] \rangle \in \mathbf{Z}/2$$

is a bordism invariant.

Proof. Since the expression $\langle w(TM), [M]\rangle$ is additive with respect to the sum in the bordism group (disjoint union), it suffices to show that if M is null-bordant, i.e. $M = \partial W$, then $\langle w(TM), [M]\rangle = 0$.

The tangent bundle of W and M are related by $TM \oplus \epsilon = TW|_M$. Hence if $i : M \subset W$ denotes the inclusion,

$$\begin{aligned}\langle w(TM), [M]\rangle &= \langle w(TM \oplus \epsilon), [M]\rangle \\ &= \langle w(i^*(TW)), [M]\rangle \\ &= \langle i^*(w(TW)), [M]\rangle \\ &= i^*(w(TW)) \cap [M] \\ &= w(TW) \cap i_*[M] = 0,\end{aligned}$$

since in the sequence $H_{n+1}(W, M) \xrightarrow{\partial} H_n(M) \xrightarrow{i_*} H_n(W)$ the map labelled ∂ takes the generator $[W, M]$ to $[M]$. □

Definition 10.45. A *partition* of the positive integer n is an n–tuple of non-negative integers $(i_1, \cdots, i_n)$ so that $i_1 + 2i_2 + \cdots + ni_n = n$.

Given a partition $(i_1, \cdots, i_n)$ of n, the number

$$\langle w_1^{i_1} \cdots w_n^{i_n}(TM), [M]\rangle \in \mathbf{Z}/2$$

is called the *Stiefel–Whitney number* associated to the partition $(i_1, \cdots, i_n)$.

Thus to each partition $\alpha = (i_1, \cdots, i_n)$ of n we have associated the bordism invariant whose value on the manifold M is

$$w_\alpha(M) = \langle w_1^{i_1} \cdots w_n^{i_n}(TM), [M]\rangle.$$

Thom's theorem is the following.

Theorem 10.46. *Let S denote the set of partitions of n so that $i_k = 0$ whenever k has the form $2^\ell - 1$. Then the map taking a manifold to its Stiefel–Whitney numbers induces an isomorphism*

$$\oplus_{\alpha \in S}\, w_\alpha : \Omega_n^O \to \oplus_{\alpha \in S}\, \mathbf{Z}/2.$$

In other words, the unoriented bordism class of a manifold is determined by its Stiefel–Whitney numbers, and given any partition $\alpha \in S$ there exists an n-manifold M with $w_\alpha(M) \neq 0$, but $w_\beta(M) = 0$ for $\beta \neq \alpha$. □

Thom proves this theorem by a method analogous to the example of $BU_{(0)}$ we gave in the previous section. First, Thom finds sufficiently many examples of manifolds with the appropriate Stiefel–Whitney numbers, and then he uses these to define a map from the Thom spectrum to a product of Eilenberg-MacLane spectra $\mathbf{K}(\mathbf{Z}/2)$. He shows this map induces an isomorphism on homology, using the Thom isomorphism to compute the cohomology of the Thom spectrum as a module over the mod 2 Steenrod algebra. The Whitehead theorem then implies that the map is a homotopy equivalence, and so the Stiefel–Whitney numbers classify bordism.

10.11. Projects for Chapter 10

10.11.1. Unstable homotopy theory. Unstable homotopy theory is significantly harder than the stable theory, essentially because $\pi_n(X, A) \not\cong \pi_n(X/A)$. There are nevertheless some useful results; you should lecture on some or all of these.

Since S^m has a cell structure with only one 0-cell and one m-cell, the product $S^k \times S^n$ has a cell structure with 4 cells, a 0-cell $e^0 \times e^0$, a k-cell $e^k \times e^0$, an n-cell $e^0 \times e^n$, and a $(k+n)$-cell $e^k \times e^n$. Removing the top cell leaves the wedge

$$S^k \times S^n - (e^k \times e^n) = S^k \vee S^n.$$

Let $a : S^{k+n-1} \to S^k \vee S^n$ denote the attaching map for the (top) $(k+n)$-cell of $S^k \times S^n$.

The map a can be used to construct interesting elements in $\pi_n X$.

Definition 10.47. Given $f \in \pi_k X$ and $g \in \pi_n X$, define the *Whitehead product* $[f, g] \in \pi_{k+n-1} X$ to be the (homotopy class of) the composite

$$S^{k+n-1} \xrightarrow{a} S^k \vee S^n \xrightarrow{f \vee g} X.$$

For example, if $k = n = 1$, the attaching map for the 2-cell of a torus represents the commutator of the two generators, and hence the Whitehead product $\pi_1 X \times \pi_1 X \to \pi_1 X$ takes a pair of loops to their commutator. Since $\pi_2(X)$ is abelian, the suspension map $s_* : \pi_1 X \to \pi_1(\Omega SX) = \pi_2(X)$ takes any commutator to zero. More generally show (or look up) the following fact.

Proposition 10.48. *The suspension of the attaching map for the top cell of $S^k \times S^n$,*

$$Sa : S(S^{k+n-1}) = S^{k+n} \to S(S^k \vee S^n),$$

is nullhomotopic. Hence $s_*([f, g]) = 0$ for any $f \in \pi_k X, g \in \pi_n X$; i.e. the Whitehead product $[f, g]$ is in the kernel of the suspension homomorphism $\pi_{n+k-1}(X) \to \pi_{n+k} SX$. □

Thus Whitehead products produce decidedly unstable elements in $\pi_m X$. The map $s : X \to \Omega SX$ (defined in Section 10.3) induces the suspension homomorphism $s_* : \pi_\ell(X) \to \pi_\ell(\Omega SX) = \pi_{\ell+1} SX$. It can be studied in a large range (the "metastable range") by using the EHP sequence:

Theorem 10.49. *If X is an $(n-1)$-connected space, there is an exact sequence*

$$\pi_{3n-2}(X) \xrightarrow{s_*} \pi_{3n-1}(SX) \to \pi_{3n-1}(SX \wedge X) \to \pi_{3n-3}(X) \to \cdots$$
$$\cdots \to \pi_k(X) \to \pi_{k+1}(SX) \to \pi_{k+1}(SX \wedge X) \to \cdots .$$

□

The map s_* is sometimes denoted "E" in the literature (from the German word "Einhängung" for suspension), the map $\pi_{k+1}(SX) \to \pi_{k+1}(SX \wedge X)$ is usually denoted by "H" since it generalizes the Hopf invariant, and the map $\pi_{k+1}(SX \wedge X) \to \pi_k(X)$ is usually denoted "P" since its image is generated by Whitehead Products. Hence the name EHP sequence.

Thus, in the range $k \leq 3n-2$, the EHP sequence gives some control over what the kernel and cokernel of the suspension map on homotopy groups are.

An important special case of the EHP sequence is obtained by setting $X = S^n$. The sequence is

$$\pi_{3n-2}S^n \to \pi_{3n-1}S^{n+1} \to \pi_{3n-1}S^{2n+1} \to \pi_{3n-3}S^n \to \cdots . \tag{10.21}$$

A proof of Theorem 10.49 can be found in [**43**] (although it is hard to reconstruct the argument since it is explained as a consequence of a more general result of James). The proof that (10.21) is exact as well as the material below on the Hopf invariant can be found in [**36**, Section 9.3].

After substituting $\pi_{2n+1}S^{2n+1} = \mathbf{Z}$ a part of the sequence (10.21) can be written:

$$\cdots \pi_{2n}S^n \to \pi_{2n+1}S^{n+1} \xrightarrow{H} \mathbf{Z} \to \pi_{2n-1}S^n \to \cdots .$$

If n is even, Theorem 10.10 implies that the map H is zero and therefore $\pi_{2n}S^n \to \pi_{2n+1}S^{n+1}$ is onto. The Freudenthal suspension theorem implies that $\pi_{2n+1}S^{n+1} \to \pi_{2n+2}S^{n+2} = \pi_n^S$ is onto, and so we conclude that for n even, $\pi_{2n}S^n \to \pi_n^S$ is onto.

If n is odd, then Exercise 166 shows that $\pi_{2n+1}S^{n+1}/\text{torsion} = \mathbf{Z}$. Since $\pi_{2n}S^n$ is finite the map $H : \pi_{2n+1}S^{n+1} \to \mathbf{Z}$ is non-zero; it is called the *Hopf invariant.*

The famous "Hopf invariant one" problem, solved by J.F. Adams, asserts that H is onto only for $n+1 = 1, 2, 4$, and 8, and in fact the Hopf fibrations are the only maps which have Hopf invariant one. The Whitehead product $[i, i] \in \pi_{2n-1}S^{n+1}$, where $i \in \pi_{n+1}S^{n+1}$ denotes the generator, has Hopf invariant 2.

There are several other definitions of the Hopf invariant for a map $f : S^{2n+1} \to S^n$, and you should lecture on some of these. Here are two.

1. Assume that f is smooth (this can always be arranged by a small homotopy) and let $x_0, x_1 \in S^{n+1}$ be two regular values of f. Let $M_0 = f^{-1}(x_0)$ and $M_1 = f^{-1}(x_1)$. Then $H(f) = \mathrm{lk}(M_0, M_1)$, where "lk" denotes the linking number.
2. Let X be the CW-complex obtained by attaching a $(2n+2)$-cell to S^{n+1} using f as the attaching map. Then $H^q(X) = \mathbf{Z}$ for $q = 0, n+1$, and $2n+2$, and is zero otherwise. Let $e_{n+1} \in H^{n+1}(X)$ and $e_{2n+2} \in H^{2n+2}(X)$ denote the generators. Then $(e_{n+1})^2 = H(f)e_{2n+2}$.

Chapter 11

Simple-Homotopy Theory

Two basic references for the material in this chapter are Cohen's book [**7**] and Milnor's article [**28**].

11.1. Introduction

Whitehead's theorem (Theorem 7.34) says that a map $f : X \to Y$ between CW-complexes is a homotopy equivalence if $f_* : \pi_n X \to \pi_n Y$ is an isomorphism for all n. Thus homotopy groups and hence the tools of homotopy theory give important information about the homotopy type of a space. However, important questions in geometric topology center around distinguishing homeomorphism types within a class of homotopy equivalent manifolds, to which the methods we have studied so far do not directly apply.

For example, suppose W is a compact manifold with two boundary components: $\partial W = M_0 \amalg M_1$. Suppose the inclusion $M_0 \hookrightarrow W$ is a homotopy equivalence. Is W homeomorphic to $M_0 \times [0,1]$, and, in particular, is M_0 homeomorphic to M_1? The answer to this question is provided by the *s-cobordism theorem* (see Section 11.7) which states that there exists a functor

$$\mathrm{Wh}\colon \{\text{Groups}\} \to \{\text{Abelian groups}\}$$

so that in the situation described above, an element $\tau(W, M_0) \in \mathrm{Wh}(\pi_1 W)$ is defined and vanishes if W is homeomorphic to $M_0 \times [0,1]$. Conversely, if the dimension of M_0 is greater than 4 and $\tau(W, M_0) = 0$, then W is homeomorphic to $M_0 \times [0,1]$.

Of course the point of this theorem is that the functor Wh has a functorial, geometric, and somewhat computable definition. (One could stupidly define $\mathrm{Wh}(\pi) = \mathbf{Z}/2$ for all groups π and define $\tau(W, M_0)$ to be 0 or 1 according to whether or not W is homeomorphic to a product.)

In this chapter we will give the complete definition of the Whitehead group $\mathrm{Wh}(\pi)$ and the Whitehead torsion τ.

Exercise 191. Use the fact that $\mathrm{Wh}(1) = 0$ to prove the generalized Poincaré conjecture for $n > 5$: a closed manifold Σ which has the homotopy type of S^n is homeomorphic to S^n. (Hint: Remove two open disks; it can be shown that the complement has the homotopy type of a CW-complex. Assume this and apply the s-cobordism theorem.)

Another collection of examples is provided by lens spaces. We will use obstruction theory to give a homotopy classification of 3-dimensional lens spaces, and then use the machinery of simple-homotopy theory to prove the following theorem.

Theorem 11.1. *The 3-dimensional lens spaces $L(p,q)$ and $L(p',q')$ are homotopy equivalent if and only if $p = p'$ and there exists an integer b so that $q \equiv \pm b^2 q' \bmod p$.*

Moreover, $L(p,q)$ is homeomorphic to $L(p,q')$ if and only if $p = p'$ and $q \equiv \pm (q')^{\pm 1} \bmod p$.

In particular $L(7,1)$ and $L(7,2)$ are closed three-manifolds which have the same homotopy type but are not homeomorphic. Whitehead torsion must be a subtle and powerful invariant to make such a distinction.

J.H.C. Whitehead developed the theory of *simple-homotopy equivalence*, a refinement of homotopy equivalence for finite CW-complexes which takes into account the cell structure. It was proven by Chapman that homeomorphic finite CW-complexes are simple-homotopy equivalent. Hence simple-homotopy theory provides a weapon by which to attack homeomorphism problems that are impervious to the homotopy theoretic machinery developed in the previous chapters.

Suppose X, Y are finite CW-complexes. Then a cellular map $f : X \to Y$ is a homotopy equivalence if and only if the mapping cylinder of f deformation retracts to X via a cellular map.

Exercise 192. Prove this using obstruction theory.

The geometric approach of simple-homotopy theory is to investigate when a pair (K, L) which admits a deformation retract of K to L admits a particular "simple" type of deformation.

In this chapter a *finite CW-pair* (K, L) will mean a CW-complex K with finitely many cells and a subcomplex $L \subset K$. Thus L is a closed subspace of K which is a union of cells. (See Definition 1.3 for the definition of a CW-complex.) We will also use infinite CW-pairs, but these will always be of the form $(\tilde{K}, \tilde{L})$ where (K, L) is a finite CW-pair, $\tilde{K} \to K$ a covering space, and $\tilde{L}$ the inverse image of L in $\tilde{K}$.

Definition 11.2. If (K, L) is a finite CW-pair, we say K *collapses to* L *by an elementary collapse*, denoted $K \searrow^e L$, if the following two conditions hold.

1. $K = L \cup e^{n-1} \cup e^n$ where the cells e^{n-1} and e^n are not in L.
2. Write $\partial D^n = S^{n-1} = D^{n-1}_+ \cup_{S^{n-2}} D^{n-1}_-$. Then there exists a characteristic map $\varphi : D^n \to K$ for e^n so that
   (a) $\varphi_{|D^{n-1}_+} : D^{n-1}_+ \to K$ is a characteristic map for e^{n-1}, and
   (b) $\varphi(D^{n-1}_-) \subset L$.

Thus K is obtained by gluing D^n to L along a map $D^{n-1}_- \longrightarrow L$, where $D^{n-1}_- \subset \partial D^n \subset D^n$.

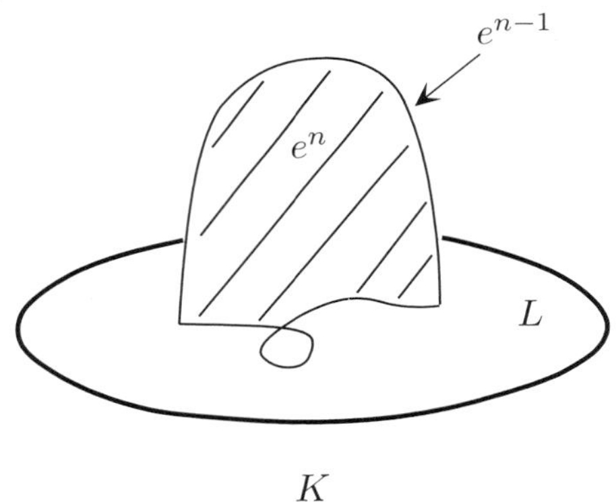

K

Note K can be viewed as the mapping cylinder of a map $D^{n-1}_- \to L$. Thus L is a deformation retract of K.

Definition 11.3.

1. One says that K *collapses to* L, or L *expands to* K, if there are subcomplexes $K = K_0 \supset K_1 \supset \cdots \supset K_n = L$ so that $K_0 \searrow^e K_1 \searrow^e \cdots \searrow^e K_n$. Write $K \searrow L$ or $L \nearrow K$.
2. A map $f : K \to L$ is called a *simple-homotopy equivalence* if there exists a finite sequence of CW-complexes $K = K_0, K_1, \ldots, K_n = L$ so that f is homotopic to a composite $K_0 \to K_1 \to K_2 \to \cdots \to K_n$ where each map $K_i \to K_{i+1}$ is either the inclusion map of an expansion, the deformation retraction of a collapse, or a cellular homeomorphism.

Exercise 193. Prove that simple-homotopy equivalence is an equivalence relation.

We now give two examples concerned with collapsing. Suppose L is a finite simplicial subcomplex of a triangulated open subset of Euclidean space. Then the *regular neighborhood* $K = N(L)$ is the union of all simplices whose closure intersects L. This is an analogue of a normal bundle, but L does not have to be a manifold. It is not difficult to see that $K \searrow L$.

The second example is where K is the "house with two rooms" pictured below. Here K is a 2-dimensional CW-complex. To get to the large room on the lower floor, you must enter the house from the top through the small cylinder on the left. Similarly, one enters the upper room via the small right cylinder. Then it is not difficult to see that K is simple-homotopy equivalent to a point, but that K does not collapse to a point; i.e. some expansions are needed.

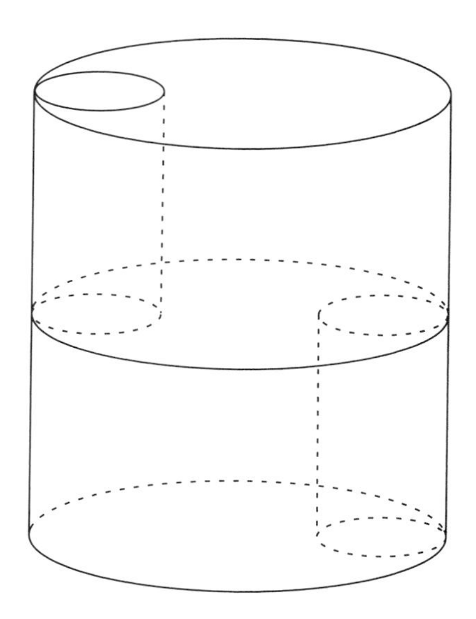

11.2. Invertible matrices and $K_1(R)$

In this section we will define the Whitehead group and in the next section define torsion. Since there are two sections of algebra coming up, we will give you some geometric motivation to help you through.

It will turn out that the question of whether a homotopy equivalence is simple can be understood in the following way. Assume $f : L \to K$ is a cellular inclusion, and a homotopy equivalence. Then if $\pi = \pi_1 L = \pi_1 K$, the relative chain complex $C_*(\tilde{K}, \tilde{L})$ (where $\tilde{K}$, $\tilde{L}$ denote universal covers) is a *free and acyclic* $\mathbf{Z}\pi$*-chain complex* and has a $\mathbf{Z}\pi$*-basis* labeled by the cells of $K - L$.

Elementary collapses and changing base points change the cellular chain complex $C_*(\tilde{K}, \tilde{L})$, and so one wants to classify acyclic, based chain complexes over $\mathbf{Z}\pi$ up to some equivalence relation, so that $C_*(\tilde{K}, \tilde{L})$ is equivalent to 0 if and only if $K \hookrightarrow L$ is a simple-homotopy equivalence. The main result will be that the chain complex $C_*(\tilde{K}, \tilde{L})$ determines an element $\tau(K, L) \in \mathrm{Wh}(\pi)$ which vanishes if and only if the map f is a simple-homotopy equivalence.

Once the machinery is set up, other useful applications will follow from considering rings R more general than the integral group ring $\mathbf{Z}\pi$. For example, if $\mathbf{Z}\pi \to R$ is a ring homomorphism, it may be easier to work with the chain complex $R \otimes_{\mathbf{Z}\pi} C_*$ than to work directly with C_*. This is especially true if R is a commutative ring or, even better, a field.

The simplest acyclic, based chain complexes are of the form:

$$0 \to C_n \xrightarrow{\partial} C_{n-1} \to 0.$$

Since this complex is based, ∂ is given by a matrix, which is invertible since (C_*, ∂) is acyclic.

Motivated by the previous discussion, we study invertible matrices over a (not necessarily commutative) ring R. We assume that all our rings are rings with 1. Unfortunately, two bizarre phenomena can arise when considering free modules over a ring R.

- It may be the case that $R^m \cong R^n$ with $m \neq n$.
- It may be the case that $M \oplus R^m \cong R^n$, but that M itself is not free. In this case we say the module M is *stably free* but not free.

Fortunately, the first problem does not occur for group rings, because there is a homomorphism $\epsilon : \mathbf{Z}\pi \to \mathbf{Z}$, $\sum a_g g \mapsto \sum a_g$. Thus $(\mathbf{Z}\pi)^m \cong (\mathbf{Z}\pi)^n$ implies

$$\mathbf{Z}^m = \mathbf{Z} \otimes_{\mathbf{Z}\pi} (\mathbf{Z}\pi)^m \cong \mathbf{Z} \otimes_{\mathbf{Z}\pi} (\mathbf{Z}\pi)^n = \mathbf{Z}^n$$

and so $m = n$. Henceforth

We assume all rings have the property that $R^m \cong R^m$ implies $m = n$.

Thus we exclude rings like the endomorphism ring of an infinite-dimensional vector space.

The second pathology does occur for certain group rings, so we cannot assume it away. It will be a thorn in our side in the next section, but we will deal with it.

Definition 11.4. Denote by $GL(n, R)$ the group of all $n \times n$ matrices over R which have a two-sided inverse. An inclusion $GL(n, R) \hookrightarrow GL(n+1, R)$

is defined by

$$A \longmapsto \begin{pmatrix} A & 0 \\ 0 & 1 \end{pmatrix}.$$

Let $GL(R) = \bigcup_n GL(n, R)$. Think of $GL(R)$ as the group of all invertible infinite matrices which are "eventually" the identity. We will always identify an invertible $n \times n$ matrix with its image in $GL(R)$. In particular, if A and B are invertible matrices, their product in $GL(R)$ makes sense even if their sizes are different.

We next define an important subgroup of $GL(R)$, the subgroup generated by elementary matrices.

Definition 11.5. $E(R) \subset GL(R)$ is the subgroup generated by the *elementary matrices*, i.e. the matrices of the form:

$$I + rE_{ij} \quad (i \neq j)$$

where I is the identity matrix, E_{ij} is the matrix with 1 in the ij spot and 0's elsewhere, and $r \in R$.

The effect of multiplying a matrix A by the elementary matrix $I + rE_{ij}$ on the right is the column operation which replaces the jth column of A by the sum of the jth column of A and r times the ith column of A. Multiplying A on the left by an elementary matrix performs the corresponding row operation.

Recall that the *commutator subgroup* of a group G is the subgroup $[G, G]$ generated by all *commutators* $ghg^{-1}h^{-1}$ where $g, h \in G$. This is the smallest normal subgroup of G such that the corresponding quotient group is abelian.

Lemma 11.6 (Whitehead lemma). *The group generated by elementary matrices equals the commutator subgroup of* $GL(R)$

$$E(R) = [GL(R), GL(R)].$$

Proof. First, $(I + rE_{ij})^{-1} = I - rE_{ij}$, and

$$E_{ij}E_{k\ell} = \begin{cases} 0 & \text{if } j \neq k \\ E_{i\ell} & \text{if } j = k. \end{cases}$$

Thus if i, j, k are *distinct*,

$$I + rE_{ik} = (I + rE_{ij})(I + E_{jk})(I + rE_{ij})^{-1}(I + E_{jk})^{-1}.$$

So any $n \times n$ elementary matrix with $n \geq 3$ can be expressed as a commutator. Hence $E(R) \subset [GL(R), GL(R)]$.

The opposite inclusion follows from the matrix identities

$$\begin{pmatrix} ABA^{-1}B^{-1} & 0 \\ 0 & I \end{pmatrix} = \begin{pmatrix} A & 0 \\ 0 & A^{-1} \end{pmatrix} \begin{pmatrix} B & 0 \\ 0 & B^{-1} \end{pmatrix} \begin{pmatrix} (BA)^{-1} & 0 \\ 0 & BA \end{pmatrix}$$

$$\begin{pmatrix} A & 0 \\ 0 & A^{-1} \end{pmatrix} = \begin{pmatrix} I & A \\ 0 & I \end{pmatrix} \begin{pmatrix} I & 0 \\ I - A^{-1} & I \end{pmatrix} \begin{pmatrix} I & -I \\ 0 & I \end{pmatrix} \begin{pmatrix} I & 0 \\ I - A & I \end{pmatrix}$$

$$\begin{pmatrix} I & X \\ 0 & I \end{pmatrix} = \prod_{i=1}^{m} \prod_{j=1}^{n} (I + x_{ij} E_{i,j+m})$$

which are valid for $A \in GL(m, R)$, $B \in GL(n, R)$, and $X = (x_{ij})$ an $m \times n$ matrix. The identities show that any commutator in $GL(n, R)$ can be expressed as a product of elementary matrices in $GL(2n, R)$. All three identities are easily checked; the last two are motivated by the elementary row operations one would do to transform $\begin{pmatrix} A & 0 \\ 0 & A^{-1} \end{pmatrix}$ and $\begin{pmatrix} I & X \\ 0 & I \end{pmatrix}$ to $\begin{pmatrix} I & 0 \\ 0 & I \end{pmatrix}$.

□

So $E(R)$ is a normal subgroup of $GL(R)$ with abelian quotient.

Definition 11.7.

$$K_1(R) = GL(R)/E(R).$$

The quotient homomorphism $GL(R) \to K_1(R)$, $A \mapsto [A]$ should be thought of as a generalized determinant function.

Exercise 194.

1. For a commutative ring R, there is a well-defined map $K_1(R) \to R^\times$, $[A] \mapsto \det A$, which is a split epimorphism. Here $R^\times$ is the group of units of R, where a unit is an element of R with a two-sided multiplicative inverse.
2. For a field F, show that $K_1(F) \cong F^\times = F - \{0\}$.
3. Show that $K_1(\mathbf{Z}) = \{[(\pm 1)]\} \cong \mathbf{Z}/2$.

Exercise 195. Show that K_1 is a *functor* from the category of rings with 1 to the category of abelian groups.

In fact, for every $n \in \mathbf{Z}$, there is a functor K_n, with the various K_n's intertwined by Künneth theorems. Composing K_n with the functor taking a group π to its integral group ring $\mathbf{Z}\pi$ defines a functor $\pi \mapsto K_n(\mathbf{Z}\pi)$ from the category of groups to the category of abelian groups.

The following equalities in $K_1(R)$ are useful in computations and applications. They are reminiscent of properties of determinants.

Theorem 11.8.

1. *Let* $A \in GL(R,m)$, $B \in GL(R,n)$, X *be an* $m \times n$ *matrix, and* Y *an* $n \times m$ *matrix. Then*
$$\left[\begin{pmatrix} A & X \\ 0 & B \end{pmatrix}\right] = \left[\begin{pmatrix} A & 0 \\ 0 & B \end{pmatrix}\right] = \left[\begin{pmatrix} AB & 0 \\ 0 & I \end{pmatrix}\right] \in K_1(R),$$
$$\left[\begin{pmatrix} A & 0 \\ Y & B \end{pmatrix}\right] = \left[\begin{pmatrix} A & 0 \\ 0 & B \end{pmatrix}\right] = \left[\begin{pmatrix} AB & 0 \\ 0 & I \end{pmatrix}\right] \in K_1(R).$$

2. *Let* $P \in GL(n,R)$ *be the permutation matrix obtained by permuting the columns of the identity matrix using the permutation* $\sigma \in S_n$. *Let* $\operatorname{sign}(\sigma) \in \{\pm 1\}$ *be the sign of the permutation. Then*
$$[P] = [(\operatorname{sign}(\sigma))] \in K_1(R).$$

Proof. Note
$$\begin{pmatrix} A & X \\ 0 & B \end{pmatrix} = \begin{pmatrix} I & XB^{-1} \\ 0 & I \end{pmatrix} \begin{pmatrix} A & 0 \\ 0 & B \end{pmatrix}$$
and the middle matrix is in $E(R)$ as in the proof of the Whitehead lemma. Likewise
$$\begin{pmatrix} A & 0 \\ 0 & B \end{pmatrix} = \begin{pmatrix} AB & 0 \\ 0 & I \end{pmatrix} \begin{pmatrix} B^{-1} & 0 \\ 0 & B \end{pmatrix}.$$
The last matrix is in $E(R)$ as in the proof of the Whitehead lemma. The first equation in Part 1 above follows. The proof of the second equation is similar.

For Part 2, note
$$\begin{pmatrix} 0 & 1 \\ 1 & 0 \end{pmatrix} = \begin{pmatrix} -1 & 0 \\ 0 & 1 \end{pmatrix} \begin{pmatrix} 1 & -1 \\ 0 & 1 \end{pmatrix} \begin{pmatrix} 1 & 0 \\ 1 & 1 \end{pmatrix} \begin{pmatrix} 1 & -1 \\ 0 & 1 \end{pmatrix}.$$
The last three matrices are in $E(R)$, so we see the assertion is true for 2×2 matrices. For a general 2-cycle σ, the same method shows that P is equivalent to a diagonal matrix with 1's down the diagonal except for a single -1. By Part 1, this is equivalent to the 1×1 matrix (-1).

Every permutation is a product of 2-cycles so the result follows. □

Theorem 11.8 shows that the group operation in the abelian group $K_1(R) = GL(R)/E(R)$ can be thought of either as matrix multiplication
$$([A],[B]) \mapsto [AB]$$
or as *block sum*
$$([A],[B]) \mapsto \left[\begin{pmatrix} A & 0 \\ 0 & B \end{pmatrix}\right].$$

The group operation in $K_1(R)$ will be written additively. Hence

$$[A] + [B] = [AB] = \left[\begin{pmatrix} A & 0 \\ 0 & B \end{pmatrix}\right].$$

Definition 11.9. Define the *reduced* K*-group*

$$\widetilde{K}_1(R) = K_1(R)/[(-1)].$$

Using this group will allow us to use unordered bases for free modules.

Exercise 196. Let $i : \mathbf{Z} \to R$ be the unique ring map from the integers to R taking the 1 of $\mathbf{Z}$ to the 1 of R. Show that $\widetilde{K}_1(R)$ is the cokernel of the induced map $K_1(\mathbf{Z}) \to K_1(R)$.

Now we switch to group rings. For a group ring, the map $i : \mathbf{Z} \to \mathbf{Z}\pi$ is split by the augmentation map $\epsilon : \mathbf{Z}\pi \to \mathbf{Z}$, $\sum a_g g \mapsto \sum a_g$. Hence $K_1(\mathbf{Z}\pi) = \mathbf{Z}/2 \oplus \widetilde{K}_1(\mathbf{Z}\pi)$.

If X is a CW-complex with fundamental group π and universal cover $\tilde{X}$, then $C_*(\tilde{X})$ is a free $\mathbf{Z}\pi$-chain complex with generators corresponding to the cells of X. However, the generators are not uniquely determined by the cells; in addition one must choose an orientation and a lift of the cell to the cover. In other words, generators are determined only up to a multiple $\pm g$ where $g \in \pi$. This helps motivate the definition of the Whitehead group.

Definition 11.10. Let E_π be the subgroup of $GL(\mathbf{Z}\pi)$ generated by $E(\mathbf{Z}\pi)$ and 1×1 matrices $(\pm g)$, where $g \in \pi$. Then the *Whitehead group of* π is

$$\mathrm{Wh}(\pi) = GL(\mathbf{Z}\pi)/E_\pi = K_1(\mathbf{Z}\pi)/\{[(\pm g)] : g \in \pi\}.$$

The elements that we mod out by are represented by matrices of the form

$$\begin{pmatrix} \pm g & & & \\ & 1 & & \\ & & 1 & \\ & & & \ddots \end{pmatrix} \text{ for } g \in \pi.$$

The Whitehead group is a functor from groups to abelian groups. There is a short exact sequence of abelian groups

$$0 \to \{\pm 1\} \times \pi^{ab} \to K_1(\mathbf{Z}\pi) \to \mathrm{Wh}(\pi) \to 0,$$

where $\pi^{ab} = \pi/[\pi, \pi]$. The reason for injectivity is that the composite of the maps

$$\{\pm 1\} \times \pi^{ab} \to K_1(\mathbf{Z}\pi) \to K_1(\mathbf{Z}[\pi^{ab}]) \xrightarrow{\det} \mathbf{Z}[\pi^{ab}]^\times$$

is the inclusion.

The elements of the subgroup $\pm\pi = \{\pm g : g \in \pi\}$ of $(\mathbf{Z}\pi)^\times$ are called the *trivial units* of $\mathbf{Z}\pi$. The ring $\mathbf{Z}\pi$ might contain other units, depending on what π is. To some extent the existence of nontrivial units is measured by the nontriviality of the Whitehead group, but the only precise statement in this direction is that if π is abelian and $\mathbf{Z}\pi$ contains nontrivial units, then the Whitehead group $\mathrm{Wh}(\pi)$ is nontrivial. This uses the fact that the determinant map $K_1(R) \to R^\times$ is a split epimorphism for a commutative ring.

Here are three interesting examples.

1. Let $\mathbf{Z}/5$ have generator t. Then in $\mathbf{Z}[\mathbf{Z}/5]$,
$$(1-t+t^2)(t+t^2-t^4) = 1.$$
Thus $1-t+t^2$ is a nontrivial unit and the Whitehead group is nontrivial. It can be shown that $\mathrm{Wh}(\mathbf{Z}/5)$ is infinite cyclic with this unit as generator.

2. It is easy to see that $\mathbf{Z}[\mathbf{Z}]$ has only trivial units (exercise!). It can be shown that $\mathrm{Wh}(\mathbf{Z}) = 0$.

3. This next example due to Whitehead is a nontrivial unit which represents the zero element of $\mathrm{Wh}(\pi)$.

   Let $\pi = \langle x, y \mid y^2 = 1\rangle = \mathbf{Z} * \mathbf{Z}/2$. Let $a = 1-y$ and $b = x(1+y)$ in $\mathbf{Z}\pi$. Notice $1-ab$ is a nontrivial unit, since $(1-ab)(1+ab) = 1$. However we will show $[(1-ab)]$ is zero in the Whitehead group. It can be shown that $\mathrm{Wh}(\pi) = 0$.

   Note that $(1-y)(x(1+y)) = x + xy - yx - yxy \neq 0$ and also $(x(1+y))(1-y) = x(1-y^2) = 0$. So $ab \neq 0$ and $ba = 0$.

   Then
$$\begin{pmatrix}1 & 0\\ b & 1\end{pmatrix}\begin{pmatrix}1 & a\\ 0 & 1\end{pmatrix}\begin{pmatrix}1 & 0\\ b & 1\end{pmatrix}^{-1}\begin{pmatrix}1 & a\\ 0 & 1\end{pmatrix}^{-1} = \begin{pmatrix}1 & 0\\ b & 1\end{pmatrix}\begin{pmatrix}1 & a\\ 0 & 1\end{pmatrix}\begin{pmatrix}1 & 0\\ -b & 1\end{pmatrix}\begin{pmatrix}1 & -a\\ 0 & 1\end{pmatrix}$$
$$= \begin{pmatrix}1-ab & 0\\ 0 & 1\end{pmatrix}.$$

   Thus one must *stabilize* (i.e. include the 1×1 matrices into the 2×2 matrices) before $1-ab$ becomes "trivial", i.e. becomes a product of elementary matrices.

The actual computation of Whitehead groups can be a difficult business, involving number theory in the case of finite groups and geometry in the case of infinite groups. We mention a result and a conjecture. The result, due to Bass-Milnor-Serre, is that $\mathrm{Wh}(\mathbf{Z}/n)$ is a free abelian group of rank $[n/2] + 1 - d(n)$ where $d(n)$ is the number of positive divisors of n [**4**]. The conjecture (proven in many cases) is that $\mathrm{Wh}(\pi) = 0$ when π is a torsion-free group.

The next lemma will enable us to remove the dependence on base points when we move to a geometric context. In particular, it shows that the assignment $X \to \mathrm{Wh}(\pi_1 X)$ gives a well-defined functor from the category of path-connected spaces to the category of abelian groups.

Lemma 11.11. *If $f : \pi \to \pi$ is the inner automorphism given by $f(g) = xgx^{-1}$ for some $x \in \pi$, then the induced map on Whitehead groups $f_* : \mathrm{Wh}(\pi) \to \mathrm{Wh}(\pi)$ is the identity map.*

Proof. The automorphism f induces $f : \mathbf{Z}\pi \to \mathbf{Z}\pi$ by the formula

$$f(\sum n_g g) = \sum n_g xgx^{-1} = x(\sum n_g g)x^{-1},$$

which in turn induces a group automorphism $f : GL(n, \mathbf{Z}\pi) \to GL(n, \mathbf{Z}\pi)$ by $A \mapsto (xI)A(x^{-1}I)$. Hence,

$$\begin{aligned} f_*[A] &= [xI \cdot A \cdot x^{-1}I] \\ &= [xI \cdot x^{-1}I \cdot A] \qquad \text{since } \mathrm{Wh}(\pi) \text{ is abelian} \\ &= [A]. \end{aligned}$$

□

This is reminiscent of the fact that an inner automorphism of π induces the identity on $H_*(\pi)$, pointing out an analogy between the two functors from groups to abelian groups.

We conclude this section with a remark about matrices over noncommutative rings. If $f : M \to M'$ is an isomorphism of R-modules and if M and M' have bases $\{b_1, \dots, b_n\}$ and $\{b'_1, \dots, b'_n\}$ respectively, we wish to define $[f] \in K_1(R)$ to be $[F]$, where F is a matrix representative of f. There are several ways to define a matrix for f, and the result depends on whether we are working with right or left R-modules.

For our main application, the modules we take are the cellular n-chains on the universal cover of a CW-complex X. These are right $\mathbf{Z}\pi$-modules. For that reason we consider right R-modules and define the *matrix* of a map $f : M \to M'$ of right based R-modules to be (f_{ij}) where

$$f(b_i) = \sum_{i=1}^{n} b'_j f_{ij}.$$

With this definition, assigning a matrix to a map of right based R-modules

$$(-) : \mathrm{Hom}_R(M, M') \to M_n(R)$$

is a homomorphism; i.e. $(f + g) = (f) + (g)$ and $(fg) = (f)(g)$ and taking the equivalence class defines a *homomorphism*

$$\mathrm{Iso}_R(M, M') \to K_1(R), \; f \mapsto [(f_{ij})].$$

We will write $[f] = [(f_{ij})] \in K_1(R)$.

11.3. Torsion for chain complexes

We next make the transition from matrices to acyclic, based chain complexes. A *based* R-module is a free, finite-dimensional R-module with a specified basis. A chain complex C_* over a ring R is *bounded* if there exists an N so that $C_n = 0$ for $|n| \leq N$, *bounded below* if there exists an N so that $C_n = 0$ for $n \geq N$, *based* if each C_n is based and C_* is bounded, *free* if each C_n is free, *projective* if each C_n is projective, *finite* if $\oplus C_n$ is finitely generated, and *acyclic* if the homology of C_* is zero. We will often write C instead of C_*. As above we assume that the ring R has the property that $R^m \cong R^n$ implies that $m = n$. For example, a group ring $\mathbf{Z}\pi$ has this property since it maps epimorphically to $\mathbf{Z}$.

Let $\widetilde{K}_1(R) = K_1(R)/[(-1)]$ where $(-1) \in GL(1, R)$. An isomorphism $f : M' \to M$ of based R-modules determines an element $[f] \in \widetilde{K}_1(R)$. (The reason that we use $\widetilde{K}_1$ rather than K_1 is that it is both messy and unnecessary for us to fuss with ordered bases.)

We wish to generalize $[f]$ in two ways. First, we wish to replace M and M' by chain complexes. Given a chain isomorphism $f : C' \to C$ between based chain complexes, define the *torsion of* f by

$$\tau(f) = \sum (-1)^n [f_n : C'_n \to C_n] \in \widetilde{K}_1(R).$$

The second way we will generalize $[f]$ is to consider $f : M' \to M$ as a acyclic, based chain complex

$$\cdots \to 0 \to M' \xrightarrow{f} M \to 0 \to \cdots .$$

Then we will have $[f] = \pm\tau(f)$.

The following theorem gives an axiomatic characterization of the torsion $\tau(C)$ of an acyclic, based chain complex. Its proof will be an easy consequence of Theorem 11.14 discussed below.

Theorem 11.12. *Let $\mathcal{C}$ be the class of acyclic, based chain complexes over R. Then there is a unique map $\mathcal{C} \to \widetilde{K}_1(R), C \mapsto \tau(C)$ satisfying the following axioms:*

1. *If $f : C \to C'$ is a chain isomorphism, then $\tau(f) = \tau(C') - \tau(C)$.*
2. $\tau(C \oplus C') = \tau(C) + \tau(C')$.
3. $\tau(0 \to C_n \xrightarrow{f} C_{n-1} \to 0) = (-1)^{n-1}[f]$.

Definition 11.13. For an R-module M and an integer n, define the *elementary chain complex* $E(M, n)$

$$E(M,n)_i = \begin{cases} 0 & \text{if } i \neq n, n-1 \\ M & \text{if } i = n, n-1, \end{cases}$$

and with all differentials zero except $\partial_n = \mathrm{Id} : E(M,n)_n \to E(M,n)_{n-1}$. A *simple chain complex* is a finite direct sum of elementary chain complexes of the form $E(R^{k_n}, n)$.

For example, if K collapses to L by an elementary collapse, then $C(\widetilde{K}, \widetilde{L})$ is an elementary chain complex with $M = R = \mathbf{Z}\pi$. If K collapses to L, then $C(\widetilde{K}, \widetilde{L})$ is a simple chain complex.

Note that a simple chain complex is an acyclic, based complex. It is of the shape pictured below.

$$\begin{array}{ccccccccc} & R^{k_n} & \xrightarrow{\mathrm{Id}} & R^{k_n} & & R^{k_{n-2}} & \xrightarrow{\mathrm{Id}} & R^{k_{n-2}} & \\ & \oplus & & \oplus & & \oplus & & \oplus & \\ \xrightarrow{\mathrm{Id}} & R^{k_{n+1}} & & R^{k_{n-1}} & \xrightarrow{\mathrm{Id}} & R^{k_{n-1}} & & R^{k_{n-3}} & \xrightarrow{\mathrm{Id}} \end{array}$$

Theorem 11.14.

1. *Let $f : C \to C'$ be an chain isomorphism between simple chain complexes. Then $\tau(f) = 0 \in \widetilde{K}_1(R)$.*
2. *Let C be a finite, free, acyclic chain complex. There are simple chain complexes E and F and a chain isomorphism $f : E \to C \oplus F$.*

Corollary 11.15. *Let C be an acyclic, based chain complex. If E, F, E', F' are simple chain complexes and if $f : E \to C \oplus F$ and $g : E' \to C \oplus F'$ are chain isomorphisms, then $\tau(f) = \tau(g)$.*

Proof. Consider the three chain isomorphisms

$$\tilde{f} = f \oplus \mathrm{Id}_{F'} : E \oplus F' \to C \oplus F \oplus F'$$
$$\tilde{p} = \mathrm{Id}_C \oplus s : C \oplus F \oplus F' \to C \oplus F' \oplus F$$
$$\tilde{g} = g \oplus \mathrm{Id}_F : E' \oplus F \to C \oplus F' \oplus F$$

where $s : F \oplus F' \to F' \oplus F$ is the obvious switch map. We then have

$$\begin{aligned} 0 &= \tau(\tilde{g}^{-1} \circ \tilde{p} \circ \tilde{f}) && \text{by Theorem 11.14, Part 1} \\ &= \tau(\tilde{g}^{-1}) + \tau(\tilde{p}) + \tau(\tilde{f}) && \\ &= -\tau(g) + \tau(f) && \text{Theorem 11.8, Part 2 shows that } \tau(\tilde{p}) = 0. \end{aligned}$$

□

We can now use the previous theorem and corollary to define torsion.

Definition 11.16. Let C be an acyclic, based complex. Define the *torsion of C* by

$$\tau(C) = \tau(f : E \to C \oplus F)$$

where E, F are simple chain complexes and f is a chain isomorphism.

Exercise 197. Prove Theorem 11.12 assuming Theorem 11.14 and Corollary 11.15.

What remains is to prove Theorem 11.14. We strongly advise you to put down this book and prove the theorem by yourself, assuming (at first) that all stably free modules are free.

Welcome back! The proof that we give for Part 1 is the same as you found, but the proof we will give for Part 2 uses the fundamental lemma of homological algebra and is much slicker and less illuminating than the inductive proof you figured out.

We separate out Part 1 as a lemma.

Lemma 11.17. *Let $f : C \to C'$ be a chain isomorphism between simple chain complexes. Then $\tau(f) = 0 \in \widetilde{K}_1(R)$.*

Proof. Write

$$C_n = E(R^{k_n}, n)_n \oplus E(R^{k_{n+1}}, n+1)_n = C'_n.$$

It is easy to see using the fact that f is a chain map that the block matrix form of $f_n : C_n \to C'_n$ is

$$\begin{pmatrix} A_n & 0 \\ B_n & A_{n+1} \end{pmatrix}.$$

Then

$$\begin{aligned} \tau(f) &= \sum (-1)^n \left[\begin{pmatrix} A_n & 0 \\ B_n & A_{n+1} \end{pmatrix} \right] && \text{definition of } \tau(f) \\ &= \sum (-1)^n ([A_n] + [A_{n+1}]) && \text{by Theorem 11.8} \\ &= 0. \end{aligned}$$

□

Before we prove Part 2 we need some preliminaries.

Definition 11.18. A *chain contraction* $s : C \to C$ for a chain complex C is a sequence of maps $s_n : C_n \to C_{n+1}$ satisfying $\partial_{n+1} s_n + s_{n-1} \partial_n = \mathrm{Id}_{C_n}$.

A chain contraction is a chain homotopy between the identity map and the zero map. If C has a chain contraction, then $H_*(C) = 0$.

Proposition 11.19. *Let $s : C \to C$ be a chain contraction. Let $B_n = \partial(C_{n+1}) \subset C_n$.*

1. $C_n = B_n \oplus s(B_{n-1})$.
2. $\partial : s(B_n) \to B_n$ *is an isomorphism with inverse* $s : B_n \to s(B_n)$.
3. C *is isomorphic to the direct sum of chain complexes* $\oplus_n E(B_n, n+1)$.

Proof. Consider the short exact sequence

$$0 \to B_n \hookrightarrow C_n \xrightarrow{\partial} B_{n-1} \to 0.$$

The formula $x = \partial s(x) + s\partial(x)$ is valid for all x. So $\partial s(x) = x - s\partial(x) = x$ for $x \in B_n$ and for $s(y)$ with $y \in B_n$, $s\partial s(y) = s(y) - s\partial\partial(y) = s(y)$. This shows that this short exact sequence is split by $s_{n-1} : B_{n-1} \to C_n$, proving the first and second assertions.

The map

$$C_n = B_n \oplus s(B_{n-1}) \to B_n \oplus B_{n-1} = E(B_n, n+1)_n \oplus E(B_{n-1}, n)_n$$
$$a \oplus b \mapsto a \oplus \partial b$$

is an isomorphism by the second assertion. It is easy to check that this is a chain map.

□

Lemma 11.20.

1. *If C is projective, acyclic, and bounded below, then C has a chain contraction.*
2. *If C is finite, free, and acyclic, the modules $s(B_n)$ are stably free for all n.*

Proof.

1. By reindexing if necessary, assume that $C_n = 0$ for n negative. Then by the fundamental lemma of homological algebra (Theorem 2.22), the identity and the zero map are chain maps from C to C inducing the same map on H_0, hence are chain homotopic.
2. By Proposition 11.19, $C_n \cong B_n \oplus B_{n-1}$. Using induction on n, one sees B_n is stably free for all n.

□

Proof of Theorem 11.14. We have already proven Part 1.

Let C be a finite, free, acyclic chain complex. Then there is a chain contraction $s : C \to C$ by Lemma 11.20, and hence C is chain isomorphic to $\oplus_n E(B_n, n+1)$ by Proposition 11.19. Now for every n there is a finitely generated free module F_n so that $B_n \oplus F_n$ is free. Give it a finite basis. Then $C \oplus (\oplus_n E(F_n, n+1))$ is chain isomorphic to the simple complex $\oplus_n E(B_n \oplus F_n, n+1)$. □

Exercise 198. Show that for an acyclic, based chain complex C with a chain contraction s, $s + \partial : C_{\text{odd}} \to C_{\text{even}}$ is an isomorphism and $[s+\partial] = \tau(C) \in \widetilde{K}_1(R)$. Here $C_{\text{odd}} = \oplus C_{2i+1}$ and $C_{\text{even}} = \oplus C_{2i}$. This is called "wrapping up" the chain complex and is the approach to torsion used in [**7**].

An isomorphism of based R-modules $f : M \to M'$ determines an element $[f] \in \widetilde{K}_1(R)$. We generalized this in two ways: to $\tau(f)$ for a chain isomorphism between based chain complexes, and to $\tau(C)$ for an acyclic, based chain complex. We wish to generalize further and define $\tau(f)$ for $f : C \to C'$ a chain homotopy equivalence between based complexes. We need some useful constructs from homological algebra.

Definition 11.21. Let $f : C \to C'$ be a chain map between chain complexes. Define the *algebraic mapping cone of f* to be the chain complex $C(f)$ where

$$C(f)_n = C_{n-1} \oplus C'_n$$

$$\partial = \begin{pmatrix} -\partial & 0 \\ f & \partial' \end{pmatrix} : C(f)_n \to C(f)_{n-1}.$$

Define the *algebraic mapping cylinder of f* to be the chain complex $M(f)$ where

$$M(f)_n = C_{n-1} \oplus C_n \oplus C'_n$$

$$\partial = \begin{pmatrix} -\partial & 0 & 0 \\ -\text{Id} & \partial & 0 \\ f & 0 & \partial' \end{pmatrix} : M(f)_n \to M(f)_{n-1}.$$

For a chain complex C, define the *cone on C*

$$\text{Cone}(C) = C(\text{Id} : C \to C),$$

the *cylinder on C*

$$\text{Cyl}(C) = M(\text{Id} : C \to C),$$

and the *suspension of C*, which is the chain complex SC where $(SC)_n = C_{n-1}$ and $\partial_{SC}(x) = -\partial_C(x)$. Note $H_n(SC) = H_{n-1}(C)$.

If the chain complexes involved are based, then $C(f)$, $M(f)$, $\text{Cone}(C)$, $\text{Cyl}(C)$, and SC have obvious bases.

All of these constructions are interrelated. There are short exact sequences of chain complexes

$$0 \to C' \to C(f) \to SC \to 0 \tag{11.1}$$

$$0 \to C \to M(f) \to C(f) \to 0 \tag{11.2}$$

$$0 \to C' \to M(f) \to \mathrm{Cone}(C) \to 0. \tag{11.3}$$

Here is some geometric motivation. If $f : X \to Y$ is a cellular map between CW-complexes, and $f_* : C_*(X) \to C_*(Y)$ is the associated cellular chain map, then the mapping cone $C(f)$, the mapping cylinder $M(f)$, the reduced cone CX, and the reduced suspension SX all have CW-structures. One can make the following identifications:

$$\begin{aligned} C(f_*) &= C_*(C(f), \mathrm{pt}) \\ M(f_*) &= C_*(M(f)) \\ \mathrm{Cone}(C_*(X)) &= C_*(CX, \mathrm{pt}) \\ \mathrm{Cyl}(C_*(X)) &= C_*(I \times X) \\ S(C_*(X)) &= C_*(SX, \mathrm{pt}). \end{aligned}$$

The exact sequence (11.1) gives a long exact sequence in homology

$$\cdots \to H_n(C) \to H_n(C') \to H_n(C(f)) \to H_{n-1}(C) \to \cdots$$

and one can check easily that the map $H_n(C) \to H_n(C')$ is just the map induced by f. In particular, if f induces an isomorphism in homology, then $C(f)$ is acyclic.

Definition 11.22. A chain map $f : C \to C'$ is a *weak homotopy equivalence* if it induces an isomorphism on homology. If f is a weak homotopy equivalence between finite, based chain complexes, then $C(f)$ is a finite, acyclic, based chain complex. Define

$$\tau(f) = \tau(C(f)).$$

Exercise 199. If $f : C \to C'$ is a chain isomorphism of finite, based complexes, we unfortunately have two different definitions of the torsion $\tau(f)$: as $\sum(-1)^n[f_n : C'_n \to C_n]$ and as $\tau(C(f))$. Show that they coincide.

The justification for the term weak homotopy equivalence is given by the following algebraic analogue of Whitehead's theorem.

Exercise 200. If $f : C \to C'$ is a weak homotopy equivalence between acyclic, projective chain complexes which are bounded below, then $C(f)$ has a chain contraction and f is a chain homotopy equivalence; i.e. there is a chain map $g : C' \to C$ so that $f \circ g$ and $g \circ f$ are homotopic to identity maps.

Theorem 11.23. *Let*

$$0 \to C' \xrightarrow{i} C \xrightarrow{p} C'' \to 0$$

be a short exact sequence of acyclic, based chain complexes. Assume the bases are compatible, which means that for every n, the basis of C_n is of the form

$$\{i(b_1'), i(b_2'), \dots, i(b_{i_j}'), c_1'', c_2'', \dots, c_{i_k}''\} \subset C_n$$

where

$$\{b_1', b_2', \dots, b_{i_j}'\} \subset C_n'$$

is the given basis for C_n' and

$$\{p(c_1''), p(c_2''), \dots, p(c_{i_k}'')\} \subset C_n''$$

is the given basis for C_n''. Then $\tau(C) = \tau(C') + \tau(C'')$.

Lemma 11.24. *Let*

$$0 \to C' \xrightarrow{i} C \xrightarrow{p} C'' \to 0$$

be a short exact sequence of free chain complexes which are bounded below. If i is a weak homotopy equivalence, there is a chain map $s : C \to C'$ which splits i. Hence $C \cong C' \oplus C''$.

Proof. The algebraic mapping cone $C(i)$ is free, acyclic, and bounded below. Hence there is a chain contraction

$$\begin{pmatrix} s_{11} & s_{21} \\ s_{12} & s_{22} \end{pmatrix} : C_n(i) = C_{n-1} \oplus C_n' \to C_{n-1}(i) = C_{n-2} \oplus C_{n-1}'.$$

It is easy to see that $s_{21} = 0$ and that $s_{12} : C \to C'$ is a chain map which splits i. □

Proof of Theorem 11.23. Let $f : C \to C' \oplus C''$ be the chain isomorphism given by Lemma 11.24. Since f is a chain isomorphism,

$$\begin{aligned}\tau(f) &= \tau(C' \oplus C'') - \tau(C) \\ &= \tau(C') + \tau(C'') - \tau(C).\end{aligned}$$

On the other hand, since the bases are compatible, then the matrix of f_n is $\begin{pmatrix} \mathrm{Id} & * \\ 0 & \mathrm{Id} \end{pmatrix}$ in block matrix form after partitioning each basis into its $'$ part and its $''$ part. But such a matrix is trivial in $\widetilde{K}_1(R)$ by Theorem 11.8. □

Corollary 11.25. *If $f : C \to C'$ is a chain map of acyclic, based chain complexes, then $\tau(f) = \tau(C') - \tau(C)$.*

This follows from Lemma 11.23 and the short exact sequence (11.1).

We wish to prove homotopy invariance and additivity of torsion. The next lemma is a key ingredient.

Lemma 11.26. *If C is a based chain complex, then* $\text{Cone}(C)$ *is an acyclic, based chain complex with trivial torsion.*

Proof. $\text{Cone}(C)$ is finite with an obvious basis; we will show $\text{Cone}(C)$ is acyclic by induction on the total rank $\sum \dim C_i$. If the total rank is 1, both assertions are clear.

Suppose C is a based chain complex of total rank $n > 1$. Let b be a basis element of C of minimal degree. Let $C' = \{\cdots \to 0 \to Rb \to 0 \cdots \to\}$ be the corresponding subcomplex of C and let $C'' = C/C'$ be the quotient complex. There is a short exact sequence of finite complexes

$$0 \to C' \to C \to C'' \to 0$$

with compatible bases. It is easy to see there is a short exact sequence of finite complexes

$$0 \to \text{Cone}(C') \to \text{Cone}(C) \to \text{Cone}(C'') \to 0$$

with compatible bases. By induction, $\text{Cone}(C')$ and $\text{Cone}(C'')$ are acyclic with trivial torsion. By the long exact sequence in homology, $\text{Cone}(C)$ is acyclic, and by Theorem 11.23, $\text{Cone}(C)$ has trivial torsion. □

Theorem 11.27. *Let C and D be based chain complexes and let $f, g : C \to D$ be weak homotopy equivalences which are chain homotopic. In symbols,*

$$f \simeq g : C \xrightarrow{\sim} D.$$

Then $\tau(f) = \tau(g)$.

Proof. Let $s = \{s_n : C_n \to D_{n+1}\}$ be a chain homotopy from f to g satisfying $s\partial + \partial s = f - g$. Then $F = \begin{pmatrix} s & f & g \end{pmatrix} : \text{Cyl}(C) \to D$ is a chain map. There is a short exact sequence

$$0 \to C(f) \to C(F) \to \text{Cone}(C) \to 0$$

of chain complexes. Then $\tau(f) = \tau(F)$ by Lemma 11.26 and Theorem 11.23. Likewise $\tau(g) = \tau(F)$. □

Exercise 201. Let X and Y be finite complexes and f and g be cellular homotopy equivalences from X to Y which are homotopic. Show that the mapping cylinder of f is simple-homotopy equivalent to the mapping cylinder to g.

Finally, there is an additivity property of torsion.

Theorem 11.28. *Let $f : C \to C'$ and $g : C' \to C''$ be weak homotopy equivalences between based chain complexes. Then*

$$\tau(g \circ f) = \tau(f) + \tau(g).$$

The idea is to convert g to an inclusion and analyze what happens. To this end we need a definition and a lemma.

Definition 11.29. A chain map $f : C \to C'$ between based complexes is a *based injection* if for all n, the map $f : C_n \to C'_n$ is an injection and f applied to the basis of C_n is a subset of the basis of C'_n.

Lemma 11.30.

1. *Let $g : C' \to C''$ be a based injection which is a weak homotopy equivalence. Then $\tau(g) = \tau(C''/C')$.*
2. *Let $f : C \to C'$ and $g : C' \to C''$ be weak homotopy equivalences between based chain complexes. If g is a based injection, then*

$$\tau(g \circ f) = \tau(f) + \tau(C''/C').$$

Proof. We prove Part 2 first. There is a based injection

$$\begin{pmatrix} \mathrm{Id} & 0 \\ 0 & g_n \end{pmatrix} : C(f)_n = C_{n-1} \oplus C'_n \to C(g \circ f) = C_{n-1} \oplus C''_n.$$

This is part of a short exact sequence of chain complexes

$$0 \to C(f) \to C(g \circ f) \to C''/C' \to 0$$

with compatible bases. The result follows by Theorem 11.23. Part 1 is a special case of Part 2 taking $f = \mathrm{Id}$ and applying Theorem 11.23 and Lemma 11.26. □

Proof of Theorem 11.28. As advertised, we convert g to an inclusion and consider chain maps

$$C \xrightarrow{g \circ f} C'' \hookrightarrow M(g).$$

Then

$$\begin{aligned} \tau(C \to M(g)) &= \tau(g \circ f) + \tau(M(g)/C'') && \text{by Lemma 11.30} \\ &= \tau(g \circ f) + \tau(\mathrm{Cone}(g)) && \text{by (11.3)} \\ &= \tau(g \circ f) && \text{by Lemma 11.26.} \end{aligned}$$

Finally, we consider the chain maps

$$C \xrightarrow{f} C' \hookrightarrow M(g),$$

and see

$$\begin{aligned} \tau(g \circ f) &= \tau(C \to M(g)) && \text{we just proved this} \\ &= \tau(f) + \tau(M(g)/C') && \text{by Lemma 11.30} \\ &= \tau(f) + \tau(C(g)) && \text{by (11.2)} \\ &= \tau(f) + \tau(g). \end{aligned}$$

□

Perhaps the homotopy invariance and additivity of torsion are analogous to the homotopy invariance and functoriality of homology.

11.4. Whitehead torsion for CW-complexes

Let K be a finite CW-complex. Assume that K is connected. Let $x_0 \in K$ and let $\pi = \pi_1(K, x_0)$. We identify π with the group of covering transformations of the universal cover $\tilde{K} \to K$ in the usual way. We have seen (in Chapter 5) that $C_*(\tilde{K})$ is a free $\mathbf{Z}\pi$-chain complex. A basis of this chain complex is obtained by choosing a lift $\tilde{e} \subset \tilde{K}$ for each cell e of K and choosing an orientation of e or, equivalently, $\tilde{e}$. The set of lifts of cells of K with the chosen orientations defines a basis over $\mathbf{Z}\pi$ for the free $\mathbf{Z}\pi$-chain complex $C_*(\tilde{K})$.

Now suppose that $f : K \to L$ is a homotopy equivalence of finite CW-complexes. We can homotop f to a cellular map $g : K \to L$, which in turn defines a weak homotopy equivalence of based $\mathbf{Z}\pi$-chain complexes

$$g_* : C_*(\tilde{K}) \to C_*(\tilde{L}).$$

Hence we have all the data needed to define torsion as in the previous section. Define

$$\tau(f) = \tau(g_*) \in \mathrm{Wh}(\pi) = \widetilde{K}_1(\mathbf{Z}\pi)/\pm\pi.$$

The main geometric result of simple-homotopy theory is the following.

Theorem 11.31. *Let $f : K \to L$ be a homotopy equivalence of finite CW-complexes. Define the torsion $\tau(f)$ as above.*

1. *The torsion $\tau(f)$ is well-defined, independent of choice of orientations, lifts, base point x_0, identification of π with the group of covering transformations, and cellular approximation g.*
2. *If f is a simple-homotopy equivalence, then $\tau(f) = 0$.*
3. *If $\tau(f) = 0$, then f is a simple-homotopy equivalence.*

Proof. We give complete proofs of Part 1 and 2, but only the vaguest of sketches for the proof of Part 3.

Changing the lift and orientation of a cell replaces $\tilde{e}$ by $\pm\gamma\tilde{e}$ for some $\gamma \in \pi$. Thus the torsion changes by the change of basis matrix

$$\left[\begin{pmatrix} 1 & & & & \\ & \ddots & & & \\ & & \pm\gamma & & \\ & & & \ddots & \\ & & & & 1 \end{pmatrix}\right] = 0 \in \mathrm{Wh}(\pi).$$

The choice of base point and the identification of π with the group of covering transformations are dealt with by Lemma 11.11, which says that conjugation in π induces the identity map on $\mathrm{Wh}(\pi)$. Independence of the choice of cellular approximation follows from Theorem 11.27, the homotopy invariance of torsion.

Next we need to show that if $f : K \to L$ is a simple-homotopy equivalence, then $\tau(f) = 0$. Now f is a simple-homotopy equivalence if f is homotopic to a composite $K_0 \to K_1 \to K_2 \to \cdots \to K_n$ where each map $K_i \to K_{i+1}$ is either the inclusion map of an elementary expansion, the deformation retraction of an elementary collapse, or a cellular homeomorphism. We analyze the pieces. A cellular homeomorphism clearly has trivial torsion. If $i : A \hookrightarrow B$ is the inclusion map of an elementary collapse, then $\tau(i) = \tau(C(i_* : C_*(\tilde{A}) \hookrightarrow C_*(\tilde{B})))$ which is $\tau(C_*(\tilde{B}, \tilde{A}))$ by Lemma 11.30. But $C_*(\tilde{B}, \tilde{A})$ is an elementary chain complex, so has trivial torsion. Finally, if $d : B \to A$ is the associated deformation retract, then

$$\begin{aligned} \tau(d) &= \tau(d_* : C_*(\tilde{B}) \to C_*(\tilde{A})) \\ &= \tau(d_* \circ i_*) - \tau(i_*) && \text{additivity of torsion} \\ &= \tau(\mathrm{Id}_{C_*(\tilde{A})}) - \tau(i_*) && \text{homotopy invariance of torsion} \\ &= 0. \end{aligned}$$

The composite $K_0 \to K_1 \to K_2 \to \cdots \to K_n$ must have trivial torsion since all the pieces do. Thus $\tau(f) = 0$.

For the proof of Part 3, suppose that $f : K \to L$ is a cellular map between finite complexes with trivial torsion. Then f factors as $K \hookrightarrow M(f) \to L$. The second map is a collapse map and hence is a simple-homotopy equivalence and has trivial torsion. By the additivity of torsion, $\tau(K \hookrightarrow M(f)) = 0$, and it suffices to show that this map is a simple-homotopy equivalence.

Recycling our notation, we will assume K is a subcomplex of L and that the torsion of the inclusion map is trivial. The first step in showing that $K \hookrightarrow L$ is a simple-homotopy equivalence is *cell-trading* [**7**, 7.3]. If e is a cell of $L - K$ of minimal dimension (say k), one constructs a simple-homotopy equivalence $L \to L'$ rel K so that L' has one less k-cell than L, one more $(k+1)$-cell, and for $i \neq k, k+1$, the number of i-cells of L and of L' is the same. By a simple-homotopy equivalence $h : L \to L'$ rel K, we mean that K is a subcomplex of both L and L' and that h restricted to K is the identity. By continuing to trade, one reduces to proving that an inclusion $i : K \hookrightarrow L$ with trivial torsion is a simple-homotopy equivalence when the cells of $L - K$ all are in two adjoining dimensions, say n and $n+1$. Then the chain complex $C(\tilde{L}, \tilde{K})$ is described by a matrix! We can stabilize the matrix, if desired, by making expansions. Since the torsion is zero, we may assume that the matrix is a product of elementary matrices. There is a technique called *cell-sliding* [**7**, 8.3] (changing the attaching map of an $(n+1)$-cell) which gives a simple-homotopy equivalence $L \to L'$ rel K so that the matrix in the chain complex is replaced by the matrix multiplied by an elementary matrix. Thus one reduces to the case where $K \hookrightarrow L$ has the chain complex

$$C(\tilde{L}, \tilde{K}) = \{\cdots \to 0 \to \mathbf{Z}[\pi]^m \xrightarrow{\text{Id}} \mathbf{Z}[\pi]^m \to 0 \to \cdots\}$$

in which case there is one last technique, *cell-cancellation* [**7**, 8.2], which says that $K \hookrightarrow L$ is a simple-homotopy equivalence. □

Two finite CW-complexes are said to have the same *simple-homotopy type* if there is a simple-homotopy equivalence between them. Homeomorphic CW-complexes could have drastically different CW-structures. Do they have the same simple-homotopy type? A fundamental theorem of Chapman implies that simple-homotopy type is a homeomorphism invariant:

Theorem 11.32 (Chapman [**6**]). *If $f : X \to Y$ is a homeomorphism between finite CW-complexes, then f is a simple-homotopy equivalence.* □

It follows that the torsion of a homotopy equivalence $f : X \to Y$ between finite CW-complexes depends only on the underlying topological spaces.

How does simple-homotopy theory apply to manifolds? Typically, a smooth manifold is given the structure of a simplicial complex (and hence a CW-complex) by constructing a triangulation. This triangulation is unique up to subdivision, and it is not difficult to show [**28**] that the identity map between a complex and a subdivision has trivial torsion. Thus compact smooth manifolds have a well-defined simple-homotopy type. A deep theorem of Kirby-Siebenmann shows that this theory also applies for topological manifolds (Hausdorff spaces which are locally Euclidean). They show that a compact topological manifold has a canonical simple-homotopy type.

In the next section we will define Reidemeister torsion for certain finite CW-complexes. This will be an interesting and computable invariant of simple-homotopy type. If two manifolds have different Reidemeister torsions, by Chapman's theorem they cannot be homeomorphic.

11.5. Reidemeister torsion

Suppose that (C, ∂) is a finite, based (with basis $\{b_i\}$) chain complex, not necessarily acyclic, over a ring S. Let $f : S \to R$ be a ring homomorphism. Then $(C \otimes_S R, \partial \otimes \mathrm{Id}_R)$ is a finite, based (with basis $\{b_i \otimes 1\}$) chain complex over S.

If (C, ∂) is acyclic with chain contraction s, then $(C \otimes_S R, \partial \otimes \mathrm{Id}_R)$ is acyclic since it has the chain contraction $s \otimes \mathrm{Id}_R$. The torsions are related by

$$\tau(C \otimes_S R) = f_* \tau(C)$$

where $f_* : \widetilde{K}_1(S) \to \widetilde{K}_1(R)$ is the induced homomorphism. However, in many interesting cases it may happen that $C \otimes_S R$ is acyclic although C is not, so that $\tau(C \otimes_S R)$ may be defined even though $\tau(C)$ is not.

Moreover, if R is a commutative ring, or better yet, a field, then the determinant defines a homomorphism (an isomorphism if R is a field) $\det : \widetilde{K}_1(R) \to R^\times / \pm 1$ (see Exercise 194).

Definition 11.33. Let C be a based chain complex over a ring S, and $f : S \to R$ be a ring homomorphism to a commutative ring. Suppose that $C \otimes_S R$ is acyclic. Then

$$\Delta_R(C) = \det(\tau(C \otimes_S R)) \in R^\times / \pm 1$$

is called the *Reidemeister torsion of C with respect to $f : S \to R$*. Since $\Delta_R(C)$ is a unit in R we use multiplicative notation for Reidemeister torsion.

Let X be a CW-complex. Let R be some ring, which for our purposes may be taken to be commutative. Suppose

$$\rho : \mathbf{Z}\pi_1 X \to R$$

is a ring homomorphism. The chain complex

$$C_*(\tilde{X}) \otimes_{\mathbf{Z}\pi_1 X} R$$

was used in Chapter 5 to define homology with twisted coefficients $H_*(X; R)$.

A choice of lifts and orientations of cells makes this a based (right) R-complex. If it happens to be acyclic, then the torsion $\tau \in \widetilde{K}_1(R)$ is defined, and so we can take the Reidemeister torsion

$$\Delta_R(X) = \Delta_R(C_*(X)) \in R^\times$$

to be the determinant of τ.

The following exercise shows how to remove the dependence of Δ_R on the choice of lifts, orientation, and ordering of cells.

Exercise 202. Let $G \subset R^\times$ be the subgroup generated by -1 and $\det(\rho(\gamma))$ for $\gamma \in \pi_1(X)$, where $\rho : \mathbf{Z}\pi_1 X \to R$ is the given ring homomorphism. Show that the Reidemeister torsion $\Delta_R(X)$, taken in $R^\times/G$, is well defined, independent of the choice of lifts, orientation, or ordering of cells.

We usually abuse notation and consider Reidemeister torsion as an element of $R^\times$, and omit mentioning that it is only well-defined up to multiplication by an element of G.

The disadvantage of Reidemeister torsion is that it requires a map ρ so that $C_*(\tilde{X}) \otimes_{\mathbf{Z}\pi} S$ is acyclic. Perhaps for a given X, no useful map exists. The advantage is that when ρ does exist, the Reidemeister torsion gives an invariant of the space. The following proposition is a corollary of Corollary 11.25.

Proposition 11.34. *Let $f : X \to Y$ be a homotopy equivalence between finite, connected CW-complexes. Let $\pi = \pi_1 X = \pi_1 Y$. Suppose $\rho : \mathbf{Z}\pi \to R$ is a ring homomorphism to a commutative ring so that $C_*(\tilde{X}) \otimes_{\mathbf{Z}\pi} R$ and $C_*(\tilde{Y}) \otimes_{\mathbf{Z}\pi} R$ are acyclic. Then*

$$\rho_*\tau(f) = \Delta_R(Y) - \Delta_R(X) \in R^\times/G.$$

Therefore Reidemeister torsion is an invariant capable of distinguishing simple-homotopy type from homotopy type and of showing that two homotopy equivalent spaces are not homeomorphic. In the next section we will apply this idea to lens spaces.

Reidemeister torsion can also lead to interesting invariants, for example, the Alexander polynomial of a knot. Let $K \subset S^3$ be a knot, that is, a smooth submanifold of S^3 diffeomorphic to S^1. Let $X = S^3 - N(K)$, where $N(K)$ is an open tubular neighborhood of K. Alexander duality implies that $H_1(X) \cong \mathbf{Z}$, and so every knot has a canonical (up to multiplication by ± 1) homomorphism $a : \pi_1 X \to H_1 X \to \mathbf{Z}$.

Let $R = \mathbf{Q}(t)$, the field of rational functions. Then define $\rho : \mathbf{Z}\pi_1 X \to \mathbf{Q}(t)$ by

$$\rho(\sum a_\gamma \gamma) = \sum a_\gamma t^{a(\gamma)}.$$

It turns out (see the exercise below) that the chain complex

$$C = C_*(\tilde{X}) \otimes_{\mathbf{Z}\pi_1 X} \mathbf{Q}(t)$$

is acyclic. Its Reidemeister torsion $\Delta_{\mathbf{Q}(t)}(X)$ is a nonzero rational function and is well defined up to sign and powers of t. Moreover $\Delta_{\mathbf{Q}(t)}(X)$ is always of the form

$$\Delta_{\mathbf{Q}(t)}(X) = (t-1)/\Delta_K(t)$$

for some polynomial $\Delta_K(t) \in \mathbf{Z}[t]$. This polynomial is called the *Alexander polynomial of* K. It is a useful invariant for distinguishing isotopy classes of knots.

Exercise 203. Let $\mu \subset X$ be a *meridian* of the knot K, that is, the boundary circle of a small embedded disk in S^3 intersecting K transversely in one point. Show that the inclusion $\mu \hookrightarrow S^3$ induces an isomorphism on integral homology. Conclude that $H_*(X, \mu; \mathbf{Z}) = 0$ and so the cellular chain complex $C_*(X, \mu)$ is acyclic.

Let $s : C_*(X, \mu) \to C_*(X, \mu)$ be a chain contraction. Lift s to a map

$$\tilde{s} : C_*(\tilde{X}, \tilde{\mu}) \otimes_{\mathbf{Z}\pi_1 X} \mathbf{Q}(t) \to C_*(\tilde{X}, \tilde{\mu}) \otimes_{\mathbf{Z}\pi_1 X} \mathbf{Q}(t)$$

and show that $\tilde{s}\partial - \partial\tilde{s}$ is a chain isomorphism. Conclude that the chain complex $C_*(\tilde{X}, \tilde{\mu}) \otimes_{\mathbf{Z}\pi_1(X)} \mathbf{Q}(t)$ is acyclic. The long exact sequence for the pair $(\tilde{X}, \tilde{\mu})$ then shows that $C_*(\tilde{X}) \otimes_{\mathbf{Z}\pi_1 X} \mathbf{Q}(t)$ is acyclic.

11.6. Torsion and lens spaces

In this section we will use Reidemeister torsion to classify 3-dimensional lens spaces up to simple-homotopy and prove Theorem 11.1. The homotopy classification differs from the simple-homotopy classification, but the simple-homotopy classification is the same as the homeomorphism classification, the diffeomorphism classification, and the isometry classification.

Let (p, q) be a pair of relatively prime integers, with $p > 0$. For convenience let r denote an inverse for q mod p, so $rq \equiv 1$ mod p. There are many descriptions of $L(p, q)$. We will give a description which makes the cell structure on its universal cover easy to see.

Let $X = S^1 \times D^2$ be the solid torus, which we parameterize as the subset $\{(z_1, z_2) |\ |z_1| = 1, |z_2| \leq 1\} \subset \mathbf{C}^2$. The quotient of X by the equivalence relation $(z_1, z_2) \sim (z_1', z_2)$ if $|z_2| = 1$ is the 3-sphere S^3. In fact the map $f : X \to S^3$ defined by

$$f(z_1, z_2) = (z_1\sqrt{1 - |z_2|^2}, z_2)$$

defines a homeomorphism from $X/\sim$ to $S^3 \subset \mathbf{C}^2$.

Write $\zeta = e^{2\pi i/p}$ and let $\mathbf{Z}/p$ act on X by

$$(z_1, z_2) \cdot g = (z_1\zeta, z_2\zeta^q)$$

where $\mathbf{Z}/p$ is written multiplicatively as $\langle g \mid g^p = 1\rangle$ (we use a right action to be consistent with the previous sections). This defines the free $\mathbf{Z}/p$-action on $S^3 = X/\sim$

$$(w_1, w_2) \cdot g = (w_1\zeta, w_2\zeta^q).$$

The map $f : X \to S^3$ is an equivariant map. By definition the quotient space $S^3/(\mathbf{Z}/p)$ is the lens space $L(p, q)$.

We use the description $X/\sim$ to construct an equivariant cell structure on S^3. Let $\tilde{e}_0$ be the image under f of the point $(1,1) \in X$. We take the 0-cells of S^3 to be the p points $\tilde{e}_0, \tilde{e}_0 g, \cdots, \tilde{e}_0 g^{p-1}$. Notice that

$$\tilde{e}_0 g^n = f((1,1)g^n) = f(\zeta^n, \zeta^{qn}) = f(1, \zeta^{qn}).$$

Next, let $\tilde{e}_1$ be the image under f of the arc $\{(1, e^{i\theta}) \mid 0 \leq \theta \leq \frac{2\pi}{p}\}$, and take as 1-cells the translates $\tilde{e}_1, \tilde{e}_1 g, \cdots, \tilde{e}_1 g^{p-1}$. Then

$$\begin{aligned} \partial \tilde{e}_1 &= f(1,\zeta) - f(1,1) \\ &= f(1,\zeta^{qr}) - f(1,1) \\ &= \tilde{e}_0(g^r - 1), \end{aligned} \tag{11.4}$$

since $qr \equiv 1 \bmod p$.

Let $\tilde{e}_2$ be $f(1 \times D^2)$. We take as 2-cells the translates of $\tilde{e}_2$ by g, $\tilde{e}_2, \tilde{e}_2 g, \cdots, \tilde{e}_2 g^{p-1}$. Then

$$\begin{aligned} \partial \tilde{e}_2 &= f(\{(1, e^{i\theta}) \mid 0 \leq \theta \leq 2\pi\}) \\ &= \tilde{e}_1 + \tilde{e}_1 g + \cdots + \tilde{e}_1 g^{p-1} \\ &= \tilde{e}_1(1 + g + \cdots + g^{p-1}). \end{aligned} \tag{11.5}$$

For the 3-cells, consider the solid cylinder

$$[0, \tfrac{2\pi}{p}] \times D^2 \cong \{e^{i\theta} \mid \theta \in [0, \tfrac{2\pi}{p}]\} \times D^2 \subset X.$$

This is homeomorphic to a closed 3-ball. Its image in S^3

$$f([0, \tfrac{2\pi}{p}] \times D^2) = ([0, \tfrac{2\pi}{p}] \times D^2)/\sim$$

is a "lens", also homeomorphic to a closed 3-ball.

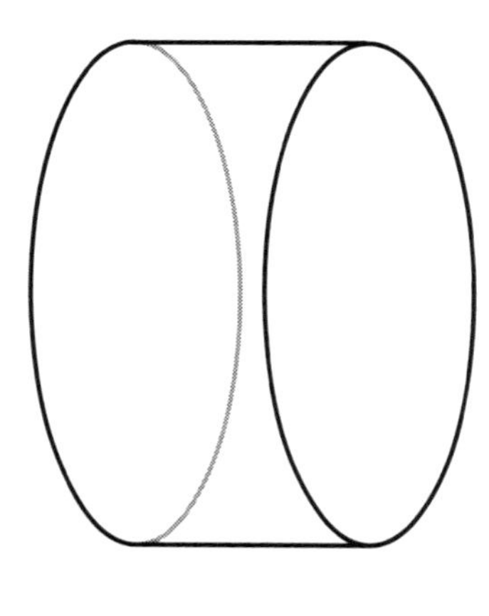

$[0, \frac{2\pi}{p}] \times D^2$

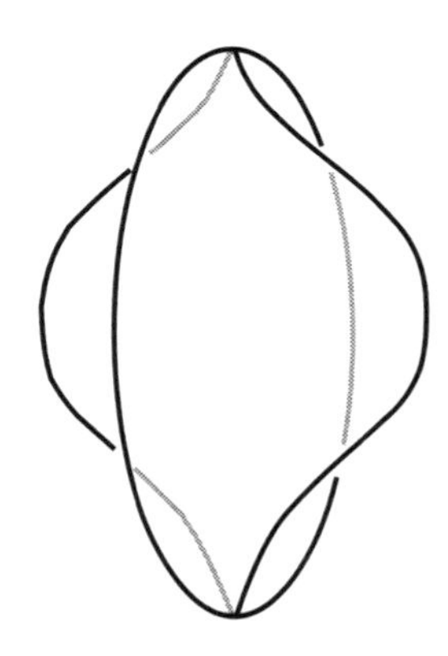

$f([0, \frac{2\pi}{p}] \times D^2)$

We let $\tilde{e}_3$ be the 3-cell $f([0, \frac{2\pi}{p}] \times D^2)$. Then

$$\begin{aligned}\partial\tilde{e}_3 &= f(\{\zeta\} \times D^2) - f(\{1\} \times D^2) \\ &= f(\{\zeta\} \times (D^2\zeta^q)) - f(\{1\} \times D^2) \\ &= f((\{1\} \times D^2)\zeta) - f(\{1\} \times D^2) \\ &= \tilde{e}_2(g-1).\end{aligned} \tag{11.6}$$

Thus we have described a $\mathbf{Z}/p$-equivariant cell structure for S^3; this defines a cell structure on $L(p,q)$ with one 0-cell e_0, one 1-cell e_1, one 2-cell e_2, and one 3-cell e_3. We calculated the $\mathbf{Z}[\pi]$-chain complex for the universal cover $\widetilde{L(p,q)} = S^3$ of $L(p,q)$ to be

$$0 \longrightarrow \mathbf{Z}\pi(\tilde{e}_3) \xrightarrow{\partial_3} \mathbf{Z}\pi(\tilde{e}_2) \xrightarrow{\partial_2} \mathbf{Z}\pi(\tilde{e}_1) \xrightarrow{\partial_1} \mathbf{Z}\pi(\tilde{e}_0) \longrightarrow 0 \tag{11.7}$$

with $\partial_3 = g-1$, $\partial_2 = 1 + g + g^2 + \cdots + g^{p-1}$, and $\partial_1 = g^r - 1$.

Notice that the map from the lens to $L(p,q)$ is a quotient map which is a homeomorphism on its interior. Therefore $L(p,q)$ can be described as the identification space of the lens, where the left 2-disk in the boundary of the lens is identified with the right 2-disk by a $\frac{2\pi q}{p}$-twist.

Exercise 204. Draw a picture of this cell structure on S^3 for $p = 5$ and $q = 2$ by thinking of S^3 as $\mathbf{R}^3 - \{\infty\}$. Label the i-cells $\tilde{e}^i, g\tilde{e}^i, \cdots, g^4\tilde{e}^i$.

Exercise 205. Show that $L(2,1)$ is the real projective plane. Show that $L(p,q)$ is the union of two solid tori.

Since $\pi_1(L(p,q)) = \mathbf{Z}/p$, if $L(p,q)$ is homotopy equivalent to $L(p',q')$, then $p = p'$. The following theorem gives the homotopy classification of 3-dimensional lens spaces.

Theorem 11.35. *Suppose $f : L(p,q) \to L(p,q')$ takes g to $(g')^a$, where g, g' are the generators of $\pi_1(L(p,q)), \pi_1(L(p,q'))$ as above. Assume $(a,p) = 1$, so that f induces an isomorphism on fundamental groups. Then*

1. $q \deg(f) \equiv q'a^2 \bmod p$.
2. *f is a homotopy equivalence if and only if* $\deg(f) = \pm 1$.

Moreover, if there exists an integer a so that $a^2q' \equiv \pm q \bmod p$, then there is a homotopy equivalence $f : L(p,q) \to L(p,q')$ whose induced map on fundamental groups takes g to $(g')^a$.

Proof. Let $L = L(p,q)$, $L' = L(p,q')$. First note that $f_* : \pi_1 L \to \pi_1 L'$ is an isomorphism since $(a,p) = 1$ implies $(g')^a$ generates $\mathbf{Z}/p = \pi_1 L'$. Using the cellular approximation theorem and the homotopy extension property, we may assume that f is cellular. Denote the cells of L' by e'.

Recall that the group of covering transformations of a universal cover $p : \tilde{X} \to X$ is identified with the fundamental group of X by taking the covering transformation $h : \tilde{X} \to \tilde{X}$ to the loop $p(\alpha)$, where α is a path in $\tilde{X}$ from $\tilde{x}_0$ to $h(\tilde{x}_0)$. Since $\tilde{e}_1$ is a path in S^3 from $\tilde{e}_1$ to $\tilde{e}_1 g^r$ by (11.4), e_1 (which is a loop since there is only one 0-cell in L) represents g^r in $\pi_1 L$. Similarly e_1' represents $(g')^{r'}$ in $\pi_1 L'$.

Because f is cellular and takes g to $(g')^a$ on fundamental groups, it follows that the loop $f(e_1)$ represents $(g')^{ar} = (g')^{r'q'ar}$, and so the chain map $f_* : C_1(L) \to C_1(L')$ takes e_1 to $(q'ar)e_1'$.

Lift f to $\tilde{f} : \tilde{L} \to \tilde{L}'$, so that $\tilde{f}(\tilde{e}_0) = \tilde{e}_0'$. Then since $f(e_1)$ wraps $q'ar$ times around e_1', $f(\tilde{e}_1)$ lifts to a sum of $q'ar$ translates of $\tilde{e}_1'$. Precisely,

$$\begin{aligned} \tilde{f}(\tilde{e}_1) &= \tilde{e}_1' + \tilde{e}_1'(g')^{r'} + \cdots + \tilde{e}_1'(g')^{r'(q'ar-1)} \\ &= \tilde{e}_1'(1 + (g')^{r'} + \cdots + (g')^{r'(q'ar-1)}). \end{aligned} \tag{11.8}$$

To avoid being confused by isomorphic rings, write

$$\Lambda = \mathbf{Z}[\mathbf{Z}/p] = \mathbf{Z}[t]/(t^p - 1).$$

Identify $\mathbf{Z}[\pi_1 L]$ with Λ using the isomorphism determined by $g \mapsto t^a$, and identify $\mathbf{Z}[\pi_1 L']$ with Λ via $g' \mapsto t$. With these identifications, the equivariant chain complexes of $\widetilde{L(p,q)}$ and $\widetilde{L(p,q')}$ and the chain map between them are given by the diagram

$$\begin{array}{ccccccccccc} 0 & \longrightarrow & \Lambda & \xrightarrow{\partial_3} & \Lambda & \xrightarrow{\partial_2} & \Lambda & \xrightarrow{\partial_1} & \Lambda & \longrightarrow & 0 \\ & & \downarrow f_3 & & \downarrow f_2 & & \downarrow f_1 & & \downarrow f_0 & & \\ 0 & \longrightarrow & \Lambda & \xrightarrow[\partial_3']{} & \Lambda & \xrightarrow[\partial_2']{} & \Lambda & \xrightarrow[\partial_1']{} & \Lambda & \longrightarrow & 0 \end{array}$$

where the differentials in two chain complexes are given by multiplication by an element in Λ as follows:

$$\begin{aligned} \partial_3 &= t^a - 1 \\ \partial_2 &= 1 + t^a + \cdots + (t^a)^{p-1} = 1 + t + \cdots + t^{p-1} \\ \partial_1 &= t^{ar} - 1 \\ \partial_3' &= t - 1 \\ \partial_2' &= 1 + t + \cdots + t^{p-1} \\ \partial_1' &= t^{r'} - 1. \end{aligned}$$

These equations follow from Equations (11.4), (11.5), and (11.6) and the identifications of Λ with $\mathbf{Z}[\pi_1 L]$ and $\mathbf{Z}[\pi_1 L']$.

Since $\tilde{f}(\tilde{e}_0) = \tilde{e}_0'$ and f takes g to $(g')^a$, it follows that $f_0 =$Id. From Equation (11.8) we conclude that

$$f_1 = 1 + t^{r'} + \cdots + t^{r'(q'ar-1)}.$$

Now $f_1\partial_2 = \partial_2' f_2$, i.e. $(1+t+\cdots+t^{p-1})f_1 = (1+t+\cdots+t^{p-1})f_2$. This implies that

$$f_2 = f_1 + \xi(1-t) \text{ for some } \xi \in \Lambda. \tag{11.9}$$

Similarly we have that $f_2\partial_3 = \partial_3' f_3$; i.e. $f_2(t^a-1) = (t-1)f_3$, and therefore $(t-1)(t^{a-1}+\cdots+1)f_2 = (t-1)f_3$. Hence

$$f_3 = (t^{a-1}+\cdots+1)f_2 + \beta(1+t+\cdots+t^{p-1}) \text{ for some } \beta \in \Lambda. \tag{11.10}$$

Let $\varepsilon : \Lambda \to \mathbf{Z}$ be the augmentation map defined by map $\sum n_i t^i \mapsto \sum n_i$. Then $H_3\tilde{L} = \ker\ \partial_3 = \operatorname{span}(1+t+\cdots+t^{p-1}) \cong \mathbf{Z}$. The isomorphism is given by

$$(1+t+\cdots+t^{p-1})\cdot\alpha \mapsto \varepsilon(\alpha).$$

(Check that this indeed gives an isomorphism. Facts like this come from the identity $(1+t+\cdots+t^{p-1})(1-t) = 0$ in $\mathbf{Z}[\mathbf{Z}/p]$.)

Similarly

$$H_3\tilde{L}' \cong \ker\partial_3' = \operatorname{span}(1+t+\cdots+t^{p-1}) \cong \mathbf{Z}.$$

Thus, $\deg\tilde{f} = n$ if and only if $f_3(1+t+\cdots+t^{p-1}) = n(1+t+\cdots+t^{p-1})$. Now

$$f_3(1+t+\cdots+t^{p-1}) = \varepsilon(f_3)(1+t+\cdots+t^{p-1})$$

and, using the computations above,

$$\begin{aligned}
\varepsilon(f_3) &= \varepsilon((t^{a-1}+t+\cdots+1)f_2 + \beta\cdot(1+t+\cdots+t^{p-1})) \\
&= a\varepsilon(f_2) + \varepsilon(\beta)\cdot p \\
&= a\varepsilon(f_1 + \xi(1-t)) + \varepsilon(\beta)\cdot p \\
&= a\varepsilon(f_1) + \varepsilon(\beta)\cdot p \\
&= a\cdot q'ar + \varepsilon(\beta)\cdot p.
\end{aligned}$$

In these equations, ξ and β are the elements defined in Equations (11.9) and (11.10). Thus the degree of $\tilde{f}$ equals $a^2\cdot q'r + \varepsilon(\beta)\cdot p$, and in particular the degree of $\tilde{f}$ is congruent to $a^2q'r \bmod p$.

The covers $\tilde{L} \to L$, $\tilde{L}' \to L'$ both have degree p. Since the degree multiplies under composition of maps between oriented manifolds, it follows that $p\deg f = p\deg\tilde{f}$, and so $\deg f = \deg\tilde{f} = a^2q'r + \varepsilon(\beta)p$, proving the second assertion.

If $\deg f = \deg\tilde{f} = \pm 1$, then since $\tilde{L} \cong S^3 \cong \tilde{L}'$, the map $\tilde{f} : S^3 \to S^3$ is a homotopy equivalence. Thus $f : L \to L'$ induces an isomorphism on all homotopy groups and is therefore a homotopy equivalence by Theorem 6.71.

It remains to prove the last assertion of Theorem 11.35. Define a map on the 1 skeleton $f^{(1)} : L^{(1)} \to L'$ so that the induced map on fundamental groups takes g to $(g')^a$. Since $((g')^a)^p = 1$, $f^{(1)}$ extends over the 2 skeleton.

The obstruction to extending over the 3-skeleton lies in $H^3(L;\pi_2 L')$. Notice that $\pi_1 L'$ acts trivially on $\pi_k L'$ for all k, since the covering transformations $S^3 \to S^3$ have degree 1 and so are homotopic to the identity. Thus the results of obstruction theory (Chapter 7) apply in this situation.

Since $\pi_2 L' = \pi_2 S^3 = 0$, $H^3(L;\pi_2 L') = 0$, and so this obstruction vanishes. Hence we can extend over the 3-skeleton; since this does not alter the map on the 1-skeleton, we obtain a map $f : L \to L'$ so that $f_*(g) = (g')^a$.

If $a^2 q' r \equiv \pm 1 \mod p$, then $\tilde{f}$ has a degree $\pm 1 \mod p$. We assert that f can be modified so that the resulting lift $\tilde{f}$ is replaced by another equivariant map $\tilde{f}'$ such that deg $\tilde{f}' =$ deg $\tilde{f} \pm p$. This is a formal consequence of the technique used to extend f over the 3-skeleton in obstruction theory; we outline the construction in this specific case.

Let $x \in L$, and redefine f on a neighborhood of x as indicated in the following figure.

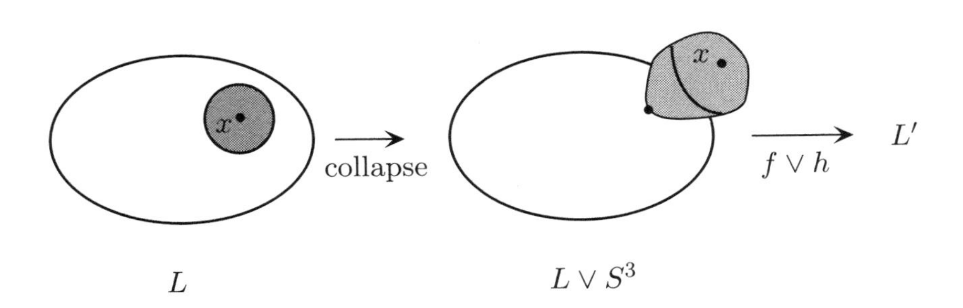

Here $h : S^3 \to L'$ is a degree $\pm p$ map (e.g. take h to be the universal covering). Denote by f' this composition of the collapsing map $L \to L \vee S^3$ and $f \vee h$. There are many ways to see that the degree of f' equals the $\deg(f) \pm p$. Notice that since f is only modified on a 3-cell, f and f' induce the same map on fundamental groups.

If $a^2 q' r \equiv \pm 1 \mod p$, then repeating this modification as needed we can arrange that $\deg f = \pm 1$, and so f is a homotopy equivalence. This completes the proof of Theorem 11.35. □

Exercise 206.

1. Show that $L(5,1)$ and $L(5,2)$ have the same homotopy and homology groups, but are not homotopy equivalent.
2. Show that $L(7,1)$ and $L(7,2)$ are homotopy equivalent. Show that any homotopy equivalence is orientation preserving, i.e. has degree 1. Is this true for any pair of homotopy equivalent lens spaces?

Having completed the homotopy classification, we turn now to the simple-homotopy classification. This is accomplished using Reidemeister torsion. The chain complex of the universal covers of a lens space $L = L(p,q)$ is not acyclic, since it has the homology of S^3. We will tensor with $\mathbf{C}$ to turn them into acyclic complexes and compute the corresponding Reidemeister torsion. Thus we need a ring map $\mathbf{Z}[\mathbf{Z}/p] \to \mathbf{C}$.

Let $\zeta = e^{2\pi i/p} \in \mathbf{C}$. Let $\rho : \mathbf{Z}[\mathbf{Z}/p] \to \mathbf{C}$ be the ring homomorphism defined by

$$h(t) = \zeta.$$

Note that $h(1 + t + \cdots + t^{p-1}) = 0$, and that if $p \nmid a$, then $\zeta^a \neq 1$.

In this notation we assumed $\mathbf{Z}/p$ had the generator t. Let g be the generator of $\pi_1 L$, and choose an isomorphism $\pi_1 L \cong \mathbf{Z}/p$ so that g corresponds to t^a where $(a,p) = 1$. Let $D_* = C_*(\tilde{L}) \otimes_{\pi_1 L} \mathbf{C}$. This is a based complex over the complex numbers with basis of the form $e \otimes 1$ where e is an (oriented) cell of L. Since L has one cell in each dimension $d = 0, 1, 2, 3$,

$$D_n = \begin{cases} \mathbf{C} & \text{if } n = 0, 1, 2, 3, \\ 0 & \text{otherwise.} \end{cases}$$

Moreover, from the chain complex (11.7) one easily sees that

$$D_* = \{\ 0 \longrightarrow \mathbf{C} \xrightarrow{\zeta^a - 1} \mathbf{C} \xrightarrow{0} \mathbf{C} \xrightarrow{\zeta^{ar} - 1} \mathbf{C} \longrightarrow 0\ \}$$

and hence D_* is acyclic.

The following diagram exhibits a chain isomorphism of D_* with an elementary complex

$$\begin{array}{ccccccccccc} 0 & \longrightarrow & \mathbf{C} & \xrightarrow{\zeta^a - 1} & \mathbf{C} & \xrightarrow{0} & \mathbf{C} & \xrightarrow{\zeta^{ar} - 1} & \mathbf{C} & \longrightarrow & 0 \\ & & \downarrow{\scriptstyle \zeta^a - 1} & & \downarrow{\scriptstyle \mathrm{Id}} & & \downarrow{\scriptstyle \zeta^{ar} - 1} & & \downarrow{\scriptstyle \mathrm{Id}} & & \\ 0 & \longrightarrow & \mathbf{C} & \xrightarrow{\mathrm{Id}} & \mathbf{C} & \xrightarrow{0} & \mathbf{C} & \xrightarrow{\mathrm{Id}} & \mathbf{C} & \longrightarrow & 0 \end{array}$$

and so the Reidemeister torsion is

$$\Delta_{\mathbf{C}}(L) = (\zeta^a - 1)(\zeta^{ar} - 1)$$

(recall that we are using multiplicative notation). Notice that the Reidemeister torsion takes its values in $\mathbf{C}^\times / \pm\{1, \zeta, \cdots, \zeta^{p-1}\}$.

Now suppose $f : L(p,q) \to L(p,q')$ is a homotopy equivalence which takes g to $(g')^a$ for some a so that $(a,p) = 1$. Then if we choose the isomorphism $\pi_1 L' \cong \mathbf{Z}/p$ so that g' corresponds to t, we have

$$\Delta_{\mathbf{C}}(L') = (\zeta - 1)(\zeta^{r'} - 1).$$

If f is a simple-homotopy equivalence, $\tau(f) = 0$; by Proposition 11.34 $\Delta_{\mathbf{C}}(L)$ and $\Delta_{\mathbf{C}}(L')$ are equal in the quotient $\mathbf{C}^\times / \pm\{1, \zeta, \cdots, \zeta^{p-1}\}$. We summarize this conclusion (and the conclusion of Theorem 11.35 for convenience) in the following proposition.

Proposition 11.36. *If $f : L(p,q) \to L(p,q')$ is a simple-homotopy equivalence which takes the generator $g \in \pi_1 L$ to $(g')^a \in \pi_1 L'$, then*

1. $a^2 q' \equiv \pm q \bmod p$.
2. *For each pth root of unity $\zeta \neq 1$, there exists an $s \in \mathbf{Z}/p$ so that*
$$(\zeta^a - 1)(\zeta^{ar} - 1) = \pm \zeta^s (\zeta - 1)(\zeta^{r'} - 1)$$
where r and r' are determined by the equations $rq \equiv 1 \bmod p$ and $r'q' \equiv 1 \bmod p$.

□

Example of L(7,1) and L(7,2). Let $L(7,1) = L$ and $L(7,2) = L'$. Then the equation $a^2 = \pm 2 \bmod 7$ has only the solutions $a = 3,\ a = 4$. We have $r = 1$ and $r' = 4$.

Exercise 206 shows that L and L' are homotopy equivalent. Suppose that L and L' were simple-homotopy equivalent.

1. If $a = 3$, then for each seventh root of unity ζ there exists an $s \in \mathbf{Z}$ with
$$(\zeta^3 - 1)^2 = \pm \zeta^s (\zeta - 1)(\zeta^4 - 1).$$
This implies that
$$|\zeta^3 - 1|^2 = |\zeta - 1||\zeta^4 - 1|.$$
Note that $|\zeta^3 - 1| = |\zeta^4 - 1|$. But we leave it to you as an exercise to show $|\zeta^b - 1| = 2\cos(b\pi/p)$, and we will allow the use of a calculator to show $|\zeta - 1| \neq |\zeta^3 - 1|$.
2. If $a = 4$, the equation reads:
$$(\zeta^4 - 1)^2 = \pm \zeta^s (\zeta - 1)(\zeta^4 - 1).$$
For similar reasons as in the first case this is impossible.

Thus $L(7,1)$ is homotopy equivalent but not simple-homotopy equivalent to $L(7,2)$. Theorem 11.32 implies that $L(7,1)$ and $L(7,2)$ are not homeomorphic.

We will now give the simple-homotopy classification for 3-dimensional lens spaces. The proof will rely on a number-theoretic result about roots of unity.

Theorem 11.37. *If $f : L(p,q) \to L(p,q')$ is a simple-homotopy equivalence, with $f_*(g) = (g')^a$, then $L(p,q)$ is homeomorphic to $L(p,q')$ and either*
$$a = \pm 1 \text{ and } q \equiv \pm q' \bmod p$$

or

$$a = \pm q \textit{ and } q \equiv \pm(q')^{-1} \bmod p.$$

In particular, $q \equiv \pm(q')^{-1} \bmod p$.

Proof. Suppose that $f : L(p,q) \to L(p,q')$ is a simple-homotopy equivalence. Proposition 11.36 shows that for any p^{th} root of unity $\zeta \neq 1$ there exists an s with

$$(\zeta^a - 1)(\zeta^{ar} - 1) = \pm\zeta^s(\zeta - 1)(\zeta^{r'} - 1).$$

Note that $|\zeta^s|^2 = 1$ and for any x, $|\zeta^x - 1|^2 = (\zeta^x - 1)(\zeta^{-x} - 1)$. Thus

$$1 = (\zeta^a - 1)(\zeta^{-a} - 1)(\zeta^{ar} - 1)(\zeta^{-ar} - 1)(\zeta - 1)(\zeta^{-1} - 1)(\zeta^{r'} - 1)(\zeta^{-r'} - 1).$$

For each j so that $0 < j < p$ and $(j,p) = 1$ define

$$m_j = \#\{x \in \{a, -a, ar, -ar\} | x \equiv j \bmod p\}$$

and

$$n_j = \#\{x \in \{1, -1, r', -r'\} | x \equiv j \bmod p\}.$$

Then clearly $\sum_j m_j = 4 = \sum_j n_j$.

Let $a_j = m_j - n_j$. Then

(a) $\sum a_j = 0$.

(b) $a_j = m_j - n_j = m_{p-j} - n_{p-j} = a_{p-j}$.

(c) $\prod_j (\zeta^j - 1)^{a_j} = 1$.

A theorem of Franz (for a proof see [**8**]) says that if a_j is a sequence of integers so that (a), (b), and (c) hold for all p^{th} roots of unity $\zeta \neq 1$, then $a_j = 0$ for all j.

Thus $m_j = n_j$ for each j. It follows that either

1. $a \equiv \varepsilon_1$ and $ar \equiv \varepsilon_2 r' \bmod p$ for some $\varepsilon_i \in \{\pm 1\}$, or
2. $a \equiv \varepsilon_1 r'$ and $ar \equiv \varepsilon_2 \bmod p$ for some $\varepsilon_i \in \{\pm 1\}$.

The first part of the theorem follows from this and the facts that $rq \equiv 1 \bmod p$ and $r'q' \equiv 1 \bmod p$.

The homeomorphism $h : S^3 \to S^3$ taking (w_1, w_2) to $(w_1, \bar{w}_2)$ is equivariant with respect to the actions $(w_1, w_2)g = (w_1\zeta, w_2\zeta^q)$ and $(w_1, w_2)g = (w_1\zeta, w_2\zeta^{-q})$. This implies that $L(p,q)$ and $L(p,-q)$ are homeomorphic.

The homeomorphism $k : S^3 \to S^3$ taking (w_1, w_2) to (w_2, w_1) is equivariant with respect to the actions $(w_1, w_2)g = (w_1\zeta, w_2\zeta^q)$ and $(w_1, w_2)g = (w_1\zeta^q, w_2\zeta)$. The quotient space of the $\mathbf{Z}/p$ action $(w_1, w_2)g = (w_1\zeta^q, w_2\zeta)$ is the same as the quotient space of the $\mathbf{Z}/p$ action $(w_1, w_2)g = (w_1\zeta, w_2\zeta^r)$, since p and q are relatively prime and $(w_1\zeta^q, w_2\zeta) = (w_1\zeta^q, w_2(\zeta^q)^r)$. This implies that $L(p,q)$ and $L(p,r)$ are homeomorphic (where $r = q^{-1} \bmod$

p). Thus if $L(p,q)$ and $L(p,q')$ are simple-homotopy equivalent, they are homeomorphic. □

Proof of Theorem 11.1. The homotopy classification was obtained in Theorem 11.35. Theorem 11.37 shows that if $q' = \pm q^{\pm 1}$, then $L(p,q)$ and $L(p,q')$ are homeomorphic.

If $q' \neq \pm q^{\pm 1}$, then $L(p,q)$ and $L(p,q')$ are not simple-homotopy equivalent, and so by Chapman's theorem (Theorem 11.32) $L(p,q)$ and $L(p,q')$ are not homeomorphic. □

11.7. The s-cobordism theorem

Finally, we end this chapter with the statement of the s-cobordism theorem, a fundamental result of geometric topology.

Theorem 11.38 (s-cobordism theorem). *Let W be a smooth (respectively piecewise-linear, topological) compact manifold of dimension 6 or more whose boundary consists of two path components M_0 and M_1. Suppose that the inclusions $M_0 \hookrightarrow W$ and $M_1 \hookrightarrow W$ are homotopy equivalences. Let $\tau(W, M_0) \in$ Wh(π) denote the Whitehead torsion of the acyclic, based $\mathbf{Z}[\pi_1 W]$-complex $C_*(\tilde{W}, \tilde{M}_0)$.*

Then W is diffeomorphic (respectively PL-homeomorphic, homeomorphic) to $M_0 \times [0,1]$ if and only if $\tau(W, M_0) = 0$ vanishes. □

A good exposition of the proof in the smooth case is given in [**19**] and in the PL-case in [**34**]. The topological case is much harder and is based on the breakthroughs of Kirby and Siebenmann [**20**] for topological manifolds. The theorem is false in the smooth case if W has dimension 5 by results of Donaldson [**10**], and is true in the topological case in dimension 5 for many fundamental groups (e.g. $\pi_1 W$ finite) by work of Freedman-Quinn [**12**].

The method of proof of the s-cobordism theorem is to develop *handlebody structures* on manifolds. A handlebody structure is an enhanced analogue of a CW-decomposition. Provided the dimension of the manifold is high enough, then handles can be manipulated in a manner similar to the way cells are manipulated in the proof of Theorem 11.31, and the proof of the s-cobordism theorem proceeds using handle-trading, handle-sliding, and handle-cancellation.

11.8. Projects for Chapter 11

11.8.1. Handlebody theory and torsion for manifolds.

Discuss handlebody theory for smooth (or PL) manifolds and use it to indicate how torsion can be useful in the study of diffeomorphism (or PL homeomorphism) problems for manifolds. In particular, discuss how handlebody structures relate Theorems 11.31 and 11.38.

Bibliography

[1] J. F. Adams, *Algebraic Topology: A Student's Guide.* London Mathematical Society Lecture Notes Series 4, Cambridge University Press, 1972.

[2] J. F. Adams, *Stable Homotopy and Generalized Cohomology.* Chicago Lectures in Mathematics, University of Chicago Press, 1974.

[3] M. Atiyah, *K-Theory*, Notes by D. W. Anderson, Second edition. Advanced Book Classics, Addison-Wesley Publishing Company, 1989.

[4] H. Bass, J. Milnor, and J.P. Serre, "Solution of the congruence subgroup problem for SL_n $(n \geq 3)$ and Sp_{2n} $(n \geq 2)$." Inst. Hautes Études Sci. Publ. Math. No. 33, 1967, 59–137.

[5] G. Bredon, *Introduction to Compact Transformation Groups.* Pure and Applied Mathematics, Volume 46, Academic Press, 1972.

[6] T. A. Chapman, "Topological invariance of Whitehead torsion." Amer. J. Math. 96, 1974, 488–497.

[7] M. Cohen, *A Course in Simple Homotopy Theory.* Graduate Texts in Mathematics No. 10, Springer–Verlag, 1973.

[8] G. de Rham, S. Maumary, and M. A. Kervaire, *Torsion et type simple d'homotopie.* Lecture Notes in Mathematics No. 48, Springer-Verlag, 1967.

[9] A. Dold, *Lectures on Algebraic Topology.* Die Grundlehren der mathematischen Wissenschaften, Band 200, Springer–Verlag, 1972.

[10] S. K. Donaldson, "Irrationality and the h-cobordism conjecture." J. Differential Geom. 26, 1987.

[11] J. Dugundjii, *Topology.* Allyn and Bacon, 1965.

[12] M. H. Freedman and F. Quinn, *Topology of 4-manifolds.* Princeton Mathematical Series, 39, Princeton University Press, 1990.

[13] R. Fritsch and R. A. Piccinini, *Cellular Structures in Topology.* Cambridge Studies in Advanced Mathematics, 19, Cambridge University Press, 1990.

[14] M. Greenberg and J. Harper, *Algebraic Topology, a First Course.* Advanced Book Program, Benjamin/Cummins, 1981.

[15] P. A. Griffiths and J. W. Morgan, *Rational Homotopy Theory and Differential Forms.* Progress in Mathematics, 16, Birkhäuser, 1981.

[16] V. Guillemin and A. Pollack, *Differential Topology.* Prentice Hall, 1974.

[17] M. Hirsch, *Differential Topology.* Graduate Texts in Mathematics No. 33, Springer-Verlag, 1976.

[18] D. Husemoller, *Fibre Bundles.* Graduate Texts in Mathematics No. 20, Springer Verlag, 1974.

[19] M. Kervaire, "Le théorème de Barden-Mazur-Stallings." Comment. Math. Helv. 40, 1965, 31–42.

[20] R. C. Kirby and L. C. Siebenmann, *Foundational Essays on Topological Manifolds, Smoothings, and Triangulations*, Annals of Mathematical Studies No. 88, Princeton Univ. Press, 1977.

[21] S. Lang, *Algebra.* Addison–Wesley Publishing Company, 1965.

[22] S. MacLane, *Homology.* Die Grundlehren der Mathematischen Wissenschaften, Band 114, Springer-Verlag, 1975.

[23] W. S. Massey, *A Basic Course in Algebraic Topology.* Graduate Texts in Mathematics No. 127, Springer-Verlag, 1991.

[24] J. Milnor, "On spaces having the homotopy type of CW-complex." Trans. Amer. Math. Soc. 90, 1959, 272–280.

[25] J. Milnor, "On axiomatic homology theory." Pacific J. Math. 12, 1962, 337–341.

[26] J. Milnor, "The geometric realization of a semi-simplicial complex." Ann. of Math. (2) 65, 1957, 357–362.

[27] J. Milnor, *Topology from the Differentiable Viewpoint.* The University Press of Virginia, 1965.

[28] J. Milnor, "Whitehead torsion." Bull. Amer. Math. Soc. 72, 1966, 358–426.

[29] J. Milnor and D. Husemoller, *Symmetric Bilinear Forms.* Ergebnisse der Mathematik und ihrer Grenzgebiete, Band 73, Springer-Verlag, 1973.

[30] J. Milnor and J. Stasheff, *Characteristic Classes.* Annals of Mathematics Studies No. 56, Princeton University Press and University of Tokyo Press, 1974.

[31] R. Mosher and M. Tangora, *Cohomology Operations and Applications in Homotopy Theory.* Harper & Row, Publishers, 1968.

[32] D. Ravenel, *Complex Cobordism and Stable Homotopy Groups of Spheres.* Pure and Applied Mathematics, 121, Academic Press, Inc. [Harcourt Brace Jovanovich, Publishers], 1986.

[33] J. Rotman, *An Introduction to Homological Algebra.* Pure and Applied Mathematics, 85, Academic Press, Inc. [Harcourt Brace Jovanovich, Publishers], 1979.

[34] C. P. Rourke and B. J. Sanderson, *Introduction to piecewise-linear topology*, Ergebnisse der Mathematik und ihrer Grenzgebiete, Band 69, Springer-Verlag, 1972.

[35] E. H. Spanier, "Duality and S-theory." Bull. Amer. Math. Soc. 62, 1956, 194–203.

[36] E. H. Spanier, *Algebraic Topology.* McGraw-Hill Series in Higher Mathematics, McGraw-Hill, 1966.

[37] N. Steenrod, *The Topology of Fiber Bundles.* Princeton Mathematical Series No. 14, Princeton University Press, 1951.

[38] N. Steenrod, "A convenient category of topological spaces." Michigan Math. J. 14, 1967, 133–152.

[39] R. M. Switzer, *Algebraic Topology–Homotopy and Homology.* Die Grundlehren der Mathematischen Wissenschaften, Band 212, Springer-Verlag, 1975.

[40] R. Thom, "Quelques propriétés des variétés différentiables." Comm. Math. Helvetici 28, 1954, 17–86.

[41] J. Vick, *Homology Theory, an Introduction to Algebraic Topology*, Second edition. Graduate Texts in Mathematics No. 145, Springer-Verlag, 1994.

[42] C. A. Weibel, *An introduction to homological algebra*, Cambridge Studies in Advanced Mathematics, 38, Cambridge University Press, 1994.

[43] G. Whitehead, *Elements of Homotopy Theory*. Graduate Texts in Mathematics No. 61, Springer-Verlag, 1978.

[44] G. Whitehead, "Fifty years of homotopy theory." Bull. Amer. Math. Soc. 8, 1983, 1–29.

Index